Microbiome of Finfish and Shellfish

Arvind Diwan • Sanjay N. Harke •
Archana Panche
Editors

Microbiome of Finfish and Shellfish

Springer

Editors
Arvind Diwan
Institute of Biosciences and Technology
MGM University
Aurangabad, Maharashtra, India

Sanjay N. Harke
Institute of Biosciences and Technology
MGM University
Aurangabad, Maharashtra, India

Archana Panche
Centre for Biosustanability
Technical University of Denmark
Copenhagen, Denmark

ISBN 978-981-99-0851-6 ISBN 978-981-99-0852-3 (eBook)
https://doi.org/10.1007/978-981-99-0852-3

This Springer imprint is published by the registered company Springer Nature Singapore Pte Ltd.
The registered company address is: 152 Beach Road, #21-01/04 Gateway East, Singapore 189721, Singapore

Foreword

The microbiome—bacteria, archaea, microeukaryotes and fungi, and viruses—is a key factor in the health status and biological activity of most organisms. Changes in the composition of the microbiome, particularly in the gut tract, have been found to be associated with the regulation of several disorders. Scientists around the world are studying therapies that manipulate the microbiome, including probiotics, prebiotics, antibiotics, and transplants of microbes from healthy organisms. In recent years the microbiome of a number of vertebrate animal species has been sequenced including livestock and wildlife species such as the Tasmanian devil, red panda, giant panda, black howler monkey, and koala. The microbiomes of some terrestrial and aquatic invertebrates have also been described, but much remains to be discovered and investigated.

Microbiomes of finfish and shellfish contribute to digestion but also have impacts on growth, reproduction, immune system, and susceptibility of the host to diseases. Understanding the interactions between the microbiomes of the animals themselves—not limited to gut microbiota—and their surrounding environments is important for the effective management of farmed animals, biosecurity, minimizing environmental impacts, and food safety. Therefore, the study of such microbial communities is highly relevant for the development of the bio-industry, including aquaculture.

Increasing efforts are being made to discover the diversity and constituents of microbiome diversity in finfish and shellfish, and how these relate to their ecology, health, and suitability for consumption by humans. Recent advances in molecular biology technologies offer substantial and diverse opportunities for these areas of research. High-throughput sequencing methods, including PCR amplicon, metagenomic, and metatranscriptomic shotgun sequencing, provide high-resolution data on microbiome structure, functions, and host interactions. Increasingly adaptable and sophisticated bioinformatic tools enable these data to be analyzed at the level of the individual, and between populations, species, and ecological systems.

The present book brings together wide-ranging and comprehensive information on finfish and shellfish microbiomes, their identification, diversity, and functional properties, relevance to diseases, health status, data on species-specific metagenomics, probiotic research, bioinformatics skills, and other relevant topics.

The importance of microbiome research for the growth of biology-related industries including aquaculture has also been covered and discussed in detail.

Dr. A D Diwan and his team, Dr. S N Harke, Dr. A N Panche, and all of the authors are to be congratulated for the significant achievement of editing and producing this important reference book. I believe that their contribution will be useful not only to the aqua-farming industry but also to research institutes and teaching faculties, and to both professionals and students involved in microbiome research across many different organizations.

Centre for Environment, Fisheries and Aquaculture Science (Cefas)
Lowestoft, UK
David Bass

Preface

The microbiome is a virgin science and in recent years has gained a lot of importance because of its pivotal role in the protection of the health of all organisms including human beings. Microbiome actually deals with microorganisms that are associated with indigenous body parts and the entire gut system in all animals. These microbes are linked with different roles involving hereditary traits, defense against diseases, and strengthening overall immunity, which determine the health status of animals. Scientists have actually nicknamed the gut microbiome the "second brain" of the organism because it has the power to change the mood and is the core of the immune system. Considerable efforts have been made to discover the microbiome diversity and their taxonomic identification in finfish and shellfish. Further microbiome's importance has been correlated with various physiological functions and activities. In the recent past, due to the availability of advanced molecular tools, some efforts have been made on DNA sequencing of these microbes to understand the environmental impact and other stress factors on their genomic structural profile. There are also reports on the usage of next-generation sequencing (NGS) technology including amplicon and shotgun approaches, and associated bioinformatics tools to count and classify commensal microbiome at the species level. The microbiome present in the whole body, particularly in the gut systems of finfish and shellfish, not only contributes to digestion but also has an impact on nutrition, growth, reproduction, immune system, and vulnerability of the host fish to diseases. Therefore, the study of such microbial communities is highly relevant for the development of the bio-industry including aquaculture. In recent years, attempts have been made to discover the chemical ingredients present in these microbes in the form of biomolecules/bioactive compounds with their functions and usefulness for various health benefits, particularly for the treatment of different types of disorders, and therefore, it has been speculated that microbiomes hold great promise not only as a cure for ailments but also can be used for preventive measure. This kind of exploration of new breeds of microbes with their miraculous ingredients will definitely help to grow the drug, pharmaceutical, and other biological-related industries including aquaculture with the development of innovative products. The probiotic research and bioinformatics skills will further escalate these opportunities in the sector. In the present book, sincere efforts have been made to collect comprehensive information on the finfish and shellfish microbiomes, their identification, diversity and functional properties, relationship with diseases, health status, and data on species-specific metagenomics, probiotic research, and bioinformatics skills and impact of climate change. Further, microbiome research for the growth of biological-related industries including aquaculture has also been covered and discussed in detail.

We have brought together a group of outstanding national and international contributors at the forefront of microbiome research, and together they have produced this excellent reference book that is expected to be a valuable resource for researchers, teachers, students, bio and aquaculture industries, and all those working in the fisheries and aquaculture sectors.

We extend our sincere thanks to all the authors who have contributed with dedication, persistence, and cooperation in completing their chapters in a timely manner to this book and whose cooperation has made our task as editors a pleasure. We hope that this book will be quite useful, informative, and stimulating to readers all over the world in microbiome research.

We are also grateful to our Chancellor Shri Ankushrao Kadam, Mahatma Gandhi Mission University, Aurangabad, Maharashtra, India, for his constant encouragement, support, and cooperation in our endeavors for undertaking this difficult task of writing this kind of reference book in a new virgin area of science.

Aurangabad, Maharashtra, India — Arvind Diwan
Aurangabad, Maharashtra, India — Sanjay N. Harke
Copenhagen, Denmark — Archana Panche

Contents

Part I Microbiome of Finfish

1 **Introduction to Finfish Microbiome and Its Importance** 3
Jyutika M. Rajwade, Snehal S. Kulkarni, and Janhavi Vanjari

2 **Isolation and Identification of Fish Gut Microbiome** 35
A. R. Sarika, M. S. Aishwarya, R. S. Rachanamol, A. P. Lipton, and C. R. Abhirami

3 **Gut Microbiome and Fish Health: An Overview in Finfish Aquaculture Prospective** 47
Farhana Hoque, Abhijit Das, and Jitendra Kumar Sundaray

4 **Fish Microbiome and Metagenomics** 75
Akhila Dharnappa Sannejal, Mithoor Divyashree, Deekshit Vijaya Kumar, M. S. Nithin, and Praveen Rai

5 **Ready Reckoner for Using Bioinformatics Tools in Fish and Shellfish Microbiome Analysis** 93
Pavan Gollapalli, Kattapuni Suresh Prithvisagar, and Ballamoole Krishna Kumar

6 **Microbiome Applications for the Growth and Development of the Aquaculture/Biopharma Industry** 121
Arvind Diwan, Sanjay N. Harke, and Archana Panche

Part II Microbiome of Shellfish

7 **Shellfish Microbiome** 149
Aifa Fathima, Yaser Arafath, Saqib Hassan, Aaron Lipton, Arun Ninawe, George Seghal Kiran, and Joseph Selvin

8 **Shellfish Microbiome and Its Significant Role in Immune Response and Health Management** 165
A. Panigrahi, T. N. Vinay, P. K. Patil, M. S. Nithin, and A. D. Diwan

9 **Shrimp Health and Microbiome** 181
Sanath H. Kumar, Jerusha Stephen, K. Jeena, Manjusha Lekshmi, K. Sreedharan, and Rajendran Kooloth Valappil

10 **Microbial Metagenomics and the Shellfish Microbiome** 203
Andrea M. Tarnecki

11 **Shellfish Microbiome and Probiotics: A Decade in Review** 225
Andrea M. Tarnecki and Francisca Burgos

12 **Impact of Climate Change on the Gut Microbiome of Fish and Shellfish** 255
Arvind Diwan, Sanjay N. Harke, and Archana Panche

Editors and Contributors

About the Editors

A. D. Diwan is former Assistant Director General (ADG) in the Fisheries Division of the Indian Council of Agricultural Research (ICAR), Ministry of Agriculture, Govt. of India, New Delhi. Before assuming this position, he was a Professor and Head of the Fish Genetics & Biotechnology Division and also a Dean (Faculty of Fishery Science) at the Central Institute of Fisheries Education (CIFE), a Deemed University, in Mumbai. He started his career as an Associate Professor in the Centre of Advanced Studies (CAS) in Mariculture, an ICAR/UNDP project of the Government of India at the Central Marine Fisheries Research Institute (CMFRI), Kochi.

He specializes in animal physiology and neurosciences with a vast background and experience in marine and oceanic fisheries and mariculture and published more than 140 research papers of both national and international repute. He has also edited and authored several books on physiology, biotechnology, and fisheries. He has guided several Ph.D. students and two of them have been honored with the ICAR Jawaharlal Nehru Award for excellent Post-Graduate Agricultural Research and Education. He is also a recipient of the Bharat Ratna Dr. C. Subramaniam Best Teacher National Award of ICAR, Govt. of India, New Delhi. He has served as a member of boards of many national and international committees. He is a fellow of the National Academy of Environmental Biology, the National Academy of Nature Conservator, and the Bioved Research Society. He worked as UGC Emeritus Professor at the Mahatma Gandhi Mission (MGM) University of Health Sciences, Aurangabad, Maharashtra. Under the Ministry of Science & Technology, Govt. of India, he was selected as a scientist expert for writing a specialized reference book on shrimp biotechnology. Presently, he is continuing as an emeritus professor at the MGM University of Health Sciences in the Department of Biotechnology, Aurangabad, Maharashtra, India.

Sanjay N. Harke is the founder and Director of the Institute of Biosciences & Technology under Mahatma Gandhi Mission University, Aurangabad, Maharashtra, India. He specializes in the field of biotechnology, biochemistry, molecular biology, gene technology, synthetic biology, and computational biology. He has vast experience teaching UG and PG courses in different disciplines of biotechnology and computational biology. He has published several research papers and edited a few books, and presently he is working on various research projects in the area of fungal-host cell interaction, synthetic biology, and aging.

Archana Panche presently is working as a postdoctoral researcher at the Centre for Biosustainability at the Technical University of Denmark (DTU), Copenhagen, Denmark, and Assistant Professor at the Institute of Biosciences & Technology under Mahatma Gandhi Mission University, Aurangabad, Maharashtra, India. She specializes in bioinformatics and computational biology. She has more than 10 years of experience of teaching UG and PG courses in different disciplines of biotechnology, bioinformatics, and computational biology. She has published a number of research papers in journals of national and international repute.

Contributors

C. R. Abhirami Jawaharlal Nehru Tropical Botanic Garden and Research Institute (JNTBGRI), Thiruvananthapuram, Kerala, India

M. S. Aishwarya Centre for Marine Science and Technology, Manonmaniam Sundaranar University, Rajakkamangalam, Tamil Nadu, India

Yaser Arafath Department of Microbiology, School of Life Sciences, Pondicherry University, Puducherry, India

Francisca Burgos Escuela Superior Politécnica del Litoral, ESPOL, Facultad de Ingeniería Marítima y Ciencias del Mar, Guayaquil, Ecuador

Abhijit Das Faculty of Fishery Sciences, West Bengal University of Animal and Fishery Sciences, Kolkata, India

Mithoor Divyashree Department of Environmental Health and Toxicology, Nitte University Centre for Science Education and Research (NUCSER), Nitte (Deemed to be University), Deralakatte, Mangaluru, Karnataka, India
College of Fisheries, Karnataka Veterinary, Animal and Fisheries Sciences University, Mangaluru, India

Arvind Diwan Institute of Biosciences and Technology, MGM University, Aurangabad, Maharashtra, India

Aifa Fathima Department of Microbiology, School of Life Sciences, Pondicherry University, Puducherry, India

Pavan Gollapalli Center for Bioinformatics and Biostatistics, Nitte (Deemed to Be University), Mangaluru, Karnataka, India

Sanjay N. Harke Institute of Biosciences and Technology, MGM University, Aurangabad, Maharashtra, India

Saqib Hassan Department of Microbiology, School of Life Sciences, Pondicherry University, Puducherry, India

Farhana Hoque Regional Research Station, ICAR-Central Institute of Freshwater Aquaculture (CIFA), Rahara, Kolkata, West Bengal, India

K. Jeena Fisheries Resources, Harvest & Post-Harvest Management Division, ICAR-Central Institute of Fisheries Education (CIFE), Mumbai, India
Aquatic Environment and Health Management Division, ICAR-Central Institute of Fisheries Education (CIFE), Mumbai, Maharashtra, India

George Seghal Kiran Department of Food Science and Technology, School of Life Sciences, Pondicherry University, Puducherry, India
Centre for Marine Science and Technology, Kanyakumari, India

Snehal S. Kulkarni Nanobioscience Group, Agharkar Research Institute, Pune, India

Ballamoole Krishna Kumar Department of Infectious Diseases and Microbial Genomics, Nitte University Centre for Science Education and Research, Nitte (Deemed to be University), Deralakatte, Mangaluru, Karnataka, India

Deekshit Vijaya Kumar Department of Infectious Diseases and Microbial Genomics, Nitte University Centre for Science Education and Research (NUCSER), Nitte (Deemed to be University), Deralakatte, Mangaluru, Karnataka, India
College of Fisheries, Karnataka Veterinary, Animal and Fisheries Sciences University, Mangaluru, India

Sanath H. Kumar Aquatic Environment and Health Management Division, ICAR-Central Institute of Fisheries Education (CIFE), Mumbai, India
Fisheries Resources, Harvest & Post-Harvest Management Division, ICAR-Central Institute of Fisheries Education (CIFE), Mumbai, Maharashtra, India

Manjusha Lekshmi Aquatic Environment and Health Management Division, ICAR-Central Institute of Fisheries Education (CIFE), Mumbai, India
Fisheries Resources, Harvest & Post-Harvest Management Division, ICAR-Central Institute of Fisheries Education (CIFE), Mumbai, Maharashtra, India

A. P. Lipton Centre for Marine Science and Technology, Manonmaniam Sundaranar University, Rajakkamangalam, Tamil Nadu, India

Aaron Lipton Centre for Marine Science and Technology, Kanyakumari, India

Arun Ninawe Department of Biotechnology, Ministry of Science Government of India, New Delhi, India

M. S. Nithin Division of Infectious Diseases, Nitte University Centre for Science Education and Research (NUCSER), Nitte (Deemed to be University), Mangaluru, Karnataka, India
College of Fisheries, Karnataka Veterinary, Animal and Fisheries Sciences University, Mangaluru, Karnataka, India

Archana Panche Centre for Biosustainability, Technical University of Denmark, Copenhagen, Denmark

A. Panigrahi ICAR-Central Institute of Brackishwater Aquaculture (CIBA), Chennai, Tamil Nadu, India

P. K. Patil ICAR-Central Institute of Brackishwater Aquaculture (CIBA), Chennai, Tamil Nadu, India

Kattapuni Suresh Prithvisagar Department of Infectious Diseases and Microbial Genomics, Nitte University Centre for Science Education and Research, Nitte (Deemed to be University), Deralakatte, Mangaluru, Karnataka, India

R. S. Rachanamol A. J College of Arts and Science, Thiruvananthapuram, India

Praveen Rai Department of Infectious Diseases and Microbial Genomics, Nitte University Centre for Science Education and Research (NUCSER), Nitte (Deemed to be University), Deralakatte, Mangaluru, Karnataka, India
College of Fisheries, Karnataka Veterinary, Animal and Fisheries Sciences University, Mangaluru, India

Jyutika M. Rajwade Nanobioscience Group, Agharkar Research Institute, Pune, India

Akhila Dharnappa Sannejal Department of Food Safety and Nutrition, Nitte University Centre for Science Education and Research (NUCSER), Nitte (Deemed to be University), Deralakatte, Mangaluru, Karnataka, India
College of Fisheries, Karnataka Veterinary, Animal and Fisheries Sciences University, Mangaluru, India

A. R. Sarika Kerala Biotechnology Commission, Kerala State Council for Science, Technology and Environment, Thiruvananthapuram, Kerala, India

Joseph Selvin Department of Microbiology, School of Life Sciences, Pondicherry University, Puducherry, India

K. Sreedharan ICAR-Central Institute of Fisheries Education (CIFE) Rohtak Centre, Rohtak, Haryana, India

Jerusha Stephen Aquatic Environment and Health Management Division, ICAR-Central Institute of Fisheries Education (CIFE), Mumbai, India
Fisheries Resources, Harvest & Post-Harvest Management Division, ICAR-Central Institute of Fisheries Education (CIFE), Mumbai, Maharashtra, India

Jitendra Kumar Sundaray ICAR-Central Institute of Freshwater Aquaculture (CIFA), Bhubaneswar, Odisha, India

Andrea M. Tarnecki Auburn University Shellfish Lab, School of Fisheries, Aquaculture and Aquatic Sciences, Auburn University, Dauphin Island, AL, USA

Rajendran Kooloth Valappil Aquatic Environment and Health Management Division, ICAR-Central Institute of Fisheries Education (CIFE), Mumbai, Maharashtra, India

Janhavi Vanjari Bioengineering Department, VIT Bhopal University, Sehore, Madhya Pradesh, India

T. N. Vinay ICAR-Central Institute of Brackishwater Aquaculture (CIBA), Chennai, Tamil Nadu, India

Part I

Microbiome of Finfish

Introduction to Finfish Microbiome and Its Importance

1

Jyutika M. Rajwade, Snehal S. Kulkarni, and Janhavi Vanjari

Abstract

The microbiome is a complex yet defined set of microbial species in a given habitat. The mining of the human microbiome is pursued in several research projects globally. The microbiota of other vertebrate animals also needs to be studied. Fish represent a diverse group of vertebrates and are the primary source of animal protein. The fish gut microbiota plays a crucial role in the host's metabolism and can be manipulated to maximize production. Similar to humans, the fish microbiota is studied using culture-dependent and culture-independent techniques, including next-generation sequencing, bioinformatics analysis, and omics. Various fish organs have been examined for their dominant and distinctive microbiota species, including the gills, skin, gut, kidney, liver, and spleen. Under various environmental, dietary, and other physiological situations, the gut microbiota is explored. Gnotobiotic zebrafish represent a valuable model for studying microbe–microbe and microbe–host interactions. Microbe–microbe interactions can be modulated by pre-, pro, and synbiotics, influencing the development of the fish immune system, which will eventually reduce the need for antibiotics. This chapter highlights the necessity of research to advance knowledge of fish health, immune system, metabolism, probiotic research, etc., and to extract novel biomolecules with functional features for creating new medications and pharmaceuticals.

J. M. Rajwade (✉) · S. S. Kulkarni
Nanobioscience group, Agharkar Research Institute, Pune, India
e-mail: jrajwade@aripune.org

J. Vanjari
Bioengineering Department, VIT Bhopal university, Sehore, Madhya Pradesh, India

A. Diwan et al. (eds.), *Microbiome of Finfish and Shellfish*,
https://doi.org/10.1007/978-981-99-0852-3_1

Keywords

Finfish · Gut microbiome · Next-generation sequencing · Finfish health · Finfish productivity · Immunomodulation · Host-microbiome interaction

1.1 Introduction

The microbiome is a complex yet defined set of microbial species present in a given habitat. The study of microbiomes has its origins in "microbial ecology," where environmental microbiomes were studied, and the latter was a platform for interdisciplinary work in practically all fields ranging from agriculture to personalized human medicine (Berg et al. 2020). Massive projects for mining the human microbiome are being pursued. With the rapid development in technologies for genome sequencing, the cost of analysis per sample is reduced significantly (Cullen et al. 2020). As a result, the microbiome research field provides massive information on microbes which could have immense applications. To cite a few examples, the understanding of the interactions of the gut microbial community with the hosts is increasing with time, and the emerging data can enable the development of new diagnostic techniques for diseases, or design interventional strategies to prevent diseases, etc., as part of personalized medicine. Similarly, understanding the agricultural microbiome may help keep away from the toxic chemicals used in agriculture, including aquaculture, to ensure sustainable use of environmental resources (Berg et al. 2020).

The microbiota of other vertebrate animals also needs to be studied as several species are included as dietary supplements. Fish represent a diverse group of vertebrates that constitute a major dietary protein source and a rich source of minerals, vitamins, and omega-3-fatty acids (http://nicholasinstitute.duke.edu/publication). Aquaculture and fish farming/capture fisheries are, in fact, the fastest-growing agricultural sector (FAO 2018). According to a recent report, the average per capita intake of animal protein in the form of fish is increasing (Nissa et al. 2021). To meet the increasing demand for food, the aquaculture industry focuses on intensive, high-production farming while preventing and managing diseases of fish and crustaceans. Several environmental factors affect the yield of fish. From the findings from the human gut microbiome data, there is substantial evidence that the gut microflora/microbiota influences the health of the individual, and it is realized that much can be gained from the studies on the fish microbiome. In the past few years, it has been realized that the fish gut microbiota is emerging as an essential factor in the host metabolism, and it can be manipulated to increase productivity.

To study the microbes that reside in the fish gut, fish skin, and gills and their communication with the host, new sequencing technologies, omics technologies, and bioinformatics tools can help immensely. These are briefly discussed in the following paragraphs.

1.2 Techniques Used to Study the Microbiota

1.2.1 Culture-Dependent Techniques

These represented the classical techniques and were used to document the dominant resident (autochthonous) and transient (allochthonous) microbial communities (Horsley 1977). In Dover sole, higher aerobic heterotrophic bacterial populations were observed in the juveniles as compared to adults, and the actual counts were foregut<mid gut<hindgut. The bacterial species recorded were *Acinetobacter, Alcaligenes*, *Enterobacteriaceae* representatives, *Flavobacterium, Micrococcus, Photobacterium*, *Staphylococcus,* and *Vibrio*. The bacteria degraded complex compounds such as p-nitrophenyl-β-N-acetylglucosaminidase, chitin, and collagen in the hindgut (MacDonald et al. 1986). In the early twenty-first century, traditional culture-dependent techniques were used to document the fish microbiome. Pond et al. (2006) studied the microbiota of *Oncorhynchus mykiss* using both culture-dependent and culture-independent techniques. The group identified *Aeromonas* sp. and *Carnobacterium piscicola* to be present as dominant forms over the 18-month-long experiment in laboratory conditions. The anaerobe, *Clostridium gasigenes*, was identified as the dominating intestinal microflora by molecular analytical techniques, viz., restriction fragment length polymorphism (RFLP) and 16S rRNA gene sequence analysis. Using culture-dependent techniques to assess the gut microbiota may result in biased reporting (Pond et al. 2006). Several organs in fish, such as gills, skin, gut, kidney, liver, spleen, and light-emitting organs, showed bacterial colonization. Several functions, for example, production of vitamins, and polymers, degradation of complex molecules, emission of light in deep sea fish taxa etc., were ascribed to microbes. The production of histamine (causing spoilage) in fish tissue was contributed by *Pseudomonas* (Austin 2006). *Mycoplasma, Arthrobacter, Brochothrix, Jeotgailbacillus, Ochrobactrum, Psychrobacter,* and *Sejongia* species were reported from the GI tract of different fish using culture-independent techniques (Nayak 2010). Environmental conditions, feed, and soil/ sediment influence the GI tract community. The manipulation of gut microbiota is essential for better health management, and several concepts such as probiotic strains, and pre-biotic and synbiotic concepts, have evolved (Nayak 2010). Ganguly and Prasad (2012) pointed out that the "chymous" formed after food is digested decides the abundance and the composition of the gut microbiota.

An extensive study on the isolation and characterization of culturable bacteria, including the population levels of aerobic and facultative aerobic heterotrophic autochthonous (adherent) and allochthonous (transient) bacteria present in the mid- and distal intestines of Atlantic salmon, was reported using traditional culture-based techniques (Ringø et al. 2008). Ghosh et al. (2021) reported the findings on microbiomes from three Indian Major Carps (IMCs), viz., rohu (*Labeo rohita*), catla (*Catla catla*), and mrigal (*Cirrhinus mrigala*), using culture-dependent methods. The study identified probiotic bacteria. The cultures were bacilli with *Bacillus licheniformis,* the commonest bacterium.

1.2.2 Molecular Techniques

Several studies combined traditional culture-based and molecular methods to report the fish microbiome. Pond et al. (2006) studied the microbiota of *Oncorhynchus mykiss* using both culture-dependent and culture-independent techniques. The group identified *Aeromonas* sp. and *Carnobacterium piscicola* to be present as dominant forms over the 18-month-long experiment in laboratory conditions. The importance of the fish gut microbiota was elucidated in a review by Nayak (2010). Culture-independent techniques were used to identify the microflora, which resulted in the identification of several unknown bacteria belonging to *Mycoplasma, Arthrobacter, Brochothrix, Jeotgailbacillus, Ochrobactrum, Psychrobacter,* and *Sejongia* species in the GI tract of different fish species. *Vibrio, Aeromonas, Flavobacterium, Plesiomonas, Pseudomonas, Enterobacteriaceae, Micrococcus, Acinetobacter, Clostridium, Fusarium,* and *Bacteroides* were reported from both freshwater and marine fish. Environmental conditions influence the composition of the intestinal microbiota. The anaerobe, *Clostridium gasigenes*, was identified as the dominating intestinal microflora by molecular analytical techniques, viz., restriction fragment length polymorphism (RFLP) and 16S rRNA gene sequence analysis. The techniques, viz., terminal restriction fragment length polymorphism (T-RFLP) and 16S rRNA clone library analysis, were used by Green et al. (2013) to analyze the gut microbiota of Atlantic salmon (*Salmo salar*).

The microbes native to the intestinal tract of *Notothenia coriiceps* and *Chaenocephalus aceratus* were analyzed using bacterial 16S rRNA gene sequences. Both exhibited lower diversity than other fish, but comparatively, the omnivore (*N. coriiceps*) exhibited great diversity. Microbial pathogens were also reported for the first time in Antarctic notothenioid fish (Ward et al. 2009). He et al. (2013) analyzed the fish species using the PCR-denaturing gradient gel electrophoresis (DGGE) technique, which revealed the presence of several common but unique bacteria in the gut. Based on 16S rDNA gene sequences, the bacterial diversity in the gut of grass carp (*Ctenopharyngodon idellus*) was phylogenetically closer to ingested food (Han et al. 2010). The study was based on several indices, viz., the Shannon diversity index (H), Shannon equitability index (EH), and the Coverage (Cgood) and rarefaction curves, which were calculated using molecular analysis data. The habitat microflora associated with the pond ecosystem was also studied, with the highest diversity associated in the carp gut.

The study by Li et al. (2013) employed techniques such as Metagenome-based fingerprinting and high-throughput sequencing on bacterial 16S rRNA genes to study the microbiota. Further classical biochemical and histological data were collected to understand the carbohydrate and lipid metabolism of transgenic and wild-type carps.

1.2.3 Next-Generation Sequencing

With the advent of next-generation sequencing technology, the gut microbial communities of several animals and humans are being studied for their interactions with their respective hosts. The aquatic environment serves as a continuous continuum and this makes the microbiome of aquatic fishes and other animals more dynamic (Xia et al. 2014). Advances in sequencing will add to the knowledge on biotic and abiotic factors affecting the host fish and this would provide new directions to the research in fish gut microbiota (Ghanbari et al. 2015). The data obtained are validated using culture-based techniques, histology, biochemical parameters, etc. The studies on the fish gut microbiome are possible on a large scale because of the advances in high-throughput next generation sequencing (Perry et al. 2020). 16S rRNA amplicon sequencing was used to study the hindgut microbiota composition in cod (Keating et al. 2021). Stable isotope analysis (SIA) and 16S rRNA gene sequencing techniques were used to analyze the intestinal microflora in Genetically Improved Farmed Tilapia (GIFT) (Parata et al. 2020). Rangel et al. (2022) employed high-throughput 16S rRNA gene amplicon sequencing and quantitative PCR to elucidate the consequences of insect meal diet on the microbiota composition and chitinolytic activity in European sea bass. The gut microbiota of European sea bass juveniles fed with microencapsulated organic acids was analyzed using next-generation sequencing (Busti et al. 2020). The high-throughput sequencing studies for analyzing the fish microbiome were performed on Illumina Miseq platform with the variable regions (V) within the 16S ribosomal RNA gene as the predominant marker (Perry et al. 2020). Xia et al. (2014) determined the microbial composition of Asian seabass under changing environmental conditions. The total phyla, classes, orders, and families identified were 33, 66, 130, and 278, respectively, and Proteobacteria (48.8%), Firmicutes (15.3%), and Bacteroidetes (8.2%) were the important and abundant taxa. Perry et al. (2020) reported that the gut microflora of 145 species of teleost fish is analyzed under different environmental conditions, diets, and other physiological parameters and is dominated by members belonging to Bacteroidetes and Firmicutes.

1.2.4 "Omics" Technologies

As global concerns about food availability/food security/and restrictions on water use are being realized, new technologies, especially "Omics" are required in aquaculture for diagnosing infections, ensuring good health and productivity in aquaculture. The advances in genomics, proteomics, transcriptomics, and lipidomics will help unravel the complex physiological process in animals, such as reproduction, immunity, and growth. "Targeted gene expression," "omics" technologies are capable of high-throughput analysis. These technologies help in understanding the mechanisms underlying the microbiome–fish interaction at the molecular level. These techniques allow, for example, comparisons of immunomodulatory effects between different prebiotics, providing insights into functional genomics. Prebiotics are classified as "functional feeds" which promote "healthy bacteria" and ultimately improve fish performance. These new approaches to understanding the microbiome have been recently reviewed (Porter et al. 2022). The importance of fish (and food product) proteome characterization was highlighted in a recent review (Nissa et al. 2021). Such studies are practical for comparative analysis to define molecular markers for fish health, fish growth, etc., thus ensuring the quality and safety of aquaculture-derived food. Behera and their team (Behera et al. 2021) have described the utility of sequence and function-driven metagenomics in the aquatic environment. According to the authors, such studies could enable discoveries of new genes and pathways and identify novel enzymes that can be used in the industrial, environmental, and other applications. Toxicological exposure, genome homogeneity, and aquatic management would be possible using microbes as bio-indicators. The authors further highlight the various approaches to studying microbiomes, the various computational tools, and the importance of "omics" in these studies. The massively parallel sequencing strategy was employed in teleost fish to carry out microbiome study. In zebrafish models (germ-free and biotic fish), a core microbiome was defined, and the role of the microbiome in development was studied. Microbiome ontogeny study will help in the manipulation of the microbiome to possibly increase the aquaculture yields (Llewellyn et al. 2014).

1.2.5 Use of Gnotobiotic Fish

The use of "gnotobiotic" or "germ-free" small animals such as rodents and mice is quite common, and experimental data collected on the microbiome using "gnotobiotic" zebra fish are published.

About two decades ago, gnotobiotic zebrafish (*Danio rerio*) were produced and reared through late juvenile stages (Rawls et al. 2004). The study on the colonization of the gnotobiotic zebrafish with individual bacterial species that constituted the microbiota indicated specificity to host responses. These studies used gnotobiotic mice, which proved the involvement of the gut microbiota in developing the mucosal immune system, nutrient processing and absorption, angiogenesis, and even the renewal of the epithelium. The authors state that *Danio rerio* is transparent during the juvenile and larval stages, and hence molecular mechanisms underlying the host–microbe interactions can be studied very quickly by developing suitable screening methods. The same group further demonstrated that gnotobiotic zebrafish were helpful as a model system to establish the host–microbe interaction in the vertebrate gut (Rawls et al. 2007). The group used isogenic *Pseudomonas aeruginosa* with the deletion of genes that contribute to motility and pathogenesis to colonize the zebrafish. Results indicated that host innate immune responses, which were evolutionarily conserved, were attenuated. However, colonization did not cause any change in the conserved nutrient responses. Pham et al. (2008) described a detailed protocol for generating gnotobiotic zebrafish. High sterility rates (>80%) and high survival (90%) were observed, and the utility of these model systems in the elucidation of the interactions of the animal host with resident microbes was reemphasized.

The study by Semova et al. (2012) employed gnotobiotic zebrafish to prove that the microbiota was involved in the regulation of the host energy balance. The authors used fluorescent fatty acid analogues and in vivo imaging techniques to prove this fact. The introduction of these fatty acid analogs was a dietary change that led to the enrichment of bacteria belonging to phylum Firmicutes, which increased the number of lipid droplets formed in the intestinal epithelium, probably influencing fat absorption.

A gnotobiotic *Dicentrarchus labrax* larvae (sea bass) was developed as a test system (Dierckens et al. 2009). For the development of this system, sea bass eggs were sterilized by glutaraldehyde and incubated in the presence of antibiotics (rifampicin and ampicillin). After hatching, larvae were fed with *Artemia* sp. The authors confirmed the germ-free status of larvae by culture-dependent and -independent techniques and performed a challenge test with opportunistic pathogens. The findings indicate that the system can be used to study the host–microbe interactions, which can be studied by gene expression analysis. Thus, such models are useful for pharmacological research, development of immunostimulants, etc.

Hansen and Olafsen (1989) attempted the colonization of gnotobiotic eggs with antibiotic-producing strains. However, continuous protection was not observed. The authors noted that *Pseudomonas, Alteromonas, Aeromonas*, and *Flavobacterium* were found to dominate on the surface of cod (*Gadus morhua* L.) and halibut (*Hippoglossus hippoglossus*) eggs from the stage of fertilization to hatching. These fish are commercially important. Interestingly the authors reported the presence of *Leucothrix mucor*, a filamentous bacterium in less number on halibut eggs. The bacterium heavily colonized cod eggs resulting in a hairy appearance. The studies involved the use of scanning electron microscopy. *Caulobacter* and *Seliberia* spp. were reported to be present in ovaries before spawning and hence were detected in cod ovaries.

In another study, the gnotobiotic zebrafish were used to study the effect of various microbial stimuli on spacio-temporal activation of NF-κB (Kanther et al. 2011). The transparent nature of the gnotobiotic zebrafish offers the possibility for in vivo imaging. Furthermore, these animals also allow for studies on molecular and cellular responses induced by microbial colonization.

A comparative study on intestinal cell proliferation was carried out in wild type (WT) and axin1 mutants (gnotobiotic animals) of zebrafish (Cheesman et al. 2011), proving the involvement of *Aeromonas veronii* (a resident bacterium) in the epithelial cell proliferation. The cell proliferation activity is regulated by the *Wnt* signaling pathway and microbial signaling via the Myd88 receptor. The *A. veronii* enhanced the stability of β-catenin in intestinal epithelial cells and promoted intestinal cell proliferation. Thus, intrinsic pathways and extrinsic cues are involved in cell proliferation.

1.3 Finfish Microbiome

The fish microbiome is studied for various applications. Merrifield and Rodiles (2015) reviewed the fish microbiome and stated that a core microbiome exists, providing a barrier function. The microbiome is implicated in nutrient absorption, morphology, immunity, and even the GI differentiation in finfish (Wang et al. 2018). Egerton et al. (2018) reviewed the research on the gut biota of wild fishes, its importance to fish health, and the modifications required for fish farming. However, the authors said that understanding the gut microbiome is insufficient.

A table indicating the fish species studied with reference to their microbiomes is presented (Table 1.1). The available data suggest that the gut microbiome of Atlantic cod (*Gadus morhua*), Atlantic salmon (*Salmo salar L.*), Rainbow trout (*Oncorhynchus mykiss*) is the most studied, probably because of their economic importance. In contrast, *Danio rerio* is studied due to the several characteristics it offers as a model system for experimental research.

The microbial communities colonizing the skin/mucosal surface of fish are known to be important in maintaining fish health and disease resistance. McMurtrie et al. (2022) surveyed the fish skin community in tilapia, an economically important fish. Prevalence of *Cetobacterium, Paucibacter, Pseudomonas,* and *Comamonadaceae* was observed in tilapia skin; whereas cyanobacteria (*Cyanobium, Microcystis* and/or *Synechocystis*) and the diatom *Cyclotella* were most prevalent in pond water in Malawi region. The study indicated that the skin microbiota was not much influenced by the environment. These findings are vital as a baseline study for future work.

The gills and guts in two filter-feeding fish (*Hypophthalmichthys molitrix* and *Hypophthalmichthys nobilis*) were analyzed for their communities by 16S sequencing to determine the influence of sediment and water on fish microbiota in an unfed aquaculture system (Kuang et al. 2020). The diversity in water and sediment was greater than that of the fish gills and gut. Interestingly, the host microbiota shared commonality with water, and the gill and gut microbiota possessed a unique core biota that presented predictive functions. According to the authors, these studies highlight the understanding of the host microbiota with respect to its diversity, composition, and functions with its habitat association.

The various aspects of the studies conducted on the finfish microbiome are depicted (Fig. 1.1). These can be broadly categorized into microbe–microbe interactions, microbe–host interactions, and the outcomes of these studies can result in the improvement of fish health, production, disease resistance, etc. Most studies that report on the fish microbiome include both types of interactions. The available findings are summarized in the following paragraphs with the complexity involved and limited understanding.

1.3.1 Microbe-Microbe Interactions

Several essential functions, viz. (1) mediating and stimulating gastrointestinal (GI) development, (2) aiding digestive enzyme activity, (3) maintaining mucosal tolerance, (4) immunomodulation, and (5) improving disease resistance, are attributed to the gut microbiome. The short-chain fatty acids produced by the gut microbiota have an effect on the physiology and the functions of the animals, and

Table 1.1 Studies on the fish microbiome including the techniques employed[a]

Fish species Common name (s6.cientific name)	Species isolated/microbiome related aspect studied	Method used for study	Reference
Atlantic cod (*Gadus morhua*)	Presence of endogenous chitinolytic bacteria and chitinase	Spread plate method	Lindsay and Gooday (1985)
	Brochothrix Carnobacterium, *Chryseobacterium* spp., *Psychrobacter glacincola*, *A. salmonicida* ssp., *Salmonicida,* and *V. anguillarum* tested in an in vitro test	Sequencing	Ringø et al. (2006)
	With Artemia as feed, the larval digestive tract microbiota dominated by *vibrio alginolyticus* and *V. splendidus* with dry feed	Polymerase chain reaction denaturing gradient gel electrophoresis (PCR-DGGE)	Reid et al. (2009)
	Bacteroidales, Erysipelotrichales, Clostridiales, Alteromonadales, and *Deferribacterales*	454 technology, PCR	Star et al. (2013)
	Vibrio and *Photobacterium, Lachnoclostridium*	16 S rRNA amplicon sequencing	Keating et al. (2021)
Atlantic salmon (*Salmo salar L.*)	*Lactobacillus* spp., *Lactococcus* sp., *bacillus* sp., *Photobacterium phosphoreum, Acinetobacter* sp., *pseudomonas* sp., *vibrio*	PCR and denaturing gradient gel electrophoresis (DGGE)	Hovda et al. (2007)
	Gelidibacter salicanalis, Pseudoalteromonas elyakovii, Psychrobacter aquimaris, Psychrobacter cibarius, Psychrobacter fozii, Psychrobacter maritimus, Psychrobacter okhotskensis, and Psychro-bacter psychrophilus Arthrobacter bergeri, Arthrobacter	Partial sequencing	Ringø et al. (2008)

(continued)

Table 1.1 (continued)

Fish species Common name (s6.cientific name)	Species isolated/microbiome related aspect studied	Method used for study	Reference
	psychrolactophilus, Arthrobacter rhombi, Bacillus pumilus, Bacillus subtilis, Exiguobacterium spp., *Microbacterium oxydans, Planococcus maritimus, Sporosarcina ginsengisoli, Carnobacterium inhibens, Psychrobacter cibarius*		
	Molecular analysis of microbiota along the digestive tract	Temporal temperature Gradient gel electrophoresis (TTGE)	Navarrete et al. (2009)
	Relationship of Atlantic salmon gastrointestinal tract bacteria to environmental factors	454 16S rRNA-based pyrosequencing	Green et al. (2013)
	Characterization of gastrointestinal bacteria	Pyrosequencing	Zarkasi et al. (2014)
	Lactobacillus and *Pediococcus, Isoptericola, Cellulomonas*, and *Clostridium* sensu stricto, *Acinetobacter*, and *Altererythrobacter*	PCR-TTGE	Catalán et al. (2018)
	Planctomycetes, Planctomycetia, Planctomycetales, and genus *Lactococcus*	16 S rRNA	Villasante et al. (2019)
Rainbow trout (*Oncorhynchus mykiss*)	*Aeromonas* sp., *Carnobacterium piscicola*	Restriction fragment length polymorphism (RFLP) and 16S rRNA gene sequence analysis	Pond et al. (2006)
	Intestinal and mucosal layer microbial community	Culture dependent and molecular techniques	Kim et al. (2007)

	Intestinal microbiota, mostly Firmicutes and gamma Protebacteria including *hafnia, pseudomonas, Aeromonas*	Quantitative PCR	Mansfield et al. (2010)
	Changes in the intestinal microbiome	Pyrosequencing of cpn60 PCR products and 16S rRNA DGGE	Desai et al. (2012)
	Lactococcus, Citrobatcer, Kluyvera, Obesumbacterium, Shweanella	PCR-temporal temperature gradient gel electrophoresis (TTGE)	Navarrete et al. (2012)
	Firmicutes, Proteobacteria, Bacteroidetes, Actinobacteria	High-throughput 16 S rRNA	Rimoldi et al. (2018)
	Propiobacterinae, Shewanella, and *mycoplasma*	16S rRNA gene amplicon sequencing	Terova et al. (2019)
	Buttiauxella agrestis, Budvicia aquatica, Erwinia persicinus, Yersinia mollaretii, Y. kristensenii, Y. rohdei, Moellerella wisconsensis, Obesumbacterium proteus, Pantoea cedenensis, Rahnella aquatilis, and Rahnella sp.	Partial 16S rRNA gene sequence analysis	Skrodenyte-Arbaciauskiene et al. (2006)
Zebrafish (*Danio rerio*)	Genetic analysis link bacterial motility and symbiosis in the zebrafish gut	In vivo imagining	Rawls et al. (2007)
	Aeromonas hybdophila, Aeromonas sobria, vibrio parahemolyticus, Photobatcerium damselae, Pseudomonas aeruginosa, Pseudomonas fluorescens, Pseudomonas luteola, Comamonas testosterone, Ochrobactrum anthropi, Staphylococcus cohnii, Staphylococcus epidermidis, Staphylococcus capitis, and *Staphylococcus warneri*	Traditional bacterial culture, biochemical characterization, 16S rDNA sequencing	Cantas et al. (2012)

(continued)

Table 1.1 (continued)

Fish species Common name (s6.cientific name)	Species isolated/microbiome related aspect studied	Method used for study	Reference
	Lactobacillus paracasei and *Eubacterium limosum*		Toh et al. (2013)
	Lactobacillus casei	TEM, RT-PCR	Rieu et al. (2014)
	Lactobacillus plantarum, lactobacillus fermentum	Fluorescence stereomicroscope	Russo et al. (2015)
European sea bass (*Dicentrarchus labrax* L.).	*Lactobacillus, Leuconostoc,* and *bacillus* sp.	NGS	Busti et al. (2020)
	Proteobacteria, Firmicutes, Actinobacteria, and *Bacteroidetes* phyla	16S rRNA sequencing	Pérez-Pascual et al. (2020)
	Bacillaceae, Enterococcaceae, Lachnospiraceae, and Actinobacteria	High-throughput 16S rRNA gene amplicon sequencing and quantitative PCR	Rangel et al. (2022)
Goldfish, common carp (*Cyprinus carpio*), Nile tilapia (*Oreochromis niloticus*), and ayu (*Plecoglossus altivelis)*	*Cetobacterium somerae*	16S rDNA sequencing	Tsuchiya et al. (2008)
Freshwater salmon (*Salmo salar*) and sea trout (*Salmo trutta trutta*)	Predominant group in the intestinal tract of the salmon comprised *Enterobacteriaceae* (23%), *Plesiomonas* (19.2%), and *Carnobacterium* (15.3%). Predominant microbiota in sea trout intestinal tract were *Enterobacteriaceae* (52%), *Aeromonas* (22%), and *pseudomonas* (14%)	Polymerase chain reaction (PCR), rDNA sequence analysis, fluorescence in situ hybridization and differential gradient gel electrophoresis	Skrodenytė-Arbačiauskienė et al. (2008)
Orange spotted grouper *(Epinephelus coioides)*	Gut microbiota	16S rRNA gene sequencing	Sun et al. (2009)

Coral reef fish species(*Chlorurus sordidus, Lutjanus bohar, Acanthurus nigricans)*	*Vibrio* spp. and *Photobacterium* spp. including *Vibrio ponticus* and *Photobacterium damselae*. Other *Vibrionaceae*-like sequences comprised a distinct phylogenetic group that may represent the presence of "feces-specific" bacteria	Denaturing gradient gel electrophoresis profiles of 16S rRNA genes	Smriga et al. (2010)
Cichlid (*Astatotilapia burtoni, Ophthalmotilapia ventralis*)	Gut microbiome	454 sequencing	Baldo et al. (2011)
Mirror carp (*Cyprinus carpio L.)*	Diet effects on microbiome-aerobic heterotrophic bacteria affected, no effect on lactic acid bacteria—*Lactococcus, Vagococcus*	PCR-DGGE	Kühlwein et al. (2013)
Grass carp, (*Ctenopharyngodon idella*)	Detection of grass carp reovirus (GCRV)	Reverse transcription loop-mediated isothermal amplification (RT-LAMP)	Zhang et al. (2013)
Juvenile Siberian sturgeon (*Acipenser baerii*)	Effect of prebiotics, arabinoxylan oligosaccharides (AXOS)	454 pyrosequencing	Geraylou et al. (2013)
Tropical gar (*Atractosteus tropicus*)	*Proteobacteria, fusobacteria, Firmicutes,* and *Bacteroidetes* phyla *Deinococcus Thermus* in wild-type males	16 S rDNA profiling	Pérez-Pascual et al. (2020)
Common carp (*Cyprinus carpio L.)*	Fish gut dominated by *Protebacteria, fusobacteria, Bacteroidetes,* and *Firmocutes*	Metagenome- based fingerprinting, high-throughput sequencing	Li et al. (2013)
SVCP (*Hypophthalmichthys molitrix*) GZSD (*Dorosoma cepedianum*)	*Cyanobacteria, Proteobacteria, Actinobacteria,* and *Bacteroidetes* predominant phyla	16S rRNA gene pyrosequencing	Ye et al. (2014)
Marron (*Cherax cainii*)	*Lactobacillus acidophilus* and *lactobacillus plantarum*	Illumina sequencing	Foysal et al. (2020)
Yellowtail kingfish	*Mycoplasmataceae*	16S rDNA and transcriptomic (RNA seq)	Legrand et al. (2020)

[a]Intestinal microflora is documented

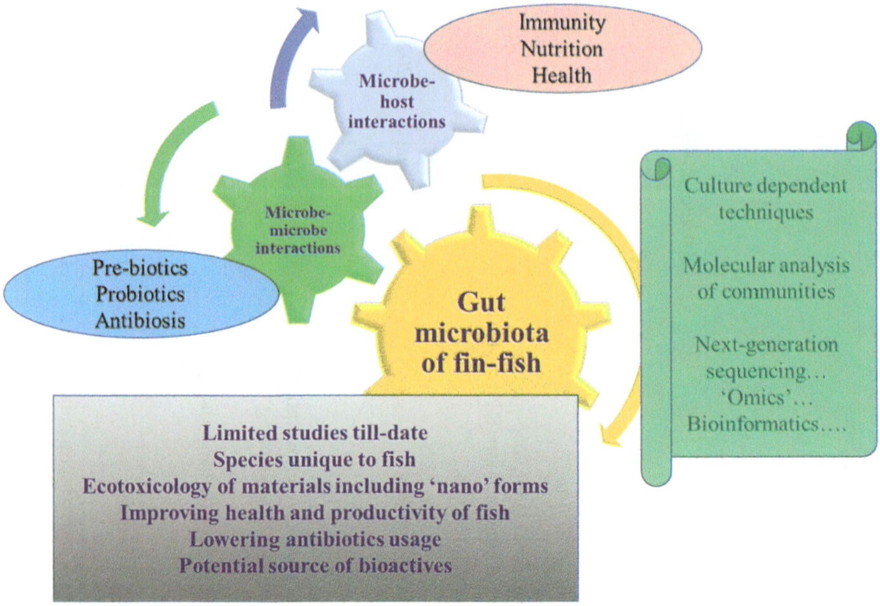

Fig. 1.1 Various aspects of the finfish microbiome

these communities can be modulated by pre-, pro-, and synbiotics (Hoseinifar et al. 2019).

1.3.1.1 Prebiotics

Fed aquaculture is practiced in order to match the need for aquatic protein production. The farmed fish are fed using a fish meal and fish oil which is sourced from wild-captured forage fish. With increased consumption of the latter, the sustainability of aquaculture is questioned. It is estimated that additional 7.4 million tons of aquafeeds will be required by 2025 (Hua et al. 2019), which means that alternate protein sources need to be explored. Thus plant-based ingredients, insect meals, fisheries by-products, biotransformed food waste, microbial and algal cells etc., need to be carefully evaluated as feed supplements in aquaculture. Such studies were initiated about a decade ago.

The comparative data on the intestinal microbiota of marine-farmed *S. salar* showed temporal variations upon feeding with experimental diets containing Soybean protein concentrate (SPC) at different levels Green et al. (2013). *Verrucomicrobia* and *Aliivibrio* were the frequently observed ribotypes in early and late summer, respectively. The intestinal population was reported to be highly complex; however, at the individual level it was simple but highly variable. *Escherichia* and *Propionibacterium* were reported in fish receiving SPC solely because of the changed diet composition. The addition of mannan-oligosaccharide as a pre-biotic led to changes in the intestinal microbiome. The diet did not cause inflammation, but the observation supports the "dysbiosis" hypothesis. The pre-biotic effect of plant-based diet in rainbow trout (*Oncorhynchus mykiss*) was elucidated (Ingerslev et al. 2014). The study showed that the plant-based diet modified the intestinal microflora, which was represented by various phyla, viz., *Firmicutes* (genus *Streptococcus, Leuconostoc,* and *Weissella* and phylum *Proteobacteria/Bacteroidetes/Actinobacteria)* and marine fed fish exhibited higher levels of IL-1b and MBL-2. During experimental infection with *Yersina ruckeri*, the survival was not different in both groups indicating the protective effect of the plant diet.

Terova et al. (2019) observed that the intestinal microbiota of *Oncorhynchus mykiss* was altered in response to feed containing insect meal. The abundance of *Mycoplasma* was noted, which produced lactic and acetic acid, and overall increased bacterial diversity was observed. Proteobacteria decreased in comparison to the fishmeal diet. These were attributed to the pre-biotic properties of the chitin in the insect meal. De Souza et al. (2020) observed that low concentrations (0.1 mg L^{-1}) of β-glucan administered in water increased tolerance to hypoxia, increased the survival rate, and regulated glucose levels in Nile tilapia fingerlings. The modulation of hematological responses under stress and increased lymphocyte count before the hypoxic challenge was observed at 0.3 mg L^{-1} of β-glucan. β-glucan, as a pre-biotic, did not affect gut diversity and growth characteristics.

Polysaccharides and oligosaccharides derived from marine algae can be used as prebiotics in aquaculture diets after proper assessment of their effects on the composition of gut biota (de Borba Gurpilhares et al. 2019). These can be extracted and treated using microbial enzymes. Production of functional dietary supplements and their use in aquaculture is being considered an environmentally safe strategy for improving fish health. Along with probiotics, the role of prebiotics is now being understood especially considering the positive response on immunity (Hoseinifar et al. 2015). Therefore, more detailed studies are warranted for intensive aquaculture in the years to come.

1.3.1.2 Probiotics

The non-pathogenic bacteria that can be exogenously introduced into the host gut, which could limit the diseases in the host, can be categorized as "probiotic" bacteria. This was largely based on the outcome of the studies on the beneficial contribution of commensal bacteria to host health and homeostasis (Rendueles et al. 2012). Probiotics are being studied for use in aquaculture specifically to improve resistance against diseases, ensuring "good health."

Eddy and Jones (2002) reported their observations on the bacteria associated with fingerling production in summer flounder (*Paralichthys dentatus*) over 3 years, corresponding to ten production cycles. The researchers observed that Artemia feeding led to the death of fish even if the pathogens were not detected. The microbiota of the live feed, the rearing water, and fish intestine were similar. Microbial succession and the microbiota stabilized after metamorphosis was observed. The dominant bacterial groups in phytoplankton, rotifers, larval fish, and tanks in early stages were *Acinetobacter, Agrobacterium, Flavobacterium, Moraxella,* and *Pseudomonas*. *Vibrio* and *Enterobacter* were reported at high levels in Artemia and older larval stages and tanks. *Bacillus* was detected in juvenile fish but not in larval stages. Such studies would help in the manipulation of the microbiota for probiotic effects.

Another extensive study on the isolation and characterization of culturable bacteria, including the population levels of aerobic and facultative aerobic heterotrophic autochthonous (adherent) and allochthonous (transient) bacteria present in the mid and distal intestines of Atlantic salmon, was reported using traditional culture-based techniques (Ringø et al. 2008). These fish were fed five different diets, and the microflora was sensitive to the dietary changes. A very high number of microbes were cultured (752) with several new species such as *Gelidibacter salicanalis, Pseudoalteromonas elyakovii, Psychrobacter aquimaris, Psychrobacter cibarius, Psychrobacter fozii, Psychrobacter maritimus, Psychrobacter okhotskensis and Psychrobacter psychrophilus, Arthrobacter bergeri, Arthrobacter psychrolactophilus, Arthrobacter rhombi, Bacillus pumilus, Bacillus subtilis, Exiguobacterium* spp., *Microbacterium oxydans, Planococcus maritimus, Sporosarcina ginsengisoli with Psychrobacter cibarius* being the dominant sp. In the GI tract. The lactic acid bacterium, viz., *Carnobacterium inhibens* was also reported. The habitat microflora associated with the pond ecosystem of *Ctenopharyngodon idellus* was studied, with the highest diversity associated in the gut of the carps representing a potential pool of probiotics to its host (Han et al. 2010).

Rendueles et al. (2012) developed a controlled co-infection model in which germ-free zebrafish were cultivated on axenic living protozoa. The group studied the effects of probiotic strains on the model zebrafish's survival under a challenge experiment with *Edwardsiella ictaluri* via the oro-intestinal route. *E. ictaluri* was a virulent bacterium that induced a robust inflammatory response resulting in rapid mortality in the zebrafish larvae. Probiotic effects were observed due to *Vibrio parahaemolyticus* and 2 *Escherichia coli* strains which were ascribed to the presence of the F-pilus, which promoted specific adhesion.

Lactobacillus casei ATCC334 was demonstrated to colonize the gut of zebrafish larvae (Rieu et al. 2014). The bacteria are known for adherence and colonization, which are the characteristics of probiotics. These bacteria exhibit anti-inflammatory properties. The supernatants from the bacterial cultures grown as biofilms suppressed TNF-α production and inhibited the NF-κB pathway probably due to GroEL. These studies were carried out using human macrophage-differentiated monocytic THP-1 cells. The study highlights the importance of biofilms of probiotic cultures in immunomodulation.

The study on the gut microbiota of economically important fishes, viz., channel catfish (*Ictalurus punctatus*), largemouth bass (*Micropterus salmoides*), and bluegill (*Lepomis macrochirus*) were significant because it established a baseline for future probiotic trials (Larsen et al. 2014). 16S rRNA pyrosequencing was performed, and statistically significant differences were observed in terms of bacterial species evenness. *Cetobacterium somerae* (phylum *Fusobacteria*) was predominant in all three fishes. Normal gut inhabitants were *Plesiomonas shigelloides*, *Fusobacterium mortiferum*, and members of the genus *Aeromonas*. According to the authors, more studies on the role of the dominant genus *Cetobacterium* in all species need to be specially studied.

The in vivo colonization ability of three probiotic lactic acid bacteria [viz., *Lactobacillus plantarum* Lp90, *L. plantarum* B2, and *Lactobacillus fermentum* PBCC11.5] was assessed in the gnotobiotic zebrafish larvae as the model system (Russo et al. 2015). These bacterial species were transformed with plasmid pRCR12, which contained the mCherry protein. Three days post-colonization bacterial displacement to the posterior intestinal tract was observed by fluorescence microscopy. The results indicated robust adhesion capability of the *L. plantarum* strains. The study provided proof of transient colonization of the intestine by probiotic microbes.

Under laboratory conditions, feeding crayfish with *Lactobacillus acidophilus* and *Lactobacillus plantarum*-containing diets did not significantly improve the animal weight but caused a shift in the bacterial taxa that were beneficial (Foysal et al. 2020). Certain parameters of the hemolymph and tail muscle composition improved. The probiotic diet caused histological changes (higher microvilli and better hepatopancreas) and enriched metabolic functions and genes associated with immune responses. Ghosh et al. (2021) reported the findings on microbiomes from three Indian Major Carps (IMCs), viz., rohu (*Labeo rohita*), catla (*Catla catla*), and mrigal (*Cirrhinus mrigala*), using culture-dependent methods. The study identified probiotic bacteria with *Bacillus licheniformis* as the commonest bacterium. Bacteria were studied for gut stability and ability to produce extracellular enzymes. Some strains, represented by *Bacillus* and *Stenotrophomonas*, showed γ-hemolysis activity but were experimentally proved as "safe" and could potentially be used as probiotics in carp culture.

The Picrust analysis of the data on the gut microbiota of European sea bass juveniles fed with microencapsulated organic acids revealed that organic acid in the feed induced pre-biotic properties that caused changes in the gut microbiota; and *Lactobacillus*, *Leuconostoc*, and *Bacillus* sp. were enriched (Busti et al. 2020). Recently, Villasante et al. (2022) proved that carbohydrate-to-protein ratio in the diet influenced growth performance, hepatic health, and dynamics of gut microbiota in Atlantic salmon (*Salmo salar*). During the fourth and eighth week, feeding carbohydrate-rich diets affected the beta diversity in the digesta of the distal intestine, which was not observed in the 12th week. According to the authors, more experiments were required to demonstrate the utility of probiotics (Lactic acid bacteria) in improving liver health.

Currently, aquaponics systems are being employed to produce vegetables and fish simultaneously. Wang et al. (2022) used 16S sequencing technology and bioinformatics analysis methods. They performed a comparative analysis of the intestinal microbiota in goldfish (*Carassius auratus* var. *Japonicus simplex*) and crucian carps (*Carassius auratus auratus*) in the presence of lettuce and water spinach cultivated in aquaponics and traditional farming. Interesting findings emerged with Proteobacteria dominating aquaponics and *Lactobacillaceae* predominant in traditional farming. In water spinach aquaponics, *Streptomyces* and *Rhizobium* were the biomarkers; and *Buttiauxella* were the probiotics in the lettuce aquaponics along with some PGPR bacteria. Experimental data indicated that aquaponics could cause changes in the structure and function of fish gut microbiota, and studies on aspects such as enhancement in fish survival rates should be undertaken on priority.

The lactic acid bacteria as probiotics in fish have been reviewed recently by Ringø et al. (2018). Another recent review considered the relevance of the host mucosal microbiota to act as probiotics to improve the natural defense systems in aquatic animals (Caipang et al. 2020). The use of allocthonous microbiota as a potential source of probiotics is not fully explored, and according to the available literature, these would be better than terrestrial microbes with probiotic characteristics. The role of probiotics in immunity, disease resistance, and adaptive immunity are well-established. The vital role in immunity parallels the fish and land vertebrates and hence needs detailed studies at a molecular level.

1.3.1.3 Antibiosis

The gut microbiota of fish plays an essential role in health. Recent findings indicate that competitive exclusion mechanisms come into play because of the commensal microbes. Furthermore, gut microbes also mediate the development and maturation of the fish immune system.

Ringø et al. (2006) showed that dietary changes affected the gut microbiota in studies using culture-based techniques supported by sequencing of the 16S rRNA genes. The GI tract of fish fed fish meal was dominated by *Brochothrix* and *Carnobacterium*; whereas *Chryseobacterium* spp., *Psychrobacter glacincola*, and *Carnobacterium* sp. dominated the digestive tract of fish receiving the soybean meal. The diet of bioprocessed soybean meal resulted in the dominance of *Psychrobacter*. Furthermore, the study identified several "new" bacterial species in the alimentary tract of Atlantic cod. Antagonistic activity of carnobacteria was observed where the pathogens (*Aeromonas salmonicida* subsp. *salmonicida* and *Vibrio anguillarum*) were inhibited. Gómez and Balcázar (2008) have extensively reviewed all studies relating to the installation of endogenous microflora and its contribution to fish health. More such studies are essential to develop therapeutic strategies to ensure disease prevention and treatment.

Tilapia hepcidin (TH)2–3 was found to function as an antimicrobial peptide and was expressed in yeast *Pichia pastoris*. This molecule was introduced as diet supplement in *Epinephelus lanceolatus* (Ting et al. 2019). With this, microbial diversity was enhanced, superoxide dismutase activity increased, and immunity improved, as evidenced by pathogen challenge. The change in the gut microbiota may have immunomodulatory effects on the grouper fish.

Juvenile *Barbonymus gonionotus* fed with diets containing chitosan at a dose of 1 g kg^{-1} feed, showed the abundance of lactic acid bacteria and inhibited the growth of pathogenic bacteria (Salam et al. 2021). Dietary chitosan supplementation improved muscle growth, nutrient, and mineral content and histological data indicated that the length of the intestinal villi increased. The kidney morphology was improved, and fat content in the liver decreased along with an improvement in muscle protein content. Thus, overall beneficial effects were observed due to the dietary inclusion of only one ingredient, chitosan.

1.3.2 Microbe–Host Interactions

1.3.2.1 Immunity

The importance of human gut microbiota in health and disease states is highlighted in several studies, and similar investigations were carried out on finfish. A detailed study involving the digestive tract of germ-free and conventionally raised zebrafish was carried out using the DNA microarray technique for comparing the expression of genes. The microbiota caused the regulation of 212 genes. Fifty-nine genes were conserved in the mouse intestine. The genes induced due to the microbiota stimulated epithelial proliferation, and nutrient metabolism was promoted and augmented the innate immune responses (Rawls et al. 2004). Pérez et al. (2010) reported the effects of bacteria on immune responses in teleost fish. The involvement of gut-associated lymphoid tissues (GALT) in discriminating pathogenic and commensal microbes is highlighted. The authors clearly stated that the intestinal microbiota positively supported the immune regulatory activities. A disturbed microbiota leads to disease development. Probiotics could help in restoring the disturbed microbiota.

In a study Galindo-Villegas and their team reported that the newly hatched zebrafish larvae were colonized by commensals, inducing neutrophils and causing the expression of pro-inflammatory and antiviral mediators, ultimately increasing the resistance to viral infection. Detailed studies indicated the involvement of TLR/MyD88 signaling pathway in microbe recognition and phagocytes as the source of the immune mediators. Furthermore, covalent modification of histone H3 at gene promoter occurred during the induction of pro-inflammatory and antiviral genes. Thus gene-specific modifications were associated with the protection of larvae (Galindo-Villegas et al. 2012). Compounds such as thymol, vanillin, and organic acids improved homeostasis and reduced inflammation. Thus, beneficial effects were observed when the sea bass juveniles were cultured under suboptimal conditions (Busti et al. 2020).

In a recirculating aquaculture system, a study on the effects of two experimental diets and a synthetic, casein-based reference diet on rainbow trout (*Oncorhynchus mykiss*) was carried out (Mansfield et al. 2010). *Carnobacterium maltaromaticum*, *Hafnia*, *Pseudomonas*, and *Aeromonas* were the most frequently observed sequences. The fish meal diet-associated microbiota consisted of *Gamma Proteobacteria* and *Firmicutes* in similar abundance, and the soybean meal diet-associated microbiota dominated by *C. maltaromaticum*. The inflammatory marker gene, viz., proliferating cell nuclear antigen (PCNA), was associated with the reference diet, and levels of interleukin-1 beta (IL-1β) indicated an association with the semi-purified reference diet. The researchers observed high variation between individuals.

The gut-associated lymphoid tissue is an essential component of the fish immune system. This tissue establishes the gut microbes and is involved in processing antigens and mounting an immune response against pathogens. These are important considerations while designing and administering oral immunoprophylactics and adjuvants in aquaculture (Lee et al. 2021). Xiong et al. (2019) have discussed fish diseases from the ecological perspective. Understanding the gut microbiota and its effect on immunity and capacity to resist evading pathogens is essential. Also, the role of commensals in resisting colonization by the pathogen and pathogen strategies to overcome the colonization are important to provide directions for future research. Challenge studies will be required in cases where the gut bacteria show antagonistic activities against fish pathogens.

1.3.2.2 Nutrition (Animal Productivity)

The review by Llewellyn et al. (2014) highlights the importance of research on microbiomes for teleost fish. This research is of utmost significance because sustainable approaches are required to improve yields in aquaculture. The global trends indicate that the demand for fish protein is increasing. Diet is one of the main factors affecting the gut microbiota, the complex community that contributes to host metabolism, nutrition, growth, and disease resistance (Rimoldi et al. 2018). Research on the effect of various diets on fish productivity, assessment of alternative diets, establishing the safety of diets, and interaction of dietary components on the microbiome is at the forefront.

The effects of plant ingredients such as peas, soybean, canola as a meal or concentrate were studied compared to the fish meal (FM) control diet. The study was carried out on rainbow trout (Desai et al. 2012). The profiles of microbes inhabiting the gut fed with the FM diet were observed to change with time, along with a reduction in the species richness and diversity. A higher *Firmicutes: Proteobacteria* ratio was observed in plant-based diets. The authors examined the microbiome using DGGE and pyrosequencing data and concluded that the intestinal microbiome contributed to the adverse health when the animals received plant food. Processing of the plant ingredients would ensure the gut microbiome stability for maintaining good health. Parma et al. (2016) carried out a next-generation sequencing-based characterization of the bacterial community in the gut of gilthead sea bream (*Sparus aurata*, L.). According to the authors, soybean meal could be successfully added to the diet at 300 g kg^{-1} and fish meal at 150 g kg^{-1}. The health in the growth stages is not affected by soybean meal, and Firmicutes dominate the gut, with *Lactobacillaceae* being the most represented family. The authors also performed histology to study diet-induced enteritidis but reported no adverse findings.

In a study, for 8 weeks, four common fishes, viz., grass carp (*Ctenopharyngodon idella*), bluntnose black bream (*Megalobrama amblycephala*), gibel carp (*Carassius gibelio*), and black carp (*Mylopharyngodon piceus*) were provided dried distiller's grain (DDG) as a supplement (He et al. 2013). This supplement is known for its high protein, fiber, and fat content. The fish species were analyzed using PCR-denaturing gradient gel electrophoresis (DGGE) technique, which revealed the presence of several common, and some unique bacteria in the gut. The supplementation of DDG altered the gut microbiota in a host-specific manner and the production (evidenced by the lower weight gain). In transgenic *Cyprinus carpio* L. (common carp) and the wild-type controls, overall dominance of *Proteobacteria, Fusobacteria, Bacteroidetes,* and *Firmicutes* was observed, but significant differences were noted. Carbohydrate metabolism was observed to increase in transgenic fish as compared to lipid metabolism. Based on the lower *Bacteroidetes: Firmicutes* ratio observed in transgenic carps, the authors observe a similarity in the gut microbiota of obese and lean individuals. The finding suggested that the presence of Firmicutes in transgenics is responsible for fast growth. However, the authors clearly state that the link between obesity and gut microbiota is very complex and beyond the ratio of *Bacteroidetes: Firmicutes* (Li et al. 2013). Since fast-growing transgenic fishes represent a food supplement, more studies on the association of gut microbiota and growth in fish are required.

The pyrosequencing of PCR-amplified 16S rRNA gene was used to assess the intestinal microbiota of European sea bass (*Dicentrarchus labrax*) fed two functional diets, which revealed *Dysgonomonas* (*Bacteroidetes*) and *Ralstonia* (*Betaproteobacteria*) as the dominant species. The fish fed with functional diets showed a decrease in the population of *Dysgonomonas* (Carda-Diéguez et al. 2014). These studies are required to determine the impact of diets on the autochthonous microbiota of the gut. Investigations on microbial ontogeny of the teleost microbiome will aid in manipulating the microbiome, ensuring greater productivity. Usually, certain substances are deliberately included in fish diets as immunostimulants. They can overcome the resistance due to antimicrobials/ chemotherapeutics used to tackle disease (Carda-Diéguez et al. 2014). The interdependence of diets providing distinct microbial communities or nutrients was analyzed in two fish, viz., stickleback and perch. The observation was that the animals fed with generalized diets were less diverse in natural and lab environments. Overall, the authors suggest that the non-additive interaction of dietary components influences gut microbial diversity (Bolnick et al. 2014).

The high-throughput 16S rRNA gene amplicon sequencing technique on MiSeq platform was utilized to study the gut microbial community profile in *Oncorhynchus mykiss* fed with poultry by-product meal (PBM). Higher *Firmicutes:Proteobacteria* ratio was obtained in plant-based food. PBM caused good growth but did not alter the richness of the gut microbial community, indicating a suitable source of dietary protein in trout (Rimoldi et al. 2018). Interpopulation variations were studied in the gut microbiomes of Atlantic salmon (Webster et al. 2018). The study carried out in seven populations revealed conserved core gut microbiota and skin microbiota (in terms of OTUs) across all populations. As expected, wild and captive populations differed with respect to their microflora, and hatchery populations had a lower diversity. According to the authors, diet contributed to a large difference in the microbiota between wild and captive fish populations. In a study, Villasante et al. (2019) reported that the effect of high dietary carbohydrate on the gut flora of Atlantic salmon was low, and the genus *Lactococcus* was present in more abundance. *Firmicutes, Actinobacteria,* and *Proteobacteria* were the major phyla determined by the high-throughput sequencing of the V4 region of the 16S rRNA gene.

In Atlantic cod, the studies on the effect of dietary supplementation with macroalgae (*Ascophyllum nodosum* or *Ulva rigida*) were carried out by Keating and group (2021), aiming to improve growth, health, and farm productivity. The macroalgae in the diet of juveniles did not cause any shift in the hindgut microbiota and caused no adverse impacts on fish health. The study also indicated the non-suitability of a diet containing 10% *Ascophyllum nodosum* in juvenile cod fish. Pérez-Pascual et al. (2020) report that 15% of fish diets can be replaced by proteins of vegetal origin, yeasts and insect larvae, cyanobacteria, or recycled processed poultry protein. Their studies proved that these are sustainable alternate sources that enhanced the growth of European sea bass (*Dicentrarchus labrax* L.). Upon supplementation, the feed conversion ratio was similar to commercial diet. The phyla *Proteobacteria, Firmicutes, Actinobacteria*, and *Bacteroidetes* were predominantly observed, and the species richness was not affected by the changed dietary composition. The study indicated adaptability to dietary change.

Wild fish-based meals are a requirement for farming carnivorous fish. However, it is realized that this practice is environmentally and economically unsustainable, and hence efforts are directed toward finding newer alternatives. Insect-based diets are considered a substitute for fish meal because of their high nutritional value and low environmental footprint (Rangel et al. 2022). In the experiments evaluating the impact of different insect meals on the composition and chitinolytic potential of the European sea bass gut and mucosa-associated bacterial communities, it was observed that the insect meals had immunomodulatory effects (Rangel et al. 2022). Feeding with black soldier fly larvae (*Hermetia illucens*, HM) and pupal exuviae (HEM) caused a shift in the microbial community, and an increase in the *Firmicutes* and *Actinobacteria* families (all taxa) was observed. These bacteria are considered beneficial for fish health. The mealworm larvae-based diet (*Tenebrio molitor*, TM) moderately impacted microbiota. The pupal exuviae (HEM) diets reported increased Paenibacillus and decreased Gammaproteobacteria and Bacteroidota members. These were associated with chitinolytic activity, as evidenced by the qPCR data. Thus, the waste product of insect rearing can be used as a functional feed in aquaculture.

The alterations in the gut physiology of wild Trinidadian guppies (*Poecilia reticulata*) fed with high- or low-quality diets were studied, and other parameters such as the gut enzyme activity, gut size, and mass were also evaluated in high- and low-predation habitats (Sullam et al. 2015). Findings of the study indicated differences in gut physiology with reference to the adaptation to habitats.

1.3.2.3 Health

Changes in Atlantic Salmon (Salmo salar) gut microbiota upon feeding fermented soybean meal were reported (Catalán et al. 2018). The authors used PCR-TTGE to study the microbiota, and gene expression analysis supported the community observation. According to the authors, the fish health and physiology improved due to the pre-biotic effects; mucin production was better, which had promotory effects on proximal gut health. The fermented diet also improved the trans-cellular uptake of water. The necessity of studies related to the various bioactive compounds derived from the feed on host health and growth were stated.

The effects of locally sourced feed supplements on genetically improved farmed tilapia (GIFT) production were studied vis-à-vis commercial diets (Parata et al. 2020). Such studies were explicitly performed to benefit the farmers with a low capital input. The fish fed with locally sourced feed was poor, as evidenced by their gut biota and other parameters. Thus, for ensuring good productivity, such studies showing the effects of diets on gut microbiota are essential.

The fish microbiome is closely related to the health of fish via the immunomodulatory effects, and it is known to be affected by various environmental stresses. Fecal samples and other non-invasive biomarker analyses can be carried out to identify the stressors rapidly. Thus, knowledge about the microbiome, its change (dysbiosis), and other biomarkers can help achieve the welfare of fish and maintain productivity (Perry et al. 2020).

1.4 Effect of Anthropogenic Substances on the Gut Microbiota of Finfish

The microbiota of finfish is responsive to several environmental factors. Xenobiotic compounds and substances of anthropogenic origin would certainly cause perturbations in the gut microbiome of fish. With the introduction of next-generation sequencing, the investigations can be carried out on a large scale, and results can lead to mitigation of the adverse effects caused and probably strategies to ensure less/no harm to the ecosystem. Various nanomaterials introduced in the environment are likely to enter the aquatic ecosystems, and the toxicological profiling of these materials can be generated by assessment of the gut microbiome of aquatic species. Recent papers have proved the consequences of engineered nanoparticles, microplastics, etc. in fish (Saggese et al. 2016; Chupani et al. 2019; Aleshina et al. 2020; Sherif et al. 2021; Chen et al. 2022).

Higuera-Llantén et al. (2018) have conclusively established that the indiscriminate use of florfenicol and oxytetracycline has caused the selection of multidrug-resistant gut microflora in the farmed *Salmo salar*. Most of the intestinal bacteria harbored antibiotic resistance genes. The MIC values reported were 2048 and 1024 μg/mL for florfenicol and oxytetracycline respectively. Antibiotics have impacted the production as well as the environment.

Anthropogenic changes affect gut microbial communities as evidenced by the findings on the gut microbiome of *Stegastes nigricans* fed with algae (Degregori et al. 2021). The hind gut and foregut microbiome had a higher alpha diversity in nutrient-enriched territories compared to controls.

The microbiota of several fishes has been sequenced, and its involvement in the fish health is established. However, the microbiota could have pathogens which could proliferate during the various stages after capture and even storage. These pathogens will pose severe problems to human consumers, so it is very pertinent to characterize the microbes associated with stored products. This work was attempted by Shehata et al. (2020) for commercially traded fillets. According to the authors, such studies are essential as they would provide information on food safety, authenticity, and traceability. In another interesting study by Thomas and group (Thomas et al. 2021) it was observed that during modified atmospheric packaging and storage, the families such as *Comamonadaceae* which were found to constitute the skin and intestine of gilthead sunbream were persistent. Storage led to a change of families from *Comamonadaceae, Enterobacteriaceae,* and *Moraxellaceae* in the skin and *Comamonadaceae, Anaplasmataceae, Bacillaceae,* and *Enterobacteriaceae* in the intestine to the dominance of *Pseudoalteromonadaceae, Psychromonadaceae*, and *Shewanellaceae* which are psychotropic and psychrophilic families. Moreover, the study indicated that the packaging retained freshness, improved quality attributes, and modified microflora contributed to the extended product shelf-life.

1.5 Importance of the Fish Microbiome

The United Nations 2030 Agenda for Sustainable Development aims to end hunger, achieve food security, and improve nutrition. All these will require substantial contribution from aquaculture, especially the capture fisheries. Fish are considered a valued protein supplement contributing to food security. The preceding sections have emphasized the strong connection between the finfish gut microbiota and its host concerning metabolism, immunity, energy utilization, and maintenance of good health (Li et al. 2013; Wang et al. 2018). Considering the vast diversity of fish that can be consumed, only a few species have been explored with respect to their gut microbiome diversity. With more information on the microbiome of teleost fish, several themes for research in aquaculture can be defined, focussing on addressing issues such as protein malnutrition by supplementation with fish protein (Llewellyn et al. 2014). The studies on the gut microbiome and host immune system, metabolism, and physiology will help manipulate the functional capacity, resulting in sustainable aquaculture (Perry et al. 2020). Diwan et al. (2022) also elucidated the importance of studying the fish and shellfish microbiome. The authors emphasized the need for such research to improve our understanding of fish health, immune system, metabolism, probiotic research, etc. Such studies also provide leads to deriving new biomolecules with functional properties, which can be helpful for developing new drugs, pharmaceuticals, etc. Luna et al. (2022) reviewed the biotechnological aspects of research on the fish microbiota. The authors mainly discussed the strategies employed to manipulate the microbiota to improve fish health, food safety, and the environmental sustainability of aquaculture.

1.6 Conclusion

More studies on the finfish microbiome are required because fishes are the most diverse vertebrates. Given the diversity of habitats, detailed investigations on the composition, diversity, and function of microbiota associated with specific niches in fish bodies such as intestines, gills, and skin have to be carried out for all fishes and aquatic forms that are consumed. Wild fishes cannot be left unstudied as water is the continuum and several studies demonstrate its effect on the intestinal microbiota. Identifying microbes/communities specific to fish will provide cues to the development of the immune system, the ability of the host to combat pathogens, and above all, will be a potential source of hitherto unknown enzymes, metabolites, and bioactive compounds. It is realized that all the efforts at sequencing the gut microbiome have used several technologies and platforms; however, emerging data is often non-uniform and needs to be supplemented with more and more studies. Using a single technology for microbiome studies will not be possible, but collaborative efforts would be essential. More efforts are required for a functional understanding of the microbiome. The improved understanding of the probiotics, prebiotics, and immunity in the context of the fish feed/diet will eventually lead to a decrease in antibiotic usage and alleviate the burgeoning problem of drug resistance. Directed attempts at manipulating the microbiome will have to be based on reproducible data, which will ensure "healthy" fish with high productivity. Well-characterized host and their associated microbiota can be helpful as model systems to study various metabolic diseases, including obesity, diabetes, etc. In the near future, several researchers across the globe could take up concerted efforts and explore the finfish microbiome.

References

Aleshina E, Miroshnikova E, Sizova E (2020) Transformation of microbiota of fish intestines and gills against the background of molybdenum oxide nanoparticles in environment. Int J Environ Sci Technol 17(2):721–732. https://doi.org/10.1007/s13762-019-02509-x

Austin B (2006) The bacterial microflora of fish, revised. Sci World J 6:931–945. https://doi.org/10.1100/tsw.2006.181

Baldo L, Santos ME, Salzburger W (2011) Comparative transcriptomics of eastern African cichlid fishes shows signs of positive selection and a large contribution of untranslated regions to genetic diversity. Genome Biol Evol 3:443–455. https://doi.org/10.1093/gbe/evr047

Behera BK, Dehury B, Rout AK, Patra B, Mantri N, Chakraborty HJ, Sarkar DJ, Kaushik NK, Bansal V, Singh I, Das BK, Rao AR, Rai A (2021) Metagenomics study in aquatic resource management: recent trends, applied methodologies and future needs. Gene Rep 25:101372. https://doi.org/10.1016/j.genrep.2021.101372

Berg G, Rybakova D, Fischer D, Cernava T, Vergès MC, Charles T, Chen X, Cocolin L, Eversole K, Corral GH, Kazou M, Kinkel L, Lange L, Lima N, Loy A, Macklin JA, Maguin E, Mauchline T, McClure R, Mitter B, Schloter M (2020) Microbiome definition re-visited: old concepts and new challenges. Microbiome 8(1):103. https://doi.org/10.1186/s40168-020-00875-0

Bolnick DI, Snowberg LK, Hirsch PE, Lauber CL, Knight R, Caporaso JG, Svanbäck R (2014) Individuals' diet diversity influences gut microbial diversity in two freshwater fish (three spine stickleback and Eurasian perch). Ecol Lett 17(8):979–987. https://doi.org/10.1111/ele.12301

Busti S, Rossi B, Volpe E, Ciulli S, Piva A, D'Amico F, Soverini M, Candela M, Gatta PP, Bonaldo A, Grilli E, Parma L (2020) Effects of dietary organic acids and nature identical compounds on growth, immune parameters and gut microbiota of European sea bass. Sci Rep 10(1):21321. https://doi.org/10.1038/s41598-020-78441-9

Caipang CMA, Suharman I, Avillanosa AL, Bargoyo VT (2020) Host-derived probiotics for finfish aquaculture. In: IOP conference series: earth and environmental science, vol 430. IOP Publishing, Bristol, p 012026. https://doi.org/10.1088/1755-1315/430/1/012026

Cantas L, Sørby JR, Aleström P, Sørum H (2012) Culturable gut microbiota diversity in zebrafish. Zebrafish 9(1):26–37. https://doi.org/10.1089/zeb.2011.0712

Carda-Diéguez M, Mira A, Fouz B (2014) Pyrosequencing survey of intestinal microbiota diversity in cultured sea bass (*Dicentrarchus labrax*) fed functional diets. FEMS Microbiol Ecol 87(2): 451–459. https://doi.org/10.1111/1574-6941.12236

Catalán N, Villasante A, Wacyk J, Ramírez C, Romero J (2018) Fermented soybean meal increases lactic acid bacteria in gut microbiota of Atlantic salmon (*Salmo salar*). Probiotics Antimicrob Proteins 10(3):566–576. https://doi.org/10.1007/s12602-017-9366-7

Cheesman SE, Neal JT, Mittge E, Seredick BM, Guillemin K (2011) Epithelial cell proliferation in the developing zebrafish intestine is regulated by the Wnt pathway and microbial signaling via Myd88. Proc Natl Acad Sci 108(supplement_1):4570–4577. https://doi.org/10.1073/pnas.1000072107

Chen P, Huang J, Rao L, Zhu W, Yu Y, Xiao F, Yu H, Wu Y, Hu R, Liu X, He Z, Yan Q (2022) Environmental effects of nanoparticles on the ecological succession of gut microbiota across zebrafish development. Sci Total Environ 806(Pt 4):150963. https://doi.org/10.1016/j.scitotenv.2021.150963

Chupani L, Barta J, Zuskova E (2019) Effects of food-borne ZnO nanoparticles on intestinal microbiota of common carp (*Cyprinus carpio* L.). Environ Sci Pollut Res Int 26(25): 25869–25873. https://doi.org/10.1007/s11356-019-05616-x

Cullen CM, Aneja KK, Beyhan S, Cho CE, Woloszynek S, Convertino M, McCoy SJ, Zhang Y, Anderson MZ, Alvarez-Ponce D, Smirnova E, Karstens L, Dorrestein PC, Li H, Sen Gupta A, Cheung K, Powers JG, Zhao Z, Rosen GL (2020) Emerging priorities for microbiome research. Front Microbiol 11:136. https://doi.org/10.3389/fmicb.2020.00136

de Borba Gurpilhares D, Cinelli LP, Simas NK, Pessoa A Jr, Sette LD (2019) Marine prebiotics: polysaccharides and oligosaccharides obtained by using microbial enzymes. Food Chem 280: 175–186. https://doi.org/10.1016/j.foodchem.2018.12.023

de Souza FP, de Lima ECS, Pandolfi VCF, Leite NG, Furlan-Murari PJ, Leal CNS, Mainardi RM, Suphoronski SA, Favero LM, Koch JFA, Pauda UDP, Lopera-Barrero NM (2020) Effect of β-glucan in water on growth performance, blood status and intestinal microbiota in tilapia under hypoxia. Aquac Rep 17:100369. https://doi.org/10.1016/j.aqrep.2020.100369

Degregori S, Casey JM, Barber PH (2021) Nutrient pollution alters the gut microbiome of a territorial reef fish. Mar Pollut Bull 169:112522. https://doi.org/10.1016/j.marpolbul.2021.112522

Desai AR, Links MG, Collins SA, Mansfield GS, Drew MD, Van Kessel AG, Hill JE (2012) Effects of plant-based diets on the distal gut microbiome of rainbow trout (*Oncorhynchus mykiss*). Aquaculture 350:134–142. https://doi.org/10.1016/j.aquaculture.2012.04.005

Dierckens K, Rekecki A, Laureau S, Sorgeloos P, Boon N, Van den Broeck W, Bossier P (2009) Development of a bacterial challenge test for gnotobiotic sea bass (*Dicentrarchus labrax*) larvae. Environ Microbiol 11(2):526–533. https://doi.org/10.1111/j.1462-2920.2008.01794.x

Diwan AD, Harke SN, Gopalkrishna, Panche AN (2022) Aquaculture industry prospective from gut microbiome of fish and shellfish: an overview. J Anim Physiol Anim Nutr 106(2):441–469. https://doi.org/10.1111/jpn.13619

Eddy SD, Jones SH (2002) Microbiology of summer flounder *Paralichthys dentatus* fingerling production at a marine fish hatchery. Aquaculture 211(1–4):9–28. https://doi.org/10.1016/S0044-8486(01)00882-1

Egerton S, Culloty S, Whooley J, Stanton C, Ross RP (2018) The gut microbiota of marine fish. Front Microbiol 9:873. https://doi.org/10.3389/fmicb.2018.00873

FAO (2018). https://www.fao.org/3/I9553EN/i9553en.pdf

Foysal MJ, Fotedar R, Siddik M, Tay A (2020) *Lactobacillus acidophilus* and *L. plantarum* improve health status, modulate gut microbiota and innate immune response of marron (*Cherax cainii*). Sci Rep 10(1):5916. https://doi.org/10.1038/s41598-020-62655-y

Galindo-Villegas J, García-Moreno D, de Oliveira S, Meseguer J, Mulero V (2012) Regulation of immunity and disease resistance by commensal microbes and chromatin modifications during zebrafish development. Proc Natl Acad Sci U S A 109(39):E2605–E2614. https://doi.org/10.1073/pnas.1209920109

Ganguly S, Prasad A (2012) Microflora in fish digestive tract plays significant role in digestion and metabolism. Rev Fish Biol Fish 22(1):11–16. https://doi.org/10.1007/s11160-011-9214-x

Geraylou Z, Souffreau C, Rurangwa E, Maes GE, Spanier KI, Courtin CM, Delcour JA, Buyse J, Ollevier F (2013) Pre-biotic effects of arabinoxylan oligosaccharides on juvenile Siberian sturgeon (*Acipenser baerii*) with emphasis on the modulation of the gut microbiota using 454 pyrosequencing. FEMS Microbiol Ecol 86(2):357–371. https://doi.org/10.1111/1574-6941.12169

Ghanbari M, Kneifel W, Domig KJ (2015) A new view of the fish gut microbiome: advances from next-generation sequencing. Aquaculture 448:464–475. https://doi.org/10.1016/j.aquaculture.2015.06.033

Ghosh K, Mukherjee A, Dutta D, Banerjee S, Breines EM, Hareide E, Ringø E (2021) Endosymbiotic pathogen-inhibitory gut bacteria in three Indian major carps under polyculture system: a step toward making a probiotics consortium. Aquac Fish 6(2):192–204. https://doi.org/10.1016/j.aaf.2020.03.009

Gómez GD, Balcázar JL (2008) A review on the interactions between gut microbiota and innate immunity of fish. FEMS Immunol Med Microbiol 52(2):145–154. https://doi.org/10.1111/j.1574-695X.2007.00343.x

Green TJ, Smullen R, Barnes AC (2013) Dietary soybean protein concentrate-induced intestinal disorder in marine farmed Atlantic salmon, *Salmo salar* is associated with alterations in gut microbiota. Vet Microbiol 166(1–2):286–292. https://doi.org/10.1016/j.vetmic.2013.05.009

Han S, Liu Y, Zhou Z, He S, Cao Y, Shi P, Yao B, Ringø E (2010) Analysis of bacterial diversity in the intestine of grass carp (*Ctenopharyngodon idellus*) based on 16S rDNA gene sequences. Aquac Res 42(1):47–56. https://doi.org/10.1111/j.1365-2109.2010.02543.x

Hansen GH, Olafsen JA (1989) Bacterial colonization of cod (*Gadus morhua* L.) and halibut (*Hippoglossus hippoglossus*) eggs in marine aquaculture. Appl Environ Microbiol 55(6):1435–1446. https://doi.org/10.1128/aem.55.6.1435-1446.1989

He S, Wu Z, Liu Y, Wu N, Tao Y, Xu L, Zhou Z, Yao B, Ringø E (2013) Effects of dietary 60 g kg− 1 dried distiller's grains in least-cost practical diets on production and gut allochthonous bacterial composition of cage-cultured fish: comparison among fish species with different natural food habits. Aquac Nutr 19(5):765–772. https://doi.org/10.1111/anu.12023

Higuera-Llantén S, Vásquez-Ponce F, Barrientos-Espinoza B, Mardones FO, Marshall SH, Olivares-Pacheco J (2018) Extended antibiotic treatment in salmon farms select multiresistant gut bacteria with a high prevalence of antibiotic resistance genes. PLoS One 13(9):e0203641. https://doi.org/10.1371/journal.pone.0203641

Horsley R (1977) A review of the bacterial flora of teleosts and elasmobranchs, including methods for its analysis. J Fish Biol 10:529–553. https://doi.org/10.1111/j.1095-8649.1977.tb04086.x

Hoseinifar SH, Esteban MÁ, Cuesta A, Sun YZ (2015) Prebiotics and fish immune response: a review of current knowledge and future perspectives. Rev Fish Sci Aquac 23(4):315–328. https://doi.org/10.1080/23308249.2015.1052365

Hoseinifar SH, Doan HV, Dadar M, Ringø E, Harikrishnan R (2019) Feed additives, gut microbiota, and health in finfish aquaculture. In: Microbial communities in aquaculture ecosystems. Springer, Cham, pp 121–142. https://doi.org/10.1007/978-3-030-16190-3_6

Hovda MB, Lunestad BT, Fontanillas R, Rosnes JT (2007) Molecular characterisation of the intestinal microbiota of farmed Atlantic salmon (*Salmo salar* L.). Aquaculture 272(1–4):581–588. https://doi.org/10.1016/j.aquaculture.2007.08.045

Hua K, Cobcroft JM, Cole A, Condon K, Jerry DR, Mangott A, Praeger C, Vucko MJ, Zeng C, Zenger K, Strugnell JM (2019) The future of aquatic protein: implications for protein sources in aquaculture diets. One Earth 1(3):316–329. https://doi.org/10.1016/j.oneear.2019.10.018

Ingerslev HC, Strube ML, von Gersdorff Jørgensen L, Dalsgaard I, Boye M, Madsen L (2014) Diet type dictates the gut microbiota and the immune response against Yersinia ruckeri in rainbow trout (*Oncorhynchus mykiss*). Fish Shellfish Immunol 40(2):624–633. https://doi.org/10.1016/j.fsi.2014.08.021

Kanther M, Sun X, Mühlbauer M, Mackey LC, Flynn EJ, Bagnat M, Jobin C, Rawls JF (2011) Microbial colonization induces dynamic temporal and spatial patterns of NF-κB activation in the

zebrafish digestive tract. Gastroenterology 141(1):197–207. https://doi.org/10.1053/j.gastro.2011.03.042
Keating C, Bolton-Warberg M, Hinchcliffe J, Davies R, Whelan S, Wan A, Fitzgerald RD, Davies SJ, Ijaz UZ, Smith CJ (2021) Temporal changes in the gut microbiota in farmed Atlantic cod (*Gadus morhua*) outweigh the response to diet supplementation with macroalgae. Anim Microbiome 3(1):7. https://doi.org/10.1186/s42523-020-00065-1
Kim DH, Brunt J, Austin B (2007) Microbial diversity of intestinal contents and mucus in rainbow trout (Oncorhynchus mykiss). J Appl Microbiol 102(6):1654–1664. https://doi.org/10.1111/j.1365-2672.2006.03185.x
Kuang T, He A, Lin Y, Huang X, Liu L, Zhou L (2020) Comparative analysis of microbial communities associated with the gill, gut, and habitat of two filter-feeding fish. Aquac Rep 18:100501. https://doi.org/10.1016/j.aqrep.2020.100501
Kühlwein H, Emery MJ, Rawling MD, Harper GM, Merrifield DL, Davies SJ (2013) Effects of a dietary β-(1,3)(1,6)-D-glucan supplementation on intestinal microbial communities and intestinal ultrastructure of mirror carp (*Cyprinus carpio* L.). J Appl Microbiol 115(5):1091–1106. https://doi.org/10.1111/jam.12313
Larsen AM, Mohammed HH, Arias CR (2014) Characterization of the gut microbiota of three commercially valuable warm water fish species. J Appl Microbiol 116(6):1396–1404. https://doi.org/10.1111/jam.12475
Lee PT, Yamamoto FY, Low CF, Loh JY, Chong CM (2021) Gut immune system and the implications of oral-administered immunoprophylaxis in finfish aquaculture. Front Immunol 12:773193. https://doi.org/10.3389/fimmu.2021.773193
Legrand T, Wynne JW, Weyrich LS, Oxley A (2020) Investigating both mucosal immunity and microbiota in response to gut enteritis in yellowtail kingfish. Microorganisms 8(9):1267. https://doi.org/10.3390/microorganisms8091267
Li X, Yan Q, Xie S, Hu W, Yu Y, Hu Z (2013) Gut microbiota contributes to the growth of fast-growing transgenic common carp (Cyprinus carpio L.). PLoS One 8(5):e64577. https://doi.org/10.1371/journal.pone.0064577
Lindsay GJH, Gooday GW (1985) Chitinolytic enzymes and the bacterial microflora in the digestive tract of cod, *Gadus morhua*. J Fish Biol 26(3):255–265. https://doi.org/10.1111/j.1095-8649.1985.tb04264.x
Llewellyn MS, Boutin S, Hoseinifar SH, Derome N (2014) Teleost microbiomes: the state of the art in their characterization, manipulation and importance in aquaculture and fisheries. Front Microbiol 5:207. https://doi.org/10.3389/fmicb.2014.00207
Luna GM, Quero GM, Kokou F, Kormas K (2022) Time to integrate biotechnological approaches into fish gut microbiome research. Curr Opin Biotechnol 73:121–127. https://doi.org/10.1016/j.copbio.2021.07.018
MacDonald N, Stark J, Austin B (1986) Bacterial microflora in the gastrointestinal tract of Dover sole (*Solea solea* L.), with emphasis on the possible role of bacteria in the nutrition of the host. FEMS Microbiol Lett 35:107–111. https://doi.org/10.1111/j.1574-6968.1986.tb01508.x
Mansfield GS, Desai AR, Nilson SA, Van Kessel AG, Drew MD, Hill JE (2010) Characterization of rainbow trout (*Oncorhynchus mykiss*) intestinal microbiota and inflammatory marker gene expression in a recirculating aquaculture system. Aquaculture 307(1–2):95–104. https://doi.org/10.1016/j.aquaculture.2010.07.014
McMurtrie J, Alathari S, Chaput DL, Bass D, Ghambi C, Nagoli J, Deboitteville JD, Mohan CD, Cable J, Temperton B, Tyler CR (2022) Relationships between pond water and tilapia skin microbiomes in aquaculture ponds in Malawi. Aquaculture 558:738367. https://doi.org/10.1101/2021.12.06.470702
Merrifield DL, Rodiles A (2015) The fish microbiome and its interactions with mucosal tissues. In: Mucosal health in aquaculture. Academic, Cambridge, MA, pp 273–295. https://doi.org/10.1016/B978-0-12-417186-2.00010-8

Navarrete P, Espejo RT, Romero J (2009) Molecular analysis of microbiota along the digestive tract of juvenile Atlantic salmon (*Salmo salar* L.). Microb Ecol 57(3):550–561. https://doi.org/10.1007/s00248-008-9448-x

Navarrete P, Magne F, Araneda C, Fuentes P, Barros L, Opazo R, Espejo R, Romero J (2012) PCR-TTGE analysis of 16S rRNA from rainbow trout (*Oncorhynchus mykiss*) gut microbiota reveals host-specific communities of active bacteria. PLoS One 7(2):e31335. https://doi.org/10.1371/journal.pone.0031335

Nayak SK (2010) Role of gastrointestinal microbiota in fish. Aquacult Res 41:1553–1573. https://doi.org/10.1111/j.1365-2109.2010.02546.x

Nissa MU, Pinto N, Parkar H, Goswami M, Srivastava S (2021) Proteomics in fisheries and aquaculture: an approach for food security. Food Control 127:108125. https://doi.org/10.1016/j.foodcont.2021.108125

Parata L, Mazumder D, Sammut J, Egan S (2020) Diet type influences the gut microbiome and nutrient assimilation of genetically improved farmed tilapia (*Oreochromis niloticus*). PLoS One 15(8):e0237775. https://doi.org/10.1371/journal.pone.0237775

Parma L, Candela M, Soverini M, Turroni S, Consolandi C, Brigidi P, Mandrioli L, Sirri R, Fontanillas R, Gatta PP, Bonaldo A (2016) Next-generation sequencing characterization of the gut bacterial community of gilthead sea bream (*Sparus aurata* L.) fed low fishmeal based diets with increasing soybean meal levels. Anim Feed Sci Technol 222:204–216. https://doi.org/10.1016/j.anifeedsci.2016.10.022

Pérez T, Balcázar JL, Ruiz-Zarzuela I, Halaihel N, Vendrell D, De Blas I, Múzquiz JL (2010) Host–microbiota interactions within the fish intestinal ecosystem. Mucosal Immunol 3(4):355–360

Pérez-Pascual D, Estellé J, Dutto G, Rodde C, Bernardet JF, Marchand Y, Duchaud E, Przybyla C, Ghigo JM (2020) Growth performance and adaptability of European sea bass (Dicentrarchus labrax) gut microbiota to alternative diets free of fish products. Microorganisms 8(9):1346. https://doi.org/10.3390/microorganisms8091346

Perry WB, Lindsay E, Payne CJ, Brodie C, Kazlauskaite R (2020) The role of the gut microbiome in sustainable teleost aquaculture. Proc R Soc B 287(1926):20200184. https://doi.org/10.1098/rspb.2020.0184

Pham LN, Kanther M, Semova I, Rawls JF (2008) Methods for generating and colonizing gnotobiotic zebrafish. Nat Protoc 3(12):1862–1875. https://doi.org/10.1038/nprot.2008.186

Pond MJ, Stone DM, Alderman DJ (2006) Comparison of conventional and molecular techniques to investigate the intestinal microflora of rainbow trout (*Oncorhynchus mykiss*). Aquaculture 261(1):194–203. https://doi.org/10.1016/j.aquaculture.2006.06.037

Porter D, Peggs D, McGurk C, Martin S (2022) Immune responses to prebiotics in farmed salmonid fish: how transcriptomic approaches help interpret responses. Fish Shellfish Immunol 127:35–47. https://doi.org/10.1016/j.fsi.2022.05.055

Rangel F, Enes P, Gasco L, Gai F, Hausmann B, Berry D, Oliva-Teles A, Serra CR, Pereira FC (2022) Differential modulation of the european sea bass gut microbiota by distinct insect meals. Front Microbiol 13:831034. https://doi.org/10.3389/fmicb.2022.831034

Rawls JF, Samuel BS, Gordon JI (2004) Gnotobiotic zebrafish reveal evolutionarily conserved responses to the gut microbiota. Proc Natl Acad Sci U S A 101(13):4596–4601. https://doi.org/10.1073/pnas.0400706101

Rawls JF, Mahowald MA, Goodman AL, Trent CM, Gordon JI (2007) *In vivo* imaging and genetic analysis link bacterial motility and symbiosis in the zebrafish gut. Proc Natl Acad Sci U S A 104(18):7622–7627. https://doi.org/10.1073/pnas.0702386104

Reid HI, Treasurer JW, Adam B, Birkbeck TH (2009) Analysis of bacterial populations in the gut of developing cod larvae and identification of *vibrio logei*, *vibrio anguillarum* and *Vibrio splendidus* as pathogens of cod larvae. Aquaculture 288(1–2):36–43. https://doi.org/10.1016/j.aquaculture.2008.11.022

Rendueles O, Ferrières L, Frétaud M, Bégaud E, Herbomel P, Levraud JP, Ghigo JM (2012) A new zebrafish model of oro-intestinal pathogen colonization reveals a key role for adhesion in protection by probiotic bacteria. PLoS Pathog 8(7):e1002815. https://doi.org/10.1371/journal.ppat.1002815

Rieu A, Aoudia N, Jego G, Chluba J, Yousfi N, Briandet R, Deschamps J, Gasquet B, Monedero V, Garrido C, Guzzo J (2014) The biofilm mode of life boosts the anti-inflammatory properties of *lactobacillus*. Cell Microbiol 16(12):1836–1853. https://doi.org/10.1111/cmi.12331

Rimoldi S, Terova G, Ascione C, Giannico R, Brambilla F (2018) Next generation sequencing for gut microbiome characterization in rainbow trout (*Oncorhynchus mykiss*) fed animal by-product meals as an alternative to fishmeal protein sources. PLoS One 13(3):e0193652. https://doi.org/10.1371/journal.pone.0193652

Ringø E, Sperstad S, Myklebust R, Refstie S, Krogdahl Å (2006) Characterisation of the microbiota associated with intestine of Atlantic cod (*Gadus morhua* L.): the effect of fish meal, standard soybean meal and a bioprocessed soybean meal. Aquaculture 261(3):829–841. https://doi.org/10.1016/j.aquaculture.2006.06.030

Ringø E, Sperstad S, Kraugerud OF, Krogdahl Å (2008) Use of 16S rRNA gene sequencing analysis to characterize culturable intestinal bacteria in Atlantic salmon (*Salmo salar*) fed diets with cellulose or non-starch polysaccharides from soy. Aquac Res 39(10):1087–1100. https://doi.org/10.1111/j.1365-2109.2008.01972.x

Ringø E, Hoseinifar SH, Ghosh K, Doan HV, Beck BR, Song SK (2018) Lactic acid bacteria in finfish—an update. Front Microbiol 9:1818. https://doi.org/10.3389/fmicb.2018.01818

Russo P, Iturria I, Mohedano ML, Caggianiello G, Rainieri S, Fiocco D, Angel Pardo M, López P, Spano G (2015) Zebrafish gut colonization by mCherry-labelled lactic acid bacteria. Appl Microbiol Biotechnol 99(8):3479–3490. https://doi.org/10.1007/s00253-014-6351-x

Saggese I, Sarà G, Dondero F (2016) Silver nanoparticles affect functional bioenergetic traits in the invasive red sea mussel *Brachidontes pharaonis*. Biomed Res Int 2016:1872351. https://doi.org/10.1155/2016/1872351

Salam MA, Rahman MA, Paul SI, Islam F, Barman AK, Rahman Z, Shaha DC, Rahman MM, Islam T (2021) Dietary chitosan promotes the growth, biochemical composition, gut microbiota, hematological parameters and internal organ morphology of juvenile *Barbonymus gonionotus*. PLoS One 16(11):e0260192. https://doi.org/10.1371/journal.pone.0260192

Semova I, Carten JD, Stombaugh J, Mackey LC, Knight R, Farber SA, Rawls JF (2012) Microbiota regulate intestinal absorption and metabolism of fatty acids in the zebrafish. Cell Host Microbe 12(3):277–288. https://doi.org/10.1016/j.chom.2012.08.003

Shehata HR, Mitterboeck TF, Hanner R (2020) Characterization of the microbiota of commercially traded finfish fillets. Food Res Int 137:109373. https://doi.org/10.1016/j.foodres.2020.109373

Sherif, AH, El-Sharawy, MES, El-Samannoudy, SI, Adel Seida, A, Sabry, NM, Eldawoudy, M, Abdelsalam, M, Younis, NA (2021). The deleterious impacts of dietary titanium dioxide nanoparticles on the intestinal microbiota, antioxidant enzymes, diseases resistances and immune response of Nile tilapia. *Aquaculture Research*, 52(12): pp.6699–6707

Skrodenyte-Arbaciauskiene V, Sruoga A, Butkauskas D (2006) Assessment of microbial diversity in the river trout *Salmo trutta fario* L. intestinal tract identified by partial 16S rRNA gene sequence analysis. Fish Sci 72:597–602. https://doi.org/10.1111/j.1444-2906.2006.01189.x

Skrodenytė-Arbačiauskienė V, Sruoga A, Butkauskas D, Skrupskelis K (2008) Phylogenetic analysis of intestinal bacteria of freshwater salmon *Salmo salar* and sea trout *Salmo trutta trutta* and diet. Fish Sci 74(6):1307–1314. https://doi.org/10.1111/j.1444-2906.2008.01656.x

Smriga S, Sandin SA, Azam F (2010) Abundance, diversity, and activity of microbial assemblages associated with coral reef fish guts and feces. FEMS Microbiol Ecol 73(1):31–42. https://doi.org/10.1111/j.1574-6941.2010.00879.x

Star B, Haverkamp TH, Jentoft S, Jakobsen KS (2013) Next generation sequencing shows high variation of the intestinal microbial species composition in Atlantic cod caught at a single location. BMC Microbiol 13:248. https://doi.org/10.1186/1471-2180-13-248

Sullam KE, Dalton CM, Russell JA, Kilham SS, El-Sabaawi R, German DP, Flecker AS (2015) Changes in digestive traits and body nutritional composition accommodate a trophic niche shift in Trinidadian guppies. Oecologia 177(1):245–257. https://doi.org/10.1007/s00442-014-3158-5

Sun Y, Yang H, Ling Z, Chang J, Ye J (2009) Gut microbiota of fast and slow growing grouper *Epinephelus coioides*. Afr J Microbiol Res 3(11):637–640

Terova G, Rimoldi S, Ascione C, Gini E, Ceccotti C, Gasco L (2019) Rainbow trout (Oncorhynchus mykiss) gut microbiota is modulated by insect meal from Hermetia illucens prepupae in the diet. Rev Fish Biol Fish 29(2):465–486. https://doi.org/10.1007/s11160-019-09558-y

Thomas A, Konteles SJ, Ouzounis S, Papatheodorou S, Tsakni A, Houhoula D, Tsironi T (2021) Bacterial community in response to packaging conditions in farmed gilthead seabream. Aquac Fish 8:410. https://doi.org/10.1016/j.aaf.2021.09.002

Ting CH, Pan CY, Chen YC, Lin YC, Chen TY, Rajanbabu V, Chen JY (2019) Impact of tilapia hepcidin 2-3 dietary supplementation on the gut microbiota profile and immunomodulation in the grouper (*Epinephelus lanceolatus*). Sci Rep 9(1):19047. https://doi.org/10.1038/s41598-019-55509-9

Toh MC, Goodyear M, Daigneault M, Allen-Vercoe E, Van Raay TJ (2013) Colonizing the embryonic zebrafish gut with anaerobic bacteria derived from the human gastrointestinal tract. Zebrafish 10(2):194–198. https://doi.org/10.1089/zeb.2012.0814

Tsuchiya C, Sakata T, Sugita H (2008) Novel ecological niche of *Cetobacterium somerae*, an anaerobic bacterium in the intestinal tracts of freshwater fish. Lett Appl Microbiol 46(1):43–48. https://doi.org/10.1111/j.1472-765X.2007.02258.x

Villasante A, Ramírez C, Catalán N, Opazo R, Dantagnan P, Romero J (2019) Effect of dietary carbohydrate-to-protein ratio on gut microbiota in Atlantic salmon (*Salmo salar*). Animals 9(3): 89. https://doi.org/10.3390/ani9030089

Villasante A, Ramírez C, Rodríguez H, Dantagnan P, Hernández A, Figueroa E, Romero J (2022) Dietary carbohydrate-to-protein ratio influences growth performance, hepatic health and dynamic of gut microbiota in Atlantic salmon (*Salmo salar*). Anim Nutr 10:261–279. https://doi.org/10.1016/j.aninu.2022.04.003

Wang AR, Ran C, Ringø E, Zhou ZG (2018) Progress in fish gastrointestinal microbiota research. Rev Aquac 10(3):626–640. https://doi.org/10.1111/raq.12191

Wang SX, Zhang JY, Du XK, Liu DJ, Liu LX, Shen XH (2022) Comparative analysis of the intestinal microbiota in goldfish and crucian carps between different aquaponics and traditional farming. Aquac Rep 25:101240. https://doi.org/10.3389/fmicb.2021.760266

Ward NL, Steven B, Penn K, Methé BA, Detrich WH 3rd (2009) Characterization of the intestinal microbiota of two Antarctic notothenioid fish species. Extremophiles 13(4):679–685. https://doi.org/10.1007/s00792-009-0252-4

Webster TMU, Consuegra S, Hitchings M, de Leaniz CG (2018) Interpopulation variation in the Atlantic salmon microbiome reflects environmental and genetic diversity. Appl Environ Microbiol 84(16):e00691. https://doi.org/10.1128/AEM.00691-18

Xia JH, Lin G, Fu GH, Wan ZY, Lee M, Wang L, Liu XJ, Yue GH (2014) The intestinal microbiome of fish under starvation. BMC Genomics 15:266. https://doi.org/10.1186/1471-2164-15-266

Xiong JB, Nie L, Chen J (2019) Current understanding on the roles of gut microbiota in fish disease and immunity. Zool Res 40(2):70–76. https://doi.org/10.24272/j.issn.2095-8137.2018.069

Ye L, Amberg J, Chapman D, Gaikowski M, Liu WT (2014) Fish gut microbiota analysis differentiates physiology and behavior of invasive Asian carp and indigenous American fish. ISME J 8(3):541–551. https://doi.org/10.1038/ismej.2013.181

Zarkasi KZ, Abell GC, Taylor RS, Neuman C, Hatje E, Tamplin ML, Katouli M, Bowman JP (2014) Pyrosequencing-based characterization of gastrointestinal bacteria of Atlantic salmon (*Salmo salar* L.) within a commercial mariculture system. J Appl Microbiol 117(1):18–27. https://doi.org/10.1111/jam.12514

Zhang QL, Yan Y, Shen JY, Hao GJ, Shi CY, Wang QT, Liu H, Huang J (2013) Development of a reverse transcription loop-mediated isothermal amplification assay for rapid detection of grass carp reovirus. J Virol Methods 187(2):384–389. https://doi.org/10.1016/j.jviromet.2012.11.005

Isolation and Identification of Fish Gut Microbiome

2

A. R. Sarika, M. S. Aishwarya, R. S. Rachanamol, A. P. Lipton, and C. R. Abhirami

Abstract

The gut region of fish and shellfish offers a unique ecological niche where a variety of microbes resides, playing important roles such as absorption of nutrients, regulation of metabolic processes, synthesis of vitamins, and most importantly, enhancing host defense. These advantages of gut microbiome necessitate the evolution of suitable strategies and methodologies to collect samples, isolate, store, characterize, and evaluate the specific characteristics of beneficial microbes. The observations emanating from global researchers about microbial community composition, isolation, identification, and possible applications for the use of microbial cultures or consortia of strains in aquaculture have been recorded and reported. In addition, the use of different methods of sample collection, phylogenetic profiling, and bioinformatic analysis which contribute to the better understanding of fish gut microbial diversity have been described.

A. R. Sarika
Kerala Biotechnology Commission, Kerala State Council for Science, Technology and Environment, Thiruvananthapuram, Kerala, India

M. S. Aishwarya · A. P. Lipton (✉)
Centre for Marine Science and Technology, Manonmaniam Sundaranar University, Rajakkamangalam, Tamil Nadu, India

R. S. Rachanamol
A.J College of Arts and Science, Thiruvananthapuram, India

C. R. Abhirami
Jawaharlal Nehru Tropical Botanic Garden and Research Institute (JNTBGRI), Thiruvananthapuram, Kerala, India

A. Diwan et al. (eds.), *Microbiome of Finfish and Shellfish*,
https://doi.org/10.1007/978-981-99-0852-3_2

Keywords

Intestinal microbiota · Finfish and shellfish · Host habitat · Dietary ecology · GI contents · Metagenomics · NGS technologies

2.1 Introduction

Microorganisms exhibit immense diversity due to their extensive ecological adaptations, thus captivating microbiologists to research their residences within and on animals. The study on the microorganisms inhabiting humans provided insights into the bacterial communities mostly inhabiting the gastrointestinal tract. The complexity and the functional contributions of the gut microbiome were unraveled through extensive research and this paved way for the studies on gut microflora in other vertebrates too.

Within vertebrates, fishes are the group with the greatest diversity of species, with an abundance of ecological diversity and distinctive structural features within their intestines, making it a likely candidate for examining the diverse microorganisms within them. Various research studies and scientific publications have contributed to our understanding of microbiomes in fish and shellfish. The majority of these reports dealt with farmed finfish and shellfish (Diwan et al. 2022) and different body parts, including skin, gills, and digestive tracts which are the major routes of microbial entry (Seetharam et al. 2015). There are a number of factors that influence the composition of a fish gut microbiome. The gut microbiome of fish is directly influenced by the water quality and microbe communities surrounding it (Wong and Rawls 2012). In the digestive tract, microbes play a multitude of physiological roles, such as defense against diseases, modulating immunity, influencing nutrient absorption, regulating metabolic processes, and synthesizing vitamins (Wu et al. 2012; Gómez and Balcázar 2008).

Fish gut microbiota typically includes fungi, yeast, viruses, protists, bacteria, and archaea, but bacteria are the most prevalent (Rombout et al. 2011). The composition of microbiota in fish guts has been shown to be influenced by a wide range of intrinsic and extrinsic factors, including salinity, temperature, geographical location, life stage, diet, farm management practices, and antibiotic and probiotic usage (Egerton et al. 2018). In addition, the use of different methods of sample collection, phylogenetic profiling, and bioinformatics analysis may also contribute to understanding fish gut microbial diversity (Romero et al. 2014). This chapter provides an overview of the procedures for fish sample collection, isolation, and characterization of microbes from the gut and different approaches employed to determine the microbial complexity.

2.2 Sample Collection from Different Geographical Locations

Among the major determinants of fish gut microbiota is the habitat. Various animal taxa are characterized by different gut microbiota, which is shaped by environmental factors and host genetics. There is, however, uncertainty regarding the extent to which these factors affect fish microbiomes. Eichmiller et al. (2016) noted that the primary composition of the gut microbiome of invasive carp species depends on the type of ecological niche that the fish inhabit. According to Sullam et al. (2012), in their analysis of the microbiome communities in freshwater fishes' digestive tracts and their relationship to different ecological niches, similar findings were also reported. Additionally, a host's trophic level, habitat, and perhaps even phylogeny can also influence the core gut microbiome diversity. In the vast majority of fishes, niches are specialized, so habitat can be linked to host taxonomy.

The fish gut has been found to be a unique ecological niche where a variety of bacteria play important roles such as defending the pathogens, providing nutrition, improving immunity, etc. There have been different studies describing the fish intestinal microbiota in light of environmental and physiological factors. These studies, however, focused on the microbial community composition, isolation, identification, and possible applications for the use of single bacterial cultures or consortia of strains to promote fish growth.

Several researchers have carried out studies on the microbiome composition of finfish and shellfish collected from different geographical locations (Table 2.1).

In order to find whether the gut microbiota in fish is shaped by the host habitat, Kim et al. (2021) analyzed the microbiota of the intestinal contents of 227 fish from 23 different sampling spots from six different environments in Korea and found that the gut microbial community in fish was more strongly shaped by host habitat than by host taxonomy or trophic level.

A variety of factors affect the microbiome of shellfish such as mussels, oysters, and clams, such as geography, season, temperature, salinity, and other environmental factors (Liu et al. 2020; Stevick et al. 2019; Pierce and Ward 2018). Since shellfish are filter feeders, the microbiome is heavily influenced by the bacteria they consume, however, the host–microbe interaction is important as shellfish microbiome significantly differs from surrounding seawater (Vezzulli et al. 2017; Trabal et al. 2012).

Research on fish gut microbiology in the past used culture-dependent methods which vastly underestimated the diversity of these communities; only 10% of bacteria could be isolated and cultured. Given the extreme conditions that deep-sea fish endure, both their microbiome and the environment they inhabit are likely to differ (Collins et al. 2021). Molecular-based culture-independent methods have recently been applied to study microbial communities colonizing fish GI tracts in studying microbial diversity in different ecosystems (Tarnecki et al. 2017). The combination of metagenomic sequencing technology and other methods can provide technical support for better exploration of the dietary ecology of fish and enrich the theoretical basis of research on dietary ecology.

Table 2.1 The finfish and shellfish species collected from different geographical locations and the identified microbes

S. No.	Fish species	Geographical Location	Major microbes identified	References
1.	*Kyphosus sydneyanus*	Greater Hauraki Gulf area, on the northeastern coast of New Zealand	*Clostridium*	Moran et al. (2005)
2.	Farmed rainbow trout *Oncorhynchus mykiss* (Walbaum)	Local trout farm, Perthshire, UK	γ-Proteobacteria—Enterobacteriaceae, Yersinia, Serratia, Hafnia, and Obesumbacterium	Lyons et al. (2017)
3.	Five fish orders—Aulopiformes, Clupeiformes, Scorpaeniformes, Perciformes, and Pleuronectiformes	Coastal waters of Hong Kong	Proteobacteria and Firmicutes; dominant genera—*Clostridium*, *Photobacterium*, *Ralstonia*, and *Acinetobacter*	Huang et al. (2020)
4.	Wild Migratory Freshwater fish Amur bream (*Megalobrama terminalis*)	Southern China—six different sites in Pearl River, Moyang River, and Wanquan River	Microbiome varied with geographical location; Fusobacteriota predominated	Liu et al. (2022)
5.	Atlantic Salmon (*Salmo salar*)	Norway farm	Proteobacteria and Firmicutes	Gajardo et al. (2016)
6.	Yellowfin seabream, *Acanthopagrus latus*	Daya Bay	Bacterium 2013Ark19i, Bacterium 2013Arg42i, and *Acinetobacter baumanni*	Pan et al. (2021)
7.	Farmed adult turbot (*Scophthalmus maximus*)	Commercial farm in Qingdao, China	Proteobacteria and Firmicutes	Xing et al. (2013)
8.	Deep Sea Fish	Suruga Bay, Japan	Proteobacteria, Brevinema, unclassified Mycoplasmataceae	Iwatsuki et al. (2021)

A majority of laboratory studies in the past have investigated gut bacterial diversity using isolation and culturing, PCR denaturing gradient gel electrophoresis (DGGE) and terminal restriction fragment length polymorphism (T-RFLP) (Liu et al. 2016; Osborn et al. 2000). Next-generation sequencing of the 16S rRNA gene has uncovered more comprehensive and complex microbial communities. Though the gut microbial composition of several fishes were explored through high-throughput DNA sequencing, there is a scant idea in terms of the difference in the microbial composition of the reared fish from those in the natural environment (Liu et al. 2016).

2.3 Procedures for Bringing the Samples to the Laboratory for Analysis

The sample collection procedures differ depending on whether the Wild fish or the cultured fish is being employed for the study. The wild fish are collected by trawl, trap or hook, and line. Tao et al. (2018) collected the marine fish samples by a research trawler from wherever the research-related trawling work was permitted by authorities. Huang et al. (2021) in his study employed the fish collected by local fishermen using gillnets, purse seine netting, or long lines from the marine environment in Hong Kong. The fish are stored in ice or are kept in recirculating tanks for dissection. When cultured fish are employed for the study, they are reared in culture tanks with recirculating seawater.

Marine Fish Sampling: The marine fish samples were collected by a research trawler (Tao et al. 2018) wherever the research-related trawling work was permitted by authorities or by local fishermen using gillnets, purse seine netting or long lines in the study conducted by Huang et al. (2020) from the marine environment in Hong Kong. Prior to being transported to the laboratory, most of the fish were dead and kept in an ice box at −20 °C. Based on their morphological characteristics, fishes were sorted and identified.

2.4 Dissection and Removal of Different Parts of the Gut System

The microbial community in the digestive system of the fish is influenced by environmental, ecology, and evolutionary factors (Egerton et al. 2018). It is important to process the fish quickly in order to reduce the changes in gut microbial communities with time and perturbation (Clements et al. 2014). Insect pins or scalpels and forceps are used depending on the fish size to dissect fish samples after they have been treated with 70% ethanol prior to dissection. The stomach and hindgut are to be removed intact, as well as the attached organs, such as the liver. The GI contents are to be squeezed from the digestive tract and then cleaned in 70% ethanol and sterile water to prevent the samples from accumulating transient bacteria, and then stored at −20 °C or directly extracted for DNA (Huang et al. 2020).

2.5 Isolation of Microbes and Determine the Complexity of Composition

Current procedures for fish gut microbiome studies include culture-dependent and culture-independent DNA analysis obtained directly from gut contents and mucosal walls. Culture-dependent techniques rely on culture techniques and the DNA sequencing of 16S rRNA gene to identify bacteria (Talwar et al. 2018).

Culture dependent method: Isolation and Identification of Gut Bacteria—An aliquot of gut homogenate was spread aseptically with 9 mL of double-strength sterile PBS onto nutrient agar and incubated at 37 °C for 24–48 h. Colonies are counted and expressed as cfu/g and the isolated purified on nutrient agar slants. The isolated bacteria are stored in nutrient agar slants. The morphological analysis and biochemical tests are to be carried out for the species-level identification of the isolates. The strain identity is to be confirmed using 16S rRNA gene sequencing and subsequent phylogenetic analysis (Sivasubramanian et al. 2012).

Using metagenomics and next-generation sequencing techniques, researchers are now able to analyze microbial genomes directly from environmental samples (Handelsman et al. 1998; Riesenfeld et al. 2004). Through these new methods of research, researchers have been able to learn more about the connections between microbes and their hosts. Fish gut microbiome research uses a variety of next-generation sequencing (NGS) platforms such as the Illumina system, Roche 454 system, and Ion Torrent Personal Genome Machine (PGM); the Illumina system is the most common (Johny et al. 2021).

2.6 Identification Based on Molecular Gene Structure (Use of Conventional and Advanced Methods of NGS or Any Other Latest Method)

Currently, 16S rRNA gene sequencing is heavily relied on for accurate bacterial identification (Rosselló-Mora 2005). A number of early efforts involved the creation of 16S clone libraries followed by sequencing individual clones using Sanger technology (Kim et al. 2007; Wu et al. 2010). This method is very accurate in terms of sequence quality, but it is time-consuming and expensive to sequence hundreds of individual clones, which limits its usefulness. The cost per megabase of DNA sequence decreased dramatically with the arrival of next generation sequencing technologies (NGS) (Wetterstrand 2014). Intestinal microbial communities can be studied most effectively using next-generation sequencing (NGS) (Ringø et al. 2016). Using next-generation sequencing (NGS) platforms, it is possible to explore the diverse microbial communities present in the gut at a massive scale (Nakayama et al. 2013; Lyons et al. 2017). The relative abundance of microorganisms within a sample is assessed by determining the number of bacterial 16S rDNA sequences derived from the sample and assigning each to a particular genus/species (Lyons et al. 2017). The following are NGS technologies being used by researchers for the fish gut microbiome studies:

1. Illumina sequencing platforms—A larger amount of data could be generated in a more cost-effective manner using these platforms. Using NGS-based metagenomic sequencing, many samples can be combined in one sequencing run and high sequence coverage is achieved, allowing it to identify bacteria with very low abundances, which would otherwise be missed or would be too expensive to identify with other methods. The HiSeq sequencers produce the largest amount of data, with the shortest reads at lowest cost per base. MiSeq generates millions of 16S rRNA gene sequences from a single sample, revealing the gut microbial community in detail. In contrast to lower-resolution molecular and culture approaches, MiSeq allows researchers to test more complex hypotheses about gut microbial communities. While the costs per sequenced base are higher, the longer read length and the lower number of reads make it particularly suitable for amplicon sequencing. Compared to HiSeq, MiSeq, and HiSeqX, the Illumina NextSeq 500 platform performs at an intermediate scale in terms of output, read length, and costs per base.
2. The 454 platform is the most commonly used NGS technique for fish-microbiome study (Ghanbari et al. 2015). It uses "pyro sequencing" in which dNTPs are incorporated into the synthesized DNA chain and released as pyrophosphate following the addition of bases. Further improvements are made to this platform. The higher cost still remains a major limitation for its extensive use.

2.7 Differentiation of Genetic Composition and Their Interpretation of Pathogenic and Healthy Microbes Based on Their Genetic Structure

Microbiota in fish guts play an important role in regulating immunity, health, and physiology in the host, so altered microbiota can cause diseases (Xiong et al. 2019). Modifying the fish gut microbiome with pre-, pro-, and synbiotics has been shown to improve fish health (Rosado et al. 2019; Bozzi et al. 2021). A better understanding of the actions and interactions of probiotics in the gut requires basic knowledge about their genetics and physiology. A healthy gut microbiota community is actively established and maintained by understanding the factors governing the gut microbiota (Tarnecki et al. 2019). In the study conducted by Ma et al. (2019), 16S rRNA sequencing technology was used to analyze the composition of the intestinal microbiota in different health states of the Yunlong Grouper fish. PICRUSt (Phylogenetic investigation of communities by reconstruction of unobserved states), a bioinformatics software package used to predict metagenomic functional genes (Langille et al. 2013) was applied to predict gene function, evaluate changes in the intestinal microbiota function, and study the correlation with the culture environment (Ma et al. 2019). Bozzi et al. (2021) used 16S rRNA gene metabarcoding to describe the distal gut microbiota of fish affected to different degrees by an external infection as well as the secondary effects of formalin treatment.

2.8 Data Collection and Development of Microbiome Libraries Using Adequate Software

Microbiome studies produce enormous amounts of data that demand sophisticated computational tools and the technologies employed are fast evolving (Galloway-Peña and Hanson 2020). Some of the technologies are as follows.

Marker Gene Analysis: 16SrRNA gene sequencing is the targeted sequencing method used to identify bacteria and identified using taxonomic information. NGS platform is used for sequencing and machine learning methods such as RDP classifier, Bayesian classifier or databases such as SILVA, and green genes are used to assign taxonomy (Balvočiūtė and Huson 2017). Mothur (Schloss et al. 2009), DADA2 (Callahan et al. 2016), QIIME (Caporaso et al. 2010) are the standard microbiome analysis packages to assign taxonomy.

Shotgun Metagenomics: This approach makes use of untargeted sequencing techniques to capture all microbial genomes in the sample and rely on reference genomes and scientific knowledge (Sharpton 2014). The taxa, their abundance, the gene coding sequences, and the functional potential can also be identified (Galloway-Peña and Hanson 2020). The metagenomics data can be assembled using software such as MetaVelvet (Namiki et al. 2012), MetaQUAST (Mikheenko et al. 2016), Ray Meta (Boisvert et al. 2012), and Met AMOS (Treangen et al. 2013), etc.

2.8.1 Metatranscriptomics, Metaproteomics, and Metabolomics

These methods precisely capture the transcribed RNA, assessing the gene expression of the microbes. The Illumina sequencing methodologies, most often the HiSeq families of instruments are primarily employed due to the high throughput and low cost per base. Mapping of the RNA transcripts to different genomes and pathways (Kanehisa and Goto 2000) can be done to identify the taxonomy of transcriptionally active organisms as well as the function of their expressed genes. Further analysis of the metatranscriptomic data can be done through SOAPdenovoprogram (Xie et al. 2014). Microbial metabolite production is profiled and quantified in metabolomics. Metabolomics and metaproteomics use mass spectrometry to identify metabolites and proteins respectively within a microbiome (Lai et al. 2021; Zierer et al. 2018).

2.9 Use of Such Data Libraries for Undertaking Metagenomic Studies

The metagenomic approach provides precision on the genetic identity of the microbes and provides knowledge of their gene function. The approach involves either sequencing-based metagenomics which provides details on microbial diversity and functional genomics on the functionality of the genes. With the obtained data using sequencing studies, metagenomics libraries can be constructed which can help in bacterial gene manipulations. The possibility of discovering and developing novel molecules in the form of enzymes and bioactive for different applications is highlighted using functional metagenomics (Diwan et al. 2022). The advancements in molecular and sequencing techniques have created great research interest in the fish gut microbiome and are expected to provide new avenues and products in the near future.

References

Balvočiūtė M, Huson DH (2017) SILVA, RDP, Greengenes, NCBI and OTT - how do these taxonomies compare? BMC Genomics 18(Suppl 2):114. https://doi.org/10.1186/s12864-017-3501-4. PMID: 28361695; PMCID: PMC5374703

Boisvert S, Raymond F, Godzaridis É et al (2012) Ray Meta: scalable de novo metagenome assembly and profiling. Genome Biol 13:R122. https://doi.org/10.1186/gb-2012-13-12-r122

Bozzi D, Rasmussen JA, Carøe C et al (2021) Salmon gut microbiota correlates with disease infection status: potential for monitoring health in farmed animals. Anim Microbiome 3:30. https://doi.org/10.1186/s42523-021-00096-2

Callahan B, McMurdie P, Rosen M et al (2016) DADA2: high-resolution sample inference from Illumina amplicon data. Nat Methods 13:581–583. https://doi.org/10.1038/nmeth.3869

Caporaso J, Kuczynski J, Stombaugh J et al (2010) QIIME allows analysis of high-throughput community sequencing data. Nat Methods 7:335–336. https://doi.org/10.1038/nmeth.f.303

Clements KD, Angert ER, Montgomery WL, Choat JH (2014) Intestinal microbiota in fishes: what's known and what's not. Mol Ecol 23(8):1891–1898. https://doi.org/10.1111/mec.12699

Collins FWJ, Walsh CJ, Gomez-Sala B, Guijarro-García E, Stokes D, Jakobsdóttir KB, Kristjánsson K, Burns F, Cotter PD, Rea MC, Hill C, Ross RP (2021) The microbiome of deep-sea fish reveals new microbial species and a sparsity of antibiotic resistance genes. Gut Microbes 13(1):1–13. https://doi.org/10.1080/19490976.2021.1921924. PMID: 33970781; PMCID: PMC8115496

Diwan AD, Harke SN, Gopalkrishna, Panche AN (2022) Aquaculture industry prospective from gut microbiome of fish and shellfish: an overview. J Anim Physiol Anim Nutr 106:441–469. https://doi.org/10.1111/jpn.13619

Egerton S, Culloty S, Whooley J, Stanton C, Ross RP (2018) The gut microbiota of marine fish. Front Microbiol 9:873. https://doi.org/10.3389/fmicb.2018.00873

Eichmiller JJ, Hamilton MJ, Staley C, Sadowsky MJ, Sorensen PW (2016) Environment shapes the fecal microbiome of invasive carp species. Microbiome 4:44. https://doi.org/10.1186/s40168-016-0190-1

Gajardo K, Rodiles A, Kortner T et al (2016) A high-resolution map of the gut microbiota in Atlantic salmon (Salmo salar): a basis for comparative gut microbial research. Sci Rep 6:30893. https://doi.org/10.1038/srep30893

Galloway-Peña J, Hanson B (2020) Tools for analysis of the microbiome. Dig Dis Sci 65(3): 674–685. https://doi.org/10.1007/s10620-020-06091-y. PMID: 32002757; PMCID: PMC7598837

Ghanbari M, Kneifel W, Domig KJ (2015) A new view of the fish gut microbiome: advances from next-generation sequencing. Aquaculture 448:464–475. https://doi.org/10.1016/j.aquaculture.2015.06.033

Gómez GD, Balcázar JL (2008) A review on the interactions between gut microbiota and innate immunity of fish. FEMS Immunol Med Microbiol 52(2):145–154. https://doi.org/10.1111/j.1574-695X.2007.00343.x. PMID: 18081845

Handelsman J, Rondon MR, Brady SF, Clardy J, Goodman RM (1998) Molecular biological access to the chemistry of unknown soil microbes: a new frontier for natural products. Chem Biol 5(10):R245–R249. https://doi.org/10.1016/s1074-5521(98)90108-9. PMID: 9818143

Huang Q, Sham RC, Deng Y, Mao Y, Wang C, Zhang T, Leung KMY (2020) Diversity of gut microbiomes in marine fishes is shaped by host-related factors. Mol Ecol 29(24):5019–5034. https://doi.org/10.1111/mec.15699. PMID: 33084100; PMCID: PMC7756402

Huang M, Ding L, Wang J, Ding C, Tao J (2021) The impacts of climate change on fish growth: a summary of conducted studies and current knowledge. Ecol Indic 121:106976. https://doi.org/10.1016/j.ecolind.2020.106976. ISSN 1470-160X

Iwatsuki T, Kanazawa T, Ogasawara T, Hosotani K, Tsuchiya K, Watanabe S, Suzuki T, Moriuchi R, Kanesaki Y, Dohra H (2021) 16S rRNA gene amplicon sequencing of gut microbiota in three species of deep-sea fish in Suruga Bay, Japan. Microbiol Resource Announce 10(1):e01260–e01220. https://doi.org/10.1128/MRA.01260-20. PMID: 33414347; PMCID: PMC8407723

Johny TK, Puthusseri RM, Bhat SG (2021) A primer on metagenomics and next-generation sequencing in fish gut microbiome research. Aquac Res 52:4574–4600

Kanehisa M, Goto S (2000) KEGG: Kyoto encyclopedia of genes and genomes. Nucleic Acids Res 28(1):27–30. https://doi.org/10.1093/nar/28.1.27. PMID: 10592173; PMCID: PMC102409

Kim DH, Brunt J, Austin B (2007) Microbial diversity of intestinal contents and mucus in rainbow trout (Oncorhynchus mykiss). J Appl Microbiol 102(6):1654–1664. https://doi.org/10.1111/j.1365-2672.2006.03185.x

Kim PS, Shin NR, Lee JB et al (2021) Host habitat is the major determinant of the gut microbiome of fish. Microbiome 9:166. https://doi.org/10.1186/s40168-021-01113-x

Lai Y, Liu CW, Yang Y et al (2021) High-coverage metabolomics uncovers microbiota-driven biochemical landscape of interorgan transport and gut-brain communication in mice. Nat Commun 12:6000. https://doi.org/10.1038/s41467-021-26209-8

Langille MG, Zaneveld J, Caporaso JG, McDonald D, Knights D, Reyes JA, Clemente JC, Burkepile DE, Thurber RL, Knight R, Beiko RG (2013) Predictive functional profiling of microbial communities using 16S rRNA marker gene sequences. Nat Biotechnol 31(9): 814–821. https://doi.org/10.1038/nbt.2676

Liu H, Guo X, Gooneratne R, Lai R, Zeng C, Zhan F, Wang W (2016) The gut microbiome and degradation enzyme activity of wild freshwater fishes influenced by their trophic levels. Sci Rep 6:24340. https://doi.org/10.1038/srep24340

Liu X, Teixeira JS, Ner S, Ma KV, Petronella N, Banerjee S, Ronholm J (2020) Exploring the potential of the microbiome as a marker of the geographic origin of fresh seafood. Front Microbiol 11:696. https://doi.org/10.3389/fmicb.2020.00696. PMID: 32362885; PMCID: PMC7181054

Liu Y, Li Y, Li J, Zhou Q, Li X (2022) Gut Microbiome analyses of wild migratory freshwater fish (Megalobrama terminalis) through geographic isolation. Front Microbiol 13:858454. https://doi.org/10.3389/fmicb.2022.858454. PMID: 35464925; PMCID: PMC9026196

Lyons PP, Turnbull JF, Dawson KA, Crumlish M (2017) Exploring the microbial diversity of the distal intestinal lumen and mucosa of farmed rainbow trout Oncorhynchus mykiss (Walbaum) using next generation sequencing (NGS). Aquac Res 48:77–91. https://doi.org/10.1111/are.12863

Ma C, Chen C, Jia L et al (2019) Comparison of the intestinal microbiota composition and function in healthy and diseased Yunlong Grouper. AMB Express 9:187. https://doi.org/10.1186/s13568-019-0913-3

Mikheenko A, Saveliev V, Gurevich A (2016) MetaQUAST: evaluation of metagenome assemblies. Bioinformatics 32(7):1088–1090. https://doi.org/10.1093/bioinformatics/btv697. PMID: 26614127

Moran D, Turner S, Clements K (2005) Ontogenetic development of the gastrointestinal microbiota in the marine herbivorous fish Kyphosussydneyanus. Microb Ecol 49:590–597. https://doi.org/10.1007/s00248-004-0097-4

Nakayama J, Jiang J, Watanabe K, Chen K, Ninxin H, Matsuda K, Kurakawa T, Tsuji H, Sonomoto K, Lee YK (2013) Up to species-level community analysis of human gutmicrobiota by 16S rRNA amplicon pyrosequencing. Biosci Microb Food Health 32:69–76

Namiki T, Hachiya T, Tanaka H, Sakakibara Y (2012) MetaVelvet: an extension of velvet assembler to de novo metagenome assembly from short sequence reads. Nucleic Acids Res 40:e155. https://doi.org/10.1093/nar/gks678

Osborn AM, Moore ER, Timmis KN (2000) An evaluation of terminal-restriction fragment length polymorphism (T-RFLP) analysis for the study of microbial community structure and dynamics. Environ Microbiol 2(1):39–50. https://doi.org/10.1046/j.1462-2920.2000.00081.x. PMID: 11243261

Pan W, Qin C, Zuo T, Yu G, Zhu W, Ma H, Xi S (2021) Is metagenomic analysis an effective way to analyze fish feeding habits? A case of the yellowfin sea bream Acanthopagrus latus (Houttuyn) in Daya Bay. Front Mar Sci 8:634651

Pierce ML, Ward JE (2018) Microbial ecology of the Bivalvia, with an emphasis on the family Ostreidae. J Shellfish Res 37:793–806. https://doi.org/10.2983/035.037.0410

Riesenfeld CS, Schloss PD, Handelsman J (2004) Metagenomics: genomic analysis of microbial communities. Annu Rev Genet 38:525–552

Ringø E, Zhou Z, Vecino J, Wadsworth S, Romero J, Krogdahl Å, Olsen R, Dimitroglou A, Foey A, Davies S, Owen M, Lauzon H, Martinsen L, De Schryver P, Bossier P, Sperstad S, Merrifield D (2016) Effect of dietary components on the gut microbiota of aquatic animals. A never-ending story? Aquac Nutr 22:219–282. https://doi.org/10.1111/anu.12346

Rombout JH, Abelli L, Picchietti S, Scapigliati G, Kiron V (2011) Teleost intestinal immunology. Fish Shellfish Immunol 31(5):616–626. https://doi.org/10.1016/j.fsi.2010.09.001. PMID: 20832474

Romero J, Ringø E, Merrifield DL (2014) The gut microbiota of fish. In: Merrifield D, Ringø E (eds) Aquaculture nutrition: gut health, probiotics and prebiotics. Wiley Blackwell Publishing, Oxford, pp 75–100

Rosado D, Pérez-Losada M, Severino R, Cable J, Xavier R (2019) Characterization of the skin and gill microbiomes of the farmed seabass (Dicentrarchus labrax) and sea bream (Sparus aurata). Aquaculture 500:57–64. https://doi.org/10.1016/j.aquaculture.2018.09.063

Rosselló-Mora R (2005) Updating prokaryotic taxonomy. J Bacteriol 187(18):6255–6257. https://doi.org/10.1128/JB.187.18.6255-6257.2005. PMID: 16159756; PMCID: PMC1236658

Schloss PD, Westcott SL, Ryabin T, Hall JR, Hartmann M, Hollister EB, Lesniewski RA, Oakley BB, Parks DH, Robinson CJ, Sahl JW, Stres B, Thallinger GG, Van Horn DJ, Weber CF (2009) Introducing Mothur: open-source, platform-independent, community-supported software for describing and comparing microbial communities. Appl Environ Microbiol 75(23): 7537–7541. https://doi.org/10.1128/AEM.01541-09. PMID: 19801464; PMCID: PMC2786419

Seetharam AS, Emily K, Zhi QD, Max FR, Andrew JS (2015) Microbiome analyses of pacific white shrimp (Litopenaeus vannamei) collected from disparate geographical locations. Genom Data 6: 67–69. https://doi.org/10.1016/j.gdata.2015.08.009

Sharpton TJ (2014) An introduction to the analysis of shotgun metagenomic data. Front Plant Sci 5: 209. https://doi.org/10.3389/fpls.2014.00209

Sivasubramanian K, Ravichandran S, Kavitha R (2012) Isolation and characterization of gut micro biota from some estuarine fishes. Mar Sci 2(2):1–6. https://doi.org/10.5923/j.ms.20120202.01

Stevick RJ, Sohn S, Modak TH, Nelson DR, Rowley DC, Tammi K et al (2019) Bacterial community dynamics in an oyster hatchery in response to probiotic treatment. Front Microbiol 10:1060. https://doi.org/10.3389/fmicb.2019.01060

Sullam KE, Essinger SD, Lozupone CA, O'Connor MP, Rosen GL, Knight R et al (2012) Environmental and ecological factors that shape the gut bacterial communities of fish: a meta-analysis. Mol Ecol 21:3363–3378. https://doi.org/10.1111/j.1365-294X.2012.05552.x

Talwar C, Nagar S, Lal R, Negi RK (2018) Fish gut microbiome: current approaches and future perspectives. Indian J Microbiol 58(4):397–414. https://doi.org/10.1007/s12088-018-0760-y. PMID: 30262950; PMCID: PMC6141390

Tao LSR, Lui KKY, Lau ETC, Ho KKY, Mak YKY, de Mitcheson YS, Leung KMY (2018) Trawl ban in a heavily exploited marine environment: responses in population dynamics of four stomatopod species. Sci Rep 8(1):17876. https://doi.org/10.1038/s41598-018-35804-7

Tarnecki AM, Burgos FA, Ray CL, Arias CR (2017) Fish intestinal microbiome: diversity and symbiosis unravelled by metagenomics. J Appl Microbiol 123:2–17. https://doi.org/10.1111/jam.13415

Tarnecki AM, Wafapoor M, Phillips RN et al (2019) Benefits of a Bacillus probiotic to larval fish survival and transport stress resistance. Sci Rep 9:4892. https://doi.org/10.1038/s41598-019-39316-w

Trabal N, Mazón-Suástegui JM, Vázquez-Juárez R, Asencio-Valle F, Morales-Bojórquez E, Romero J (2012) Molecular analysis of bacterial microbiome associated with oysters (Crassostrea gigas and Crassostrea corteziensis) in different growth phases at two cultivation sites. Microb Ecol 64:555–569. https://doi.org/10.1007/s00248-012-0039-5

Treangen TJ, Koren S, Sommer DD et al (2013) MetAMOS: a modular and open source metagenomic assembly and analysis pipeline. Genome Biol 14:R2. https://doi.org/10.1186/gb-2013-14-1-r2

Vezzulli L, Stagnaro L, Grande C, Tassistro G, Canesi L, Pruzzo C (2017) Comparative 16SrDNA gene-based microbiome profiles of the pacific oyster (Crassostrea gigas) and the Mediterranean mussel (Mytilus galloprovincialis) from a shellfish farm (Ligurian Sea, Italy). Microb Ecol 75: 495–504. https://doi.org/10.1007/s00248-017-1051-6

Wetterstrand KA (2014) DNA sequencing costs: data from the NHGRI genome sequencing program (GSP). www.genome.gov/sequencingcosts

Wong S, Rawls JF (2012) Intestinal microbiota composition in fishes is influenced by host ecology and environment. Mol Ecol 21(13):3100–3102. https://doi.org/10.1111/j.1365-294x.2012.05646.x. PMID: 22916346; PMCID: PMC4846280

Wu S, Gao T, Zheng Y, Wang W, Cheng Y, Wang G (2010) Microbial diversity of intestinal contents and mucus in yellow catfish (Pelteobagrus fulvidraco). Aquaculture 303(1–4):1–7

Wu S, Wang G, Angert ER, Wang W, Li W, Zou H (2012) Composition, diversity, and origin of the bacterial community in grass carp intestine. PLoS One 7:e30440. https://doi.org/10.1371/journal.pone.0030440

Xie Y, Wu G, Tang J, Luo R, Patterson J, Liu S, Huang W, He G, Gu S, Li S, Zhou X, Lam TW, Li Y, Xu X, Wong GK, Wang J (2014) SOAPdenovo-Trans: de novo transcriptome assembly with short RNA-Seq reads. Bioinformatics 30(12):1660–1666. https://doi.org/10.1093/bioinformatics/btu077. PMID: 24532719

Xing M, Hou Z, Yuan J, Liu Y, Qu Y, Liu B (2013) Taxonomic and functional metagenomic profiling of gastrointestinal tract microbiome of the farmed adult turbot (Scophthalmus maximus). FEMS Microbiol Ecol 86(3):432–443. https://doi.org/10.1111/1574-6941.12174. PMID: 23802730

Xiong JB, Nie L, Chen J (2019) Current understanding on the roles of gut microbiota in fish disease and immunity. Zool Res 40(2):70–76. https://doi.org/10.24272/j.issn.2095-8137.2018.069. PMID: 29976843; PMCID: PMC6378566

Zierer J, Jackson MA, Kastenmüller G, Mangino M, Long T, Telenti A, Mohney RP, Small KS, Bell JT, Steves CJ, Valdes AM, Spector TD, Menni C (2018) The fecal metabolome as a functional readout of the gut microbiome. Nat Genet 50(6):790–795. https://doi.org/10.1038/s41588-018-0135-7. PMID: 29808030; PMCID: PMC6104805

Gut Microbiome and Fish Health: An Overview in Finfish Aquaculture Prospective

3

Farhana Hoque, Abhijit Das, and Jitendra Kumar Sundaray

Abstract

The complex assemblages of microbes colonize in the digestive tracts of fish play a key role in health by development and modulation of the mucosal innate and adaptive immune system, exerting imperative part in outcompete opportunistic pathogens by preserving gut integrity and regulating intestinal barrier permeability. Apart, gut microbiome assists in the acquisition of nutrients, influencing the physiology, and life span. Vertebrate immune system development and proper ontogenetic development have both been shown to be significantly influenced by relations between both the host and intestinal microbiota. Understanding ontogenetic changes in the microbiota of fish is of great interest in aquaculture to assess the unpredictable mortality of cultured fish at early life stages. Although microbes are present in gut of all fish but the composition of gut microbiota of freshwater fish is attributed to multiple factors such as habitat, diets, life stages, seasons, trophic levels, etc. affecting their growth in captivity. Also, modulation of gut microbiota has revealed to be a promising means for disease bio control in farmed fish. The results of this field of study will undoubtedly lead to the creation of methods for influencing the gut microbiota of aquacultured fish species to promote better growth and production of healthy fish to promote sustainable aquaculture. This present chapter is intended to provide collective information

F. Hoque
Regional Research Station, ICAR-Central Institute of Freshwater Aquaculture, Kolkata, West Bengal, India

A. Das
Faculty of Fishery Sciences, West Bengal University of Animal and Fishery Sciences, Kolkata, India

J. K. Sundaray (✉)
ICAR-Central Institute of Freshwater Aquaculture, Bhubaneswar, Odisha, India

A. Diwan et al. (eds.), *Microbiome of Finfish and Shellfish*,
https://doi.org/10.1007/978-981-99-0852-3_3

on the ontogenetic characterization of the intestinal microbiota, gut microbiome of aquacultured finfishes with special reference to the improvement of immune response and disease resistance, modulation of fish gut microbiota for improved health.

Keywords

Gut microbiome · Health · Immunity · Ontogeny · Disease resistance

3.1 Introduction

The microorganisms in the gut, which cover the full spectrum of metabolic processes and incite the immune system of the host organism, are known as the gut microbiota (Gomez and Balcaazar 2008; Hanning and Diaz-Sanchez 2015). The importance of this framework can be better understood with more knowledge of the circumstances surrounding this taxonomy and metagenomic research. A combination of the genomes of different microorganisms found in the environment is referred to as a microbiota or microbiome (Yukgehnaish et al. 2020).

The population of bacteria living in the gastrointestinal (GI) tract of vertebrates (the gut microbiome) is now widely acknowledged to play a crucial role in host development, physiology, and health (Mueller et al. 2012; Llewellyn et al. 2014). The majority of our study of microbiome structure and function comes from research on mammals, which comprise less than 10% of the diversity of all vertebrates (Sullam et al. 2012). Despite encompassing nearly half of vertebrate species (Nelson 2006), few studies have examined the gut microbiome of fishes (Llewellyn et al. 2014; Ringø et al. 2015). Less than 10% of bacteria can be identified and cultivated in a lab, therefore early investigations of fish gut microbiology used methods that significantly underestimated the diversity of these communities (Amann et al. 1995). More recently, molecular-based culture-independent approaches have been used to examine the microbial communities that colonized the GI of fish, in line with previous initiatives aimed at studying microbial diversity in diverse environments. By promoting immune system growth, assisting in nutrition uptake, and preventing off opportunistic pathogens, the gut microbiota of vertebrates plays a crucial role in host health (Tarnecki et al. 2017).

The present knowledge of fish microbiome composition is actually obtained from multiple research articles as well as various studies conducted by different researchers. According to these findings, farmed fish, particularly salmonids, are the focus of microbiome research. There have been numerous attempts to examine the microbiome associated with finfish and shellfish, and early microbial research primarily focused on various body parts such the skin, gills, and gastrointestinal tract (Harrison 1929). This is because the main entry points for microorganisms into fish and shellfish are these body parts (Bøgwald and Dalmo 2014; Chen et al. 2017; Cornejo-Granados et al. 2017; Ringø et al. 2007, 2008; Seetharam et al. 2015). According to Ringø et al. (2015), the microbial communities linked to various body regions, including the intestinal tract, have an impact on fish health. The importance of mucosal adhesion, particularly in the digestive system, for bacterial infection during the early stages of fish development is underlined. While studying the function of fish's gastrointestinal microbiota, Nayak (2010) compiled a list of the most often encountered bacteria in salmon fish. The focus of current research, however, is on the microbes that exist in the gastrointestinal tract and are connected to numerous physiological functions, including disease defence, immune response modulation, nutrient absorption, regulation of metabolic processes, and vitamin synthesis (Cornejo-Granados et al. 2017; Gomez and Balcaazar 2008; Romero et al. 2014; Sullam et al. 2012). There are other additional factors that contribute to an organism's ability to sustain its overall health and well-being. Without a healthy gut microbiome, it would be impossible to digest and break down the complex food components that comprise compound molecules. Additionally, the production of the energy that species require throughout the day and the absorption of vital nutrients are both crucial functions of gut microorganisms. Additionally, the metabolic activity of gut microbiome also influences diet requirements, food cravings and determines the nutritional pattern. Phenotypic and biochemical features were previously applied as traditional approaches for identifying microbiome species, and a variety of culture conditions were also used to count the number of bacteria present. Recent advancements in genetic engineering techniques, such as next-generation sequencing (NGS) technology, have allowed for more thorough and cost-effective analyses of the microbiome population (Tarnecki et al. 2017). As a result, numerous studies on the genome structural composition of microbial species in fish have been conducted. This type of genetic knowledge on the microbes will make it simpler to manipulate them in any way to improve their desired functional qualities, such as feed efficiency and disease resistance. Establishing a healthy microbiome environment in cultivable species and in aquaculture farming systems both heavily depend on the knowledge of host–microbe relationships and the identification of microbial community functions (Tarnecki et al. 2017). While analysing the fish gut microbiome, Talwar et al. (2018) conducted a thorough investigation on the diversity and functional characteristics of the various bacteria. They also highlighted the prospect of increasing aquaculture production on a sustainable basis by examining functional features of the microbiome. Recent research has also included detailed studies exploring these topics by Legrand et al. (2019), Perry et al. (2020), and Villamil et al. (2020). Legrand et al. (2019) reported functional involvement of the microbiota with the host and their interplay mechanisms while conducting in-depth research into the interaction linked among host and microbial system in fish. For fish, a variety of determinants of microbiome diversity have been identified, including nutritional, environmental, and genetic variables. Innovative techniques such as metagenomics, meta-transcriptomics, meta-proteomics, or metabolomics are crucial for understanding the functional connections and associations between bacteria and other significant microorganisms and hosts. These techniques could provide essential knowledge on how microbes and their hosts interact in a functional way.

3.2 Ontogenetic Characterization of the Intestinal Microbiota

Understanding ontogenetic changes in the microbiota of fish is of great interest in aquaculture to assess the unpredictable mortality of cultured fish at early life stages. Llewellyn et al. (2014) opined negative interactions between the environmental microbiota and the microbiota associated with fish larvae to be likely associated with the mortality. Vertebrate immune system development and appropriate ontogeny both depend on relationships between the host and the gut microbiota (Perez et al. 2010; Hooper et al. 2012; Sommer and Backhed 2013; Gomez et al. 2013). The larvae, the underlying water, and the first feed contribute to microbial colonization. The microbiota of the aquatic environment determines which bacteria catch the eggs and so have the opportunity to colonize. Sterile larvae consume the chorion-associated bacteria, the first organisms to colonize the developing gastrointestinal system, after hatching (GIT) (Egerton et al. 2018). The GIT of the newly hatched larvae tends to contain a few bacteria. Subsequent bacterial habitats are obtained in fish larvae for the initial period when they start to drink water to maintain osmoregulation, and the microbiota is then diversified by feeding (Hansen and Olafsen 1999). *Cytophaga*, *Flavobacterium*, and *Pseudomonas* were reported as the dominant species in fish eggs (Austin 1982; Yoshimizu et al. 1986) though inter-species variations were reported (Hansen and Olafsen 1999; Romero and Navarrete 2006; Mcintosh et al. 2008). Larsen et al. (2014) documented that the initial colonizing bacteria are very much species-specific, mostly due to the variation in binding glycoproteins on the egg surface. Additionally, the microbiota of the surrounding environment influences the microbial population on eggs. Chorion-associated bacteria colonize the emerging gastrointestinal tract (GIT) of freshly hatched larvae as the first colonizers. The microbiota becomes further diversified when fish larvae begin to drink and start feeding (Hansen and Olafsen 1999; Korsnes et al. 2006; Reid et al. 2009; Lauzon et al. 2010). Despite the fact that the microbial population fluctuates with life stage and climate, some species have a very stable gut microbiota within the initial 50 days of life (Mcintosh et al. 2008; Larsen et al. 2014). Only a few studies have used molecular methods to examine the gut microbiota of aquaculture species from early larval stages to later developmental stages. Studies depicted that diet plays a major role in shaping the gut microbial community and from first feeding resulting in significant diversification (Lauzon et al. 2010). The diversity of the gut microbiota generally increases as the fish diet changes from predatory to omnivorous and omnivorous to herbivorous (Liu et al. 2016). The gut colonization can be induced by stochastic (neutral assembly) or deterministic (non-neutral) model. The ultimate composition of the gut microbial community is determined by the stochastic model's random distribution of microorganisms or events that cause the microbes to enter the colon, whereas the deterministic model acquires the assembly through host selection factors, active microbe–microbe interactions, and microbe–microbe dispersal (Talwar et al. 2018). A variety of factors including changing environmental conditions such as temperature and salinity (Macfarlane and Englyst 1986), developmental stage (Romero and Navarrete 2006), digestive physiology (Cahill 1990), and feeding strategy (Uchii et al. 2006) can cause change in the gut microbial community affecting the host. Some of the gut microfloras appear to be temporary, while other bacterial floras seem to be permanent residents (Kim et al. 2007). Herbivorous fish under-goes an ontogenetic diet shift, while transitioning from carnivorous juveniles to either omnivorous or herbivorous adults (Gallagher et al. 2001). The study of Xia et al. (2014) provided the first perception of the fish gut microbiota and its changes during starvation representative of the interaction of hosts and microbes to the dynamic environment. Nikouli et al. (2020) provided evidence on adult farmed fish in the Mediterranean Sea have a divergent and species-specific gut microbiota profile that are shaped independently of the similar environmental conditions under which they grow. Proteobacteria belonging to the genus Vibrio and Photobacterium are dominant gut microbiota of the aquatic invertebrates (Rungrassamee et al. 2016; Holt et al. 2020). The gut's longitudinal axis is expected to have an impact on microbial profiles because it has diverse shapes and functions that put varied pressures on microbial selection (Holt et al. 2020).

3.3 Gut Microbiome of Aquacultured Finfishes

Freshwater fish (Bucio et al. 2006) and marine fish (Nelson et al. 2013) both have shown changes in gut microbiota composition associated to the captive state. In general, captive fish breeding and rearing involve manipulating a variety of factors, including the climate and diets (Diwan et al. 2022). To control disease outbreaks due to stress farmers use antibiotics indiscriminately in industrial farming systems, which reduces the microbial diversity in the guts of aquaculture animals (Navarrete et al. 2008). Beside this an artificial diet can also lead to alterations in the microbiota in fish GIT (Clements et al. 2014). Dhanasiri et al. (2011) described alterations in the gut microbiota of wild Atlantic cod after captive rearing, where diversity of bacterial species significantly decreased after feeding with an artificial diet for 6 weeks. However, Holben et al. (2002) found that in contrast to their native habitat, farmed Atlantic salmon exhibited a more diverse gut microbiome. *Acinetobacter junii* and a novel *Mycoplasma* phylotype were found to predominate in the guts of farmed Atlantic salmon in Scotland and Norway, respectively, when the fish were fed various diets. According to Llewellyn et al. (2016), Proteobacteria predominate in the gut of salmon throughout the life cycle, suggesting that a core microbiota exists in the gut despite many altering conditions. Roeselers et al. (2011) provided additional evidence in favour of the "core microbiota" theory. Similarities in the gut microbiota were observed in zebrafish that were taken from natural habitat and reared in laboratories for decades. Salmonoid diets that include plant-based proteins have been linked to a variety of intestinal problems as well as changes in the intestinal flora (Bakke-McKellep et al. 2007; Green et al. 2013). The increase in gut bacterial diversity was linked to soybean meal-induced enteritis in Atlantic salmon (Bakke-McKellep et al. 2007). According to Green et al. (2013), salmon fed soy protein concentrate exhibited intestinal issues and a higher bacterial diversity, all of which are not commonly associated with marine fish. Ringø et al. (2007) reported that the GIT of Atlantic cod fed soybean meal was dominated by Gram-negative bacteria *Chryseobacterium* spp. and *Psychrobacter glacincola*, as well as Gram-positive bacteria belonging to *Carnobacterium*, whereas Atlantic cod fed fishmeal were found dominated by Gram-positive bacteria of the genera *Brochothrix* and *Carnobacterium*. For comparison and a better understanding of the effects of captive rearing, more study is required on the gut microbiota of wild fish of both freshwater and marine origin.

3.4 Gut Microbiota and Fish Health

Intestinal microbiota diversity has been suggested as a biomarker of fish health (Clarke et al. 2014), and low microbiota diversity is highly associated with fish disease (He et al. 2017; Li et al. 2017). The association of gut microbiome with health status of the host is well recognized. Research to manipulate the microbiome community to enhance immunity and health of the host is prioritized now-a-days to boost up the aquaculture industry all over the world (Berry et al. 2012; Clemente et al. 2012; Subramanian et al. 2014). Olivier et al. (1981) reported that farmed fish succumbed to infection due to poor microbiome interaction in their gut. Since the natural microbiome is crucial for maintaining a healthy aquatic environment in finfish farming, it is frequently supplied artificially to fulfil several functional roles. Lai et al. (2015) and Xiong et al. (2014) opined that to develop a healthy microbiome in the gastro-intestine of finfish, regulated aquafarms must maintain the water quality of their ponds. A more diversified gut microbial community has a higher chance of possessing species with traits that are resistant to an invasive pathogen (Fargione and Tilman 2005) providing the host with more protection (De Schryver and Vadstein 2014; Johnson et al. 2008; Zhu et al. 2016). Fish in better health have been shown to have higher alpha diversity (mean species diversity at the habitat level) than fish with diseases, such as largemouth bronze gudgeon (*Coreius guichenoti*) (Liu et al. 2016), crucian carp (*Carassius auratus*) (Li et al. 2017), and ayu (*Plecoglossus altivelis*) (Nie et al. 2017), which may be due to the fact that invading pathogens outcompete the gut commensals resulting in a reduction of diversity. Oyarbide et al. (2015) also reported that gnotobiotic zebrafish (*Danio rerio*, Hamilton 1822) were more sensitive to pathogens. Similarly, antibiotic use typically results in decreased gut microbial diversity, which makes it easier for foreign pathogens to colonize the body (He et al. 2017). Gut microbiome composition has found to be closely related to the health status of shrimp also. Hence, monitoring and manipulation of microbial consortium in the aquaculture system play a significant part in maintaining soil and water quality, hence limiting the spread of pathogenic microbial diseases among aquatic species. Moreover, thorough knowledge about the microbiome of healthy and diseased aquaculture environment is very essential for manipulating them according to the need, thus reducing the use of chemicals and antibiotics in the aquaculture system. A lack of knowledge of relevant microbial interactions and the overall ecology of the system can result in hampering production.

3.4.1 Gut Microbiota and Immune Response

The relationship between gut microbiota, immunity, and disease is well understood in fish (Nie et al. 2017). Host immunity is closely linked to microbiota composition as gut microbes promote key immune tolerance in different organs (Belkaid and Hand 2014; Zheng et al. 2020). Gut bacteria directly influence innate immunity in skin (Lai et al. 2010; Wanke et al. 2011), respiratory tract (Ichinohe et al. 2011), and development of gut-associated lymphoid tissue (GALT) and regulatory T cells in the intestines (Jandhyala et al. 2015) of higher vertebrates. Symbiosis of gut microbiome causes poor host immunity (Reikvam et al. 2011) often leading to auto-immune disorders (Scher and Abramson 2011; Jangi et al. 2016). Likewise, the effect of commensal bacteria contributing to the development and homeostasis of innate immune cells (Khosravi et al. 2015; Balmer et al. 2014; Deshmukh et al. 2014) and adaptive immunity (Zhao and Elson 2018; Honda and Littman 2016) has been well documented. Contrarily, various immunological processes of the host also regulates the microbiota composition, including physical barriers (i.e., host-secreted mucus layers (Schroeder 2019; Okumura and Takeda 2018), host-produced antimicrobial compounds (Salzman 2010), recognition molecules (i.e., pattern recognition receptors [PRRs] (Chu and Mazmanian 2013) and associated signalling (Kubinak et al. 2015)) and effector responses (i.e., secretion of antibodies (Nakajima et al. 2018)). Studies show that adding amino acids and protein hydrolysate to fish feed diets can boost defences and decrease the amount of harmful bacteria in fish guts (Bui et al. 2014; Egerton et al. 2018; Khosravi et al. 2015; Kiron 2012) may be through altering the composition of the gut microbiome in cultivable finfish (Delcroix et al. 2015) and enhancing antimicrobial activity (Sila et al. 2014). Impressively, Ringø et al. (2007) and Montero et al. (2010) reported significant enhancements in gut microbiome quality and resistance to pathogenic bacteria when fish oils in fish feed were altered with plant oils as fish oil contains a high amount of marine polyunsaturated fatty acids (Merrifield and Rodiles 2015). Strong linkages between gut microbiota and immunity have been found (Raulo 2018), which is relevant for fish because they are constantly contact with water, a source of pathogenic and opportunistic commensal microbes (Ellis 2001). Additionally, fish raised intensively are frequently stocked at high densities, making it simpler for microbes to spread. In order to avoid unfavourable microbial colonization, it is crucial for aquaculture to maintain a microbial varied gut microbiome (Balcázar et al. 2006). Although the mechanisms are not fully understood, some significant processes have been found. For instance, the probiotic bacteria genera *Bacillus* and *Lactobacillus*, which are frequently employed in aquaculture, can increase phagocytic activity (Chen et al. 2019), the number of goblet cells that produce the mucus layer (Popovic 2017), and the expression of inflammatory cytokines in the fish gut (He et al. 2017). Furthermore, it has been discovered that bacteria-induced expression of myeloperoxidase, an enzyme that enables neutrophil granulocytes to carry out antimicrobial activity, is present in gnotobiotic zebrafish (*Danio rerio*) compared to conventionally grown zebrafish (Rawls et al. 2004). Colonizing microbes can also modulate host gene expression to create favourable gut environments, thereby constraining invasion by pathogens, while also promoting expression of proinflammatory and antiviral mediators genes, leading to higher viral resistance (Galindo-Villegas et al. 2012). The immune system and gut microbiome interact bilaterally, as seen in fish secretory immunoglobulins, which fish use to detect and coat intestinal bacteria to prevent them from colonizing the gut epithelium (Zhang et al. 2010). Similar to this, a causal chain between nutrition, immunity, and microbiome was found in wild three-spined sticklebacks (*Gasterosteus aculeatus*), revealing the influence of diet on fish immunity and, consequently, the microbial composition of the gut (Friberg et al. 2019). In aquaculture, where nutrition is frequently manipulated, it is essential to comprehend how microbes, hosts, and environments interact. Studies on gnotobiotic zebrafish show that the gut microbiota improves β-catenin's stability by activating Wnt signalling, favouring intestinal cell proliferation over typical ontogenesis (Cheesman and Guillemin 2007; Rawls et al. 2004). Conventionally grown zebrafish have a larger concentration of genes linked to epithelial proliferation and innate immune response than germ-free zebrafish (Rawls et al. 2004). However, NF-κB and its target genes can be robustly activated in both intestinal and extra-intestinal tissues in germ-free zebrafish that have a commensal microbiota (Kanther and Rawls 2010). Similar to this, commensal colonization in larvae increases neutrophils and promotes proinflammatory genes through the TLR/MyD88 signalling pathway and phagocytes, which can improve zebrafish disease resistance (Galindo-Villegas et al. 2012). Byup regulating pro-IL-1β, the gut microbiome specifically activates intestinal macrophages. By recruiting neutrophils, the mature form of IL-1β (triggered by pathogen infection) prepares macrophages to annihilate pathogens (Kamada et al. 2013). During healthy zebrafish development, it has been shown that the gut microbiota and the transcription of secreted immunoglobulin M (sIgM, a marker for the development of the adaptive immune system) are significantly correlated (Stephens et al. 2016). The gut microbiota of wild-type zebrafish differs from that of functional B- and T-cell receptor immune-deficient zebrafish in that it is more distinctive and its assembly is more determined (Stagaman et al. 2017). Furthermore, the fish gut microbiota is essential for direct inhibition as well as epithelium maturation and renewal, which regulate immune responses (Wang et al. 2017). Goblet cells normally release mucus, which serves as a barrier to prevent microbial migration out of the intestinal lumen (Ringø et al. 2007). Additionally, a developed gut mucosa is necessary for pattern recognition receptors (PRR) to discriminate infections from commensals (PRRs, such as toll-like receptors, RIG-I-like receptors, NOD-like receptors, and AIM2-like receptors), which regulate immunological responses by detecting bacterial antigens and activating signalling cascades (Perez et al. 2010). For instance, in addition to recognizing conserved pathogen structures, the toll-like receptor family, a prominent component of PRRs, may attract and control the inflammatory and immunological cells that trigger and regulate systemic immune responses (Fasano and Shea-Donohue 2005). Additionally, commensals might protect the host by denying resources to invasive pathogens, secreting a variety of antibacterial compounds, and occupying the niche (de Bruijn et al. 2017). On the other hand, innate immune responses such phagocytic and alternative complement pathway activity, which defend fish against infections, are correlated with probiotic colonization (Balcázar et al. 2006). Despite considerable development in this arena, preliminary knowledge regarding cross talk between fish immunity and microbiota composition is still scarce.

3.4.2 Gut Microbiota Correlates with Occurrence of Diseases in Finfish

Apart from improving the fish production, intensification in aquaculture resulted in different fish health issues and disease incidences. Increasing the expansion of the aqua sector is primarily concerned with preventing infectious diseases brought on by bacteria, viruses, and parasites. Strategies to control disease outbreak are crucial for increasing the fish production (FAO 2018). For controlling diseases in aquaculture systems and protecting fish from emerging harmful pathogens, the introduction of beneficial microbes in the aquaculture farming is the only promising approach as an alternative to antibiotics (de Bruijn et al. 2017). Li et al. (2017) while working on carps of healthy and diseased condition reported that in healthy fish *Cetobacterium*, *Cyanobacteria,* and *Clostridiaceae* were more abundant, whereas *Aeromonas*, *Vibrio,* and *Shewanella* were more abundant in diseased individuals. Similar observations were reported in Atlantic salmon (Llewellyn et al. 2014, 2016). The bacterial genus *Aeromonas* was found higher in abundance in diseased fish compared to healthy fish, indicating that the presence of a healthy microbiome in healthy fish completely inhibited the harmful expression of *Aeromonas*. Beneficial bacterial strains have been screened and characterized to treat the bacterial dysbiosis (Gagliardi et al. 2018).

Actually, pathogen incessantly interacts with the host microbiota. The interaction is dependent on the available resources and presence of antimicrobial compounds (Xiong et al. 2017). Dominance of pathogens over beneficial microbiota can create microbiota dysbiosis. Beneficial microbiome in host organisms do not cause any diseases or disorders but when the balance of the microbial community is interrupted, pathogenic diseases proliferate and can pose a health risk (Moya and Ferrer 2016; Montalban-Argues et al. 2015; Romero et al. 2014; Turner et al. 2013). A variety of factors, including changes in the environment and the usage of numerous antibiotic drugs, can cause an imbalance in the microbial community. The microbiomes or the genes of the microbial community associated with the gut, have many health benefits. The microbiota in the gut makes vitamins, aids in food digestion, stores nutrients, and supports a healthy metabolism (Merrifield and Rodiles 2015; Salinas et al. 2015). Toxins and harmful bacteria cannot enter the body through the intestinal wall because of the integrity of the cell connections that make up the intestinal lining, which is maintained by the microbiomes. Additionally, the gut microbiome and the immune system and the control of inflammation are closely related. Villamil et al. (2020) also mentioned a close relationship between aquaculture productivity and microbial composition and proposed a gnotobiotic model, which implies that changes in microbial composition of the digestive tract reflects the overall health and performance in animals (Duan et al. 2020; Villamil et al. 2020). However, the concept of such model in aquaculture system is yet to confirm (Zhang et al. 2018), but the spread of harmful microorganisms or the loss of helpful bacteria can be used as indicators to assess productivity. However, further investigation is necessary to establish this kind of concept. The dominance of a specific group of bacteria has also been reported in the gut of shellfish playing an important role in digestion and immunity (Holt et al. 2020). The microbiomes found in fish gut have the ability to produce number of compounds and these compounds offer several benefits. Considering the beneficial and vital role played by microbiome genes, many biotech and pharmaco-companies are showing a strong interest in developing these products for market release (Table 3.1).

3.5 Modulation of Fish Gut Microbiota for Improved Health: Supplementation of Synbiotics

Fish with healthy intestinal microbiota have been shown to be more resistant to disease, support in immune system development, and have overall better health. Changes in feed formulations, including additional high-quality lipids and proteins in the diet as well as the addition of probiotics and prebiotics, are the strategy used to manipulate the gut microbial community (Desai et al. 2012). Gut microbial compositions are found correlated to the source of proteins (Desai et al. 2012); their quantity (Geurden et al. 2014) and chemical structure (Kotzamanis et al. 2007) in the diet. Peptides and glycopeptides, released through protein digestion can modulate the condition and activity of the intestinal cells as well as the microbiota of the gut. Short peptides can be added to the diet to change the composition of the bacteria in the gut by giving the bacteria suitable substrates and promoting their growth (Kotzamanis et al. 2007). Some peptides produce antimicrobial activity and help to protect against pathogenic bacteria (Sila et al. 2014). Several researchers have described how dietary management alters the microbiome. It has been also confirmed that the increased lipid concentrations resulted in a more diverse gut microbial community in fish (Ringø and Birkbeck 1999). The studies of Lødemel et al. (2001) have shown that using plant oils can increase a fish's resistance to pathogenic bacteria because these natural plant oils are deficient in marine polyunsaturated fatty acids, arachidonic acid, eicosapentaenoic acid, and docosahexaenoic acid. Several studies have confirmed that the gut microbial communities can also be manipulated with the application of probiotics (Asaduzzaman et al. 2018; Siriyappagouder et al. 2018). Live beneficial microorganisms called probiotics can give the host a number of health advantages when given in sufficient doses. Probiotics alter digestion by enhancing both microbial enzymes and beneficial microbe activity (Hoseinifar et al. 2019). Additionally, they enhanced intestinal microbiological characteristics, indicating improved food digestion and microvilli morphology (Mohapatra et al. 2012). A further definition of "dietary prebiotics" was given as "a selectively fermented ingredient that results in specific changes in the composition and/or activity of the gastrointestinal microbiota, thus conferring

Table 3.1 Fish gut microbiota with health benefits

Gut microbes	Fish	Role	Characterization method	References
Obligate anaerobes (Bacteroidaceae) predominant over facultative anaerobes (Vibrionaceae and Enterobacteriaceae)	Nile tilapia (*Tilapia nilotica*), goldfish (*Carassius auratus*), carp (*Cyprinus carpio*), rainbow trout (*Salmo gairdneri*), dace (*Tribolodon hakonensis*), and Ayu (*Plecoglossus altivelis*)	Ability to ferment glucose, fructose, mannose, lactose, galactose, etc.	Isolated at random using plates of modified EG agar and NBGT-1/3S agar	Sakata et al. (1980)
Aeromonas, *Plesiomonas*, *Pseudomonas*, *Moraxella*, *Acinetobacter*, *Staphylococcus*, *Bacillus*, *Bacteroides* type A, and *Clostridium*	Japanese eel (*Anguilla japonica*), carp (*Cyprinus carpio*), goldfish (*Carassius auratus*), Ayu (*Plecoglossus altivelis*), tilapia (*Tilapia nilotica*), and channel catfish (*Ictalurus punctatus*)	Bacteroides type A, with the potential of producing vitamin B_{12}	Serial dilution and plating	Sugita et al. (1991)
Aeromonas, *Pseudomonas*, *Clostridium*, Bacteriodaceae	*Plecoglossus altivelis*, *Cyprinus carpio*, *Anguilla japonica*, *Oreochromis niloticus*, *Ictalurus punctatus*	Amylase production and helping in digestion of starch	Serial dilution and plating	Sugita et al. (1997)
Xenorhabdus sp.	*Panaque maccus*	Cellulase production	Serial dilution, repeated plating and streaking. Identification using BIOLOG identification system and Bergey's Manual of Systematic Bacteriology (1979)	Nelson et al. (1999)
Lactococcus garvieae, *Pediococcus acidilactici*, and *Enterococcus*	*Cyprinus carpio*	Intestinal lactic acid producing bacterial strains	Morphological, physiological, biochemical tests, 16S rRNA sequencing, and DNA-DNA hybridization	Cai et al. (1999)

(continued)

Table 3.1 (continued)

Gut microbes	Fish	Role	Characterization method	References
Lactococcus raffinolactis	Silver carp (*Hypophthalmichthys molitrix*), common carp (*Cyprinus carpio*), channel catfish (*Ictalurus punctatus*), and deep bodied crucian carp (*Carassius cuvieri*)	Predominant intestinal LAB	Composition analysis of intestinal bacteria	Hagi et al. (2004)
Bacillus sp.	*Oreochromis mossambicus*	Bacteriocin production	Plating in *Lactobacillus bulgaricus* agar plates	Vijaybaskar and Somasundaram (2008)
Bacteroides type A strains such as *Cetobacterium somerae*	Goldfish (*Carassius auratus*), common carp (*Cyprinus carpio*), and Mozambique tilapia (*Oreochromis mossambicus*)	High vitamin B_{12} producers and vancomycin-resistant bacteria	Conventional isolation, phenotypic and phylogenetic characterisation	Tsuchiya et al. (2008)
Bacillus sp.	*Labeo rohita*, *Catla catla*, *Cirrhinus mrigala*, *Labeo bata*, *Labeo calbasu*, *Oreochromis niloticus*, *Anabas testudineus*, *Cyprinus carpio*, *Hypophthalmichthys molitrix*, *Ctenopharyngodon idella*	Phytase production	Culture-dependent methods, 16S rRNA sequence analysis	Roy et al. (2009)
Citrobacter, *Enterobacter* sp., *Bacillus* sp.	*Labeo rohita*, *Catla catla*, *Cirrhinus mrigala*	Amylase, cellulase, and protease production	Culture-dependent isolation techniques, Quantitative enzyme assay, 16S rRNA gene sequence analysis	Ray et al. (2010)
Pediococcus acidilactici, *Enterococcus faecalis*	Rohu (*Labeo rohita*) and catla (*Catla catla*)	Capable to ferment fish processing waste	Isolated from fish waste using Mann-Rogosa-Sharpe (MRS) agar	Rai et al. (2011)

Rhodococcus sp.	*Labeo rohita*, *Catla catla*, *Hypophthalmichthys molitrix*, and *Labeo bata*	Extracellular protease, amylase, cellulase, lipase, and phytase production	Morphological, physiological, and biochemical tests	Khan et al. (2011)
Aeromonas spp., *Pseudomonas* spp., *Plesiomonas* spp., *Vibrio* spp., *Shewanella* spp., and *Cetobacterium* spp.	Zebrafish (*Danio rerio*)	Supports the study of coevolved relationships between host and microbiota	16S rRNA gene sequence analysis	Roeselers et al. (2011)
Bacillus subtilis	*Cyprinus carpio*	Probiotic and antagonistic activity against fish pathogen *Aeromonas hydrophila*	Microscopic and biochemical examination	Al-Faragi and Alsaphar (2012)
Bacillus sp. and *Acetobacter pasteurianus*	*Pterophyllum scalare*	Antagonistic activity against Aeromonas and Probiotic capabilities	Immunohistochemistry and DNA sequencing using PCR	Dosta et al. (2012)
Bacillus subtilis and *Bacillus atrophaeus*	*Labeo bata*, *Gudusia chapra*	Phytase production	Phenotypic characterization and 16S rRNA gene sequence analysis	Khan and Ghosh (2012)
Bacillus megaterium, *Serratia marcescens*, and *Pontibacter* sp.	*Labeo rohita*	Proteases production	Phylogenetic analysis of the partial 16S rRNA gene sequence	Sumathi et al. (2012)
Aeromonas hydrophila, *Shewanella putrefaciens*, *Vibrio cholerae*, *Staphylococcus* sp., *Vibrio vulnificus*, *Pantoea* sp., *Pasteurella pneumotropica*, *Corynebacterium urealyticum*, and *Micrococcus* sp.	*Cyprinus carpio* and *Clarias gariepinus*	–	Phenotypic characterization	Uddin and Al-Harbi (2012)

(continued)

Table 3.1 (continued)

Gut microbes	Fish	Role	Characterization method	References
Enterobacter asburae, *Pichia kudriavzevii*, *Candida tropicalis*, and *Candida parapsilosis*	*Catla catla*, *Hypophthalmichthys molitrix*, *Labeo rohita*, *Ctenopharyngodon idella*, *Cirrhinus mrigala*, *Cyprinus carpio*, *Labeo bata*, *Labeo calbasu*, *Oreochromis mossambicus*, *Oreochromis niloticus*	Tannase production which can overcome the antinutritional factors in the feed	16S/26S rRNA sequencing following nucleotide BLAST	Mandal and Ghosh (2013)
Cetobacterium somerae, *Plesiomonas shigelloides*, *Fusobacterium mortiferum*, and *Aeromonas* sp.	Channel catfish (*Ictalurus punctatus*), Largemouth bass (*Micropterus salmoides*), and Bluegill (*Lepomis macrochirus*)	Strong candidates for probiotics	16S rRNA pyrosequencing	Larsen et al. (2014)
Pichia kudriavzevii and Candida rugosa	Mrigal (*Cirrhinus mrigala* and Tilapia (*Oreochromis niloticus*)	Extracellular enzyme producers	18S rDNA sequence analysis	Banerjee and Ghosh (2014)
Actinobacteria	*Schizothorax zarudnyi* and *Schizocypris altidorsalis*	Potential to produce biologically active compounds	Classical cultivation techniques, repetitive sequence-based PCR (rep-PCR), partial and full 16S rDNA sequencing followed by phylogenetic analysis	Jami et al. (2015)
Bacillus sp.	*Channa striatus*, *Mystus vittatus*, *Labeo calbasu*, and *Cirrhinus mrigala*	Antibacterial activity against *Aeromonas hydrophila*, *A. salmonicida*, *A. sobria*, and *Pseudomonas fluorescens*	Partial 16S rRNA gene sequence analysis.	Nandi et al. (2017)

Dechloromonas, *Methylocaldum*, *Planctomyces*, *Rhodobacter*, *Caulobacter*, *Flavobacterium*, and *Pseudomonas*	Grass carp (*Ctenopharyngodon idellus*)	Causes changes in amino acid metabolism, xenobiotics biodegradation and metabolism, and carbohydrate metabolism	NGS-based 16S rRNA sequencing of gut microbiota of grass carp suffering from enteritis	Tran et al. (2018)
Aneurinibacillus aneurinilyticus and *Stenotrophomonas* sp.	*Labeo calbasu*	Antibacterial activity against *Aeromonas hydrophila*, probiotic abilities resulting in enhanced growth, nutrient retention, and immune response	Serial dilution, plating, Biochemical and physiological examination, Well diffusion assay, 16S rRNA sequencing, and immunohistochemistry	Bhatnagar and Dhillon (2019)
Gammaproteobacteria, *Fosobacteria*, *Actinobacteria*, *Anerolineae*, *Betaproteobacteria*, *Deinococci*, *Clostridia*, and *Deltaproteobacteria*	Freshwater pufferfish (*Tetraodon cutcutia*)	Exhibited a high prevalence of genes associated with core housekeeping functions (Synthesis of cofactors, vitamins, prosthetic groups, pigments, amino acids and its derivatives, carbohydrate, and protein metabolism)	Metagenomic approach	Deb et al. (2020)

benefit upon host health" (Gibson et al. 2010). Galactooligosaccharides, arabinoxylooligosaccharides, fructooligosaccharides, chitooligosaccharide, mannanoligosacch-arides, inulin, and levan were the common prebiotics used in aquaculture (Ringø et al. 2014). Combination of probiotics and prebiotics, termed as synbiotics, has emerged as a promising approach for disease biocontrol in farmed fish (Hoseinifar et al. 2015a, b, c, 2017a, b, c, d). Overall, providing fish synbiotic treatments had positive benefits on their immune systems, survival, growth, gut microbiota, intestinal absorption levels, and overall health status (González-Félix et al. 2018; Hassaan et al. 2018). According to reports, *Bacillus* spp. and yeasts may play a vital function as probiotics with beneficial qualities that demonstrate useful roles in the growth and health of many cultivated species (Hosseini et al. 2016; Liu et al. 2018; Safari et al. 2018). Multiple studies have shown how dietary manipulation alters the microbiota. Numerous studies have been conducted in this area regarding the significant impacts of endogenous microbiota on the health condition of fish. For instance, the colonization of harmful germs depends critically on bacterial adhesion (Hoseinifar et al. 2015a, b, c, 2017a, b, c, d). Therefore, the ability of three fish intestine-derived probionts (*Lactobacillus plantarum* CLFP 238; *Lactobacillus fermentum* CLFP 242; and *Lactobacillus lactis* CLFP 101) to inhibit the adhesion of various pathogens (*Vibrio anguillarum*, *Yersinia ruckeri*, *Aeromonas salmonicida*, and *A. hydrophila*) to intestinal mucosae (Balcázar et al. 2006). According to Balcázar et al. (2006), only *L. lactis* CLFP 101 decreased the adherence of the four fish pathogens. Additionally, antibacterial properties of *L. lactis* CLFP 101 against the four pathogens were also investigated. A preventative impact against the adherence of *A. salmonicida* and *A. hydrophila* was also seen in the instance of *L. plantarum* CLFP 238. Additionally, the administration of *L. fermentum* CLFP 242 was found to reduce the adherence of three of the four pathogens. It is significant to note that all LAB strains had the ability to survive in fish guts with high bile concentrations and low pH levels, and they all had the ability to suppress bacterial adhesion. In a different study, the benefits of dietary *L. acidophilus* on the black swordtail (*Xiphophorus helleri*) were analysed. These impacts included performance, mucosal immunity, salinity stress tolerance, and gut microbiota. The findings indicated that *L. acidophilus* had positive impacts on fish health and growth performance when consumed as part of a healthy diet (Hoseinifar et al. 2015a, b, c). Lysozyme, protease, complements, lectins, and immunoglobulins have been identified as the primary elements of skin mucosal immunology, which defends fish against infections and serves as the first line of defence (Ángeles Esteban 2012). Similar to this, red sea bream, *Pagrus major*, showed significantly improved immunological parameters and higher growth performance in an experiment with nutritional supplementation of *L. rhamnosus* or/and *L. lactis* (Dawood et al. 2016). Additionally, culture-based microbiological studies showed a considerable increase in lactic acid bacteria and total bacteria in the gut microbiota (Dawood et al. 2016). Additionally, Askarian et al. (2011) observed that probiotic (*L. casei*) administration increased the total protein level of skin mucus in Caspian white fish (*Rutilus frisiikutum*) fry. It is suggested that the improved immune functions after dietary administration of probiotics can contribute to the enhanced skin-soluble protein content and skin mucus bactericidal activity. The innate immune system of fish, which serves as an antibacterial agent, is also significantly influenced by alkaline phosphatase (ALP) activities. However, while fish is administered with a probiotic supplementation meal, it has been found that the activity of ALP dramatically increases (Hoseinifar et al. 2015a, b, c). An enhanced mucosal immune response may be linked to the increased alkaline phosphatase activity. In addition, *L. curvatus*, a dietary probiotic, has been shown to have positive effects on beluga (*Huso huso*) and Persian sturgeon (*Acipenser persicus*) fry growth, digestive enzymes, and intestinal microbiota (Askarian et al. 2011). The best outcomes were seen when fish were fed 9×10^9 CFU g^{-1} of probiotic bacteria. Similar beneficial results were seen when 2×10^9 CFU g^{-1} *L. mesenteroides* was added to the diet of Persian sturgeon, demonstrating the significant potential of *Lactobacillus* spp. to colonize the digestive tract mucosa and cause health enhancement (Askarian et al. 2011). *Bacillus subtilis* E20 has been shown to enhance growth performance and develop resistance towards *Vibrio alginolyticus* infection in parrot fish (*Oplegnathus fasciatus*). It is worth noting that increasing probiotic levels improved the immunological characteristics of *O. fasciatus* while decreasing the growth ratio. Additionally, supplementing the diet with probiotics at a dose of 10^{10} CFU kg^{-1} significantly boosted resistance to *V. alginolyticus*. In a different study, Soltan and El-Laithy (2008) demonstrated that feeding the Nile tilapia (*O. niloticus*) a diet supplemented with *B. subtilis* decreased the AST and ALT activity, demonstrating a positive influence of *B. subtilis* on the control of the gut–liver axis in addition to the composition of the intestinal microbiota (Soltan and El-Laithy 2008). Additionally, dietary *L. acidophilus* may influence the protein profile of appetite, immune and skin mucus genes in goldfish (*Carassius auratus gibelio*) by significantly upregulating TNF-1α and TNF-2α gene expression and downregulating ghrelin gene expression. The V3-V4 16S rRNA gene massive sequencing, however, revealed notable alterations in the overall transitory microbial population and diversity, demonstrating different patterns of bacterial clumping between the supplements (probiotic, prebiotic, and combination probiotic and prebiotic) and the baseline diet. Additionally, oral administration of *Agaricus bisporus* and *L. casei* significantly increased growth-associated genes (igf1 and gh), mucosal immune-related genes (lyz, tnf-alpha, and il1b), as well as antioxidant-related genes (sod, cat) and the expression of specific genes associated to growth, nutrition, mucosal immunity, and antioxidant enzymes, along with skin mucus immunological factors. Therefore, the need to assess prebiotic and probiotic safety by learning more about their effects on gut microbiota activity is the most promising part of synbiotics studies in fish aquaculture. Furthermore, improved public acceptance of probiotic and prebiotic reasoning, more precise rules for safety claims, and preparedness may be of interest in accelerating the development of a more sustainable commercial aquaculture.

3.6 Conclusion

There are many factors that affect the gut microbiota, but it is still unclear how each one affects fish behaviour and physiology. Establishing a link between the microbiome's structure and function and the host's health status, age, genetic heritage, geographical location, and other unique characteristics is a critical difficulty for making meaningful findings. Careful analyses are also required to build a comparative picture based on research using various DNA extraction techniques from various sample types within a species. There is a need for comprehensive shotgun sequencing efforts to properly study the metabolic potential of the gut microbiota because the marker gene-based approach restricts the analysis to predictions of microbiome structure and uncover functional variation with diet or host associated factors. Considerable efforts have been made to examine the microbiome of the various commercially significant fish. It is currently widely assumed that microbiome diversity is associated to a variety of animal health issues, including the gastrointestinal tract food digesting intensity, energy production and metabolism, immune system dysfunction, infectious diseases, and the effects on nutrient absorption, development, and reproduction. Additionally, it has been suggested that the gut microbiome's composition can serve as an indicator of fish health. In this scenario, more research studies are required to determine the relationship between certain taxa of microbial species and fish disease, which may allow us to change the gut microbiota's composition to reduce fish disease. However, in such specialized fields of science, more positive research studies are needed. Moreover, additional study is required on the diversity of the microbiome, how it relates to different parameters, and how to identify these microorganisms down to the species level in different fish. It is now possible to quantify and categorize commensal microbiomes at the species level because of the availability of gene-based molecular techniques such as DNA sequencing and NGS technologies, including amplicon and shotgun techniques, and related bioinformatics expertise. The development of these molecular approaches inspired a number of researchers to study different functional aspects of bacteria, and in recent years, the discipline of microbiome metagenomics has grown massively. In order to change the gene structure and give microbes the necessary qualities, efforts are being undertaken to study the various roles of the different microbial species at the genomic levels. The bio-industries, especially aquaculture businesses, are looking for such cutting-edge research to create new solutions for preventing the spread of diseases in fish and to create aquaculture farming systems that are wholesome and sustainable. Probiotics are increasingly used instead of drugs and antibiotics to maintain the health state of aquaculture species. Additionally, it has been proposed that using probiotics encourages the growth and reproduction of cultivable organisms while also preventing the spread of disease. The probiotics are made from beneficial microbial species, and healthy microbial growth in the animal's digestive system normally maintains it healthy. Disease control is one of the biggest challenges in aquaculture, and several researchers have advised using bioinformatics skills to better understand the fundamental mechanisms by which infectious diseases harm their hosts. To minimize not only disease management but also other physiological abnormalities in cultivable species, much study is still required in this field. The bio-pharma and aquaculture industries are seeking for such creative research not only for their future possibilities, but also to provide assistance for the development of healthy and sustainable aquaculture farming.

Acknowledgment Authors are thankful to Director, ICAR Central Institute of Freshwater Aquaculture, Bhubaneswar, Odisha, India for providing facilities to operate DBNT Twining project. Financial support to FH and JKS from DBT TWNIING (BT/PR25141/NER/95/1039/2017) to execute the project is well acknowledged.

Conflict of Interest Authors declare no conflict of interest.

Author's Credit FH and JKS conceptualised the Manuscript. FH and AD wrote the Manuscript. JKS edited the full manuscript.

References

Al-faragi JK, Alsaphar SA (2012) Isolation of lactic acid bacteria (LAB) species as probiotic from intestinal contents of common carp Cyprinus carpio L. J Biotechnol Res Center 6(1):88–95

Amann RI, Ludwig W, Schleifer KH (1995) Phylogenetic identification and in situ detection of individual microbial cells without cultivation. Microbiol Rev 59:143–169

Ángeles Esteban M (2012) An overview of the immunological defenses in fish skin. Int Sch Res Notices 2012:853470

Asaduzzaman M, Sofia E, Shakil A, Haque NF, Khan MNA (2018) Host gut-derived probiotic bacteria promote hypertrophic muscle progression and up regulate growth-related gene expression of slow-growing Malaysian Mahseer Tor tambroides. Aquacult Rep 9:37–45. https://doi.org/10.1016/j.aqrep.2017.12.001

Askarian F, Kousha A, Salma W, Ringø E (2011) The effect of lactic acid bacteria administration on growth, digestive enzyme activity and gut microbiota in Persian sturgeon (Acipenser persicus) and beluga (Huso huso) fry. Aquac Nutr 17:488–497

Austin B (1982) Taxonomy of bacteria isolated from a coastal, marine fish-rearing unit. J Appl Bacteriol 53(2):253–268

Bakke-McKellep AM, Penn MH, Salas PM, Refstie S, Sperstad S, Landsverk T, Ringø E, Krogdahl A (2007) Effects of dietary soybean meal, inulin and oxytetracycline on intestinal microbiota and epithelial cell stress, apoptosis and proliferation in the teleost Atlantic salmon (Salmo salar L.). Br J Nutr 97:699–713

Balcázar JL, Decamp O, Vendrell D, De Blas I, Ruiz-Zarzuela I (2006) Health and nutritional properties of probiotics in fish and shellfish. Microb Ecol Health Dis 18:65–70. https://doi.org/10.1080/08910600600799497

Balmer ML, Schurch CM, Saito Y, Geuking MB, Li H, Cuenca M, Kovtonyuk LV, McCoy KD, Hapfelmeier S, Ochsenbein AF, Manz MG, Slack E, Macpherson AJ (2014) Microbiota-derived compounds drive steady-state granulopoiesis via MyD88/TICAM signaling. J Immunol 193: 5273–5283

Banerjee S, Ghosh K (2014) Enumeration of gut associated extracellular enzyme-producing yeasts in some freshwater fishes. J Appl Ichthyol 30(5):986–993. https://doi.org/10.1111/jai.12457

Belkaid Y, Hand TW (2014) Role of the microbiota in immunity and inflammation. Cell 157:121–141

Berry D, Schwab C, Milinovich G, Reichert J, Mahfoudh KB, Decker T, Engel M, Hai B, Hainzl E, Heider S, Kenner L, Muller M, Rauch I, Strobl B, Wagner M, Schleper C, Urich T, Loy A

(2012) Phylotype-level 16S rRNA analysis reveals new bacterial indicators of health state in acute murine colitis. ISME J 6:2091–2106. https://doi.org/10.1038/ismej.2012.39
Bhatnagar A, Dhillon O (2019) Characterization, screening, and application of bacteria with probiotic properties isolated from the gut of Labeo calbasu (Hamilton). Fisheries Aquat Life 27(4):178–189
Bøgwald J, Dalmo RA (2014) Gastrointestinal pathogenesis in aquatic animals. In: Merrifield DL, Ringo E (eds) Aquaculture nutrition: gut health, probiotics and prebiotics. Wiley, Hoboken, pp 53–74
Bucio A, Hartemink R, Schrama JW, Verreth J, Rombouts FM (2006) Presence of lactobacilli in the intestinal content of freshwater fish from a river and from a farm with a recirculation system. Food Microbiol 23:476–482. https://doi.org/10.1016/j.fm.2005.06.001
Bui HTD, Khosravi S, Fournier V, Herault M, Lee KJ (2014) Growth performance, feed utilization, innate immunity, digestibility and disease resistance of juvenile red sea bream (Pagrus major) fed diets supplemented with protein hydrolysates. Aquaculture 418:11–16. https://doi.org/10.1016/j.aquaculture.2013.09.046
Cahill MM (1990) Bacterial flora of fishes: a review. Microb Ecol 19(1):21–41. https://doi.org/10.1007/BF02015051
Cai Y, Suyanandana P, Saman P, Benno Y (1999) Classification and characterization of lactic acid bacteria isolated from the intestines of common carp and freshwater prawns. J Gen Appl Microbiol 45(4):177–184
Cheesman SE, Guillemin K (2007) We know you are in there: conversing with the indigenous gut microbiota. Res Microbiol 158:2–9. https://doi.org/10.1016/j.resmic.2006.10.005
Chen CY, Chen PC, Weng FC, Shaw GT, Wang D (2017) Habitat and indigenous gut microbes contribute to the plasticity of gut microbiome in oriental river prawn during rapid environmental change. PLoS One 12:e0181427. https://doi.org/10.1371/journal.pone.0181427
Chen SW, Liu CH, Hu SY (2019) Dietary administration of probiotic Paenibacillus ehimensis NPUST1 with bacteriocin-like activity improves growth performance and immunity against Aeromonas hydrophila and Streptococcus iniae in Nile tilapia (Oreochromis niloticus). Fish Shellfish Immunol 84:59. https://doi.org/10.1016/j.fsi.2018.10.059
Chu H, Mazmanian SK (2013) Innate immune recognition of the microbiota promotes host-microbial symbiosis. Nat Immunol 14:668–675
Clarke SF, Murphy EF, O'Sullivan O, Lucey AJ, Humphreys M, Hogan A, Hayes P, O'Reilly M, Jeffery IB, Wood-Martin R, Kerins DM, Quigley E, Ross RP, O'Toole PW, Molloy MG, Falvey E, Shanahan F, Cotter PD (2014) Exercise and associated dietary extremes impact on gut microbial diversity. Gut 63(12):1913–1920
Clemente JC, Ursell LK, Parfrey LW, Knight R (2012) The impact of the gut microbiota on human health: an integrative view. Cell 148:1258–1270. https://doi.org/10.1016/j.cell.2012.01.035
Clements KD, Angert ER, Montgomery WL, Choat JH (2014) Intestinal microbiota in fishes: what's known and what's not. Mol Ecol 23:1891–1898
Cornejo-Granados F, Lopez-Zavala AA, Gallardo-Becerra L, Mendoza-Vargas A, Sánchez F, Vichido R, Brieba LG, Viana MT, Sotelo-Mundo RR, Ochoa-Leyva A (2017) Microbiome of Pacific White leg shrimp reveals differential bacterial community composition between wild, aquacultured and AHPND/EMS outbreak conditions. Sci Rep 7:11783. https://doi.org/10.1038/s41598-017-11805-w
Dawood MA, Koshio S, Ishikawa M, Yokoyama S, El Basuini MF, Hossain MS, Nhu TH, Dossou S, Moss AS (2016) Effects of dietary supplementation of Lactobacillus rhamnosus or/and Lactococcus lactis on the growth, gut microbiota and immune responses of red sea bream, Pagrus major. Fish Shellfish Immunol 49:275–285
de Bruijn I, Liu Y, Wiegertjes GF, Raaijmakers JM (2017) Exploring fish microbial communities to mitigate emerging diseases in aquaculture. FEMS Microbiol Ecol 94(1):161
De Schryver P, Vadstein O (2014) Ecological theory as a foundation to control pathogenic invasion in aquaculture. ISME J 8(12):2360–2368. https://doi.org/10.1038/ismej.2014.84

Deb S, Das L, Das SK (2020) Composition and functional characterization of the gut microbiome of freshwater pufferfish (Tetraodon cutcutia). Arch Microbiol 202(10):2761–2770. https://doi.org/10.1007/s00203-020-01997-7

Delcroix J, Gatesoupe FJ, Desbruyeres E, Huelvan C, le Delliou H, le Gall MM, Quazuguel P, Mazurais D, Zambonino-Infante JL (2015) The effects of dietary marine protein hydrolysates on the development of sea bass larvae, Dicentrarchus labrax, and associated microbiota. Aquac Nutr 21:98–104

Desai AR, Links MG, Collins SA, Mansfield GS, Drew MD, Van Kessel AG, Hill JE (2012) Effects of plant-based diets on the distal gut microbiome of rainbow trout (Oncorhynchus mykiss). Aquaculture 350–353:134–142. https://doi.org/10.1016/j.aquaculture.2012.04.005

Deshmukh HS, Liu Y, Menkiti OR, Mei J, Dai N, O'Leary CE, Oliver PM, Kolls JK, Weiser JN, Worthen GS (2014) The microbiota regulates neutrophil homeostasis and host resistance to Escherichia coli K1 sepsis in neonatal mice. Nat Med 20:524–530

Dhanasiri AK, Brunvold L, Brinchmann MF, Korsnes K, Bergh Ø, Kiron V (2011) Changes in the intestinal microbiota of wild Atlantic cod Gadus morhua L. upon captive rearing. Microb Ecol 61(1):20–30

Diwan AD, Harke SN, Panche AN (2022) Aquaculture industry prospective from gut microbiome of fish and shellfish: an overview. J Anim Physiol Anim Nutr 106(2):441–469

Dosta MDCM, Barrera TC, Fernandez FJ, Perrino LMR, Gutierrez HH, Surez SC (2012) Bacteria with probiotic capabilities isolated from the digestive tract of the ornamental fish Pterophyllum scalare. Probiot Anim 231:45954

Duan Y, Huang J, Wang Y, Zhang J (2020) Characterization of bacterial community in intestinal and rearing water of Penaeus monodon differing growth performances in outdoor and indoor ponds. Aquac Res 51:4279–4289

Egerton S, Culloty S, Whooley J, Stanton C, Ross RP (2018) The gut microbiota of marine fish. Front Microbiol 9:873. https://doi.org/10.3389/fmicb.2018.00873

Ellis AE (2001) Innate host defense mechanisms of fish against viruses and bacteria. Dev Comp Immunol 25:827–839. https://doi.org/10.1016/S0145-305X(01)00038-6

FAO (2018) The State of World Fisheries and Aquaculture 2018 – Meeting the sustainable development goals

Fargione JE, Tilman D (2005) Diversity decreases invasion via both sampling and complementarity effects. Ecol Lett 8:604–611

Fasano A, Shea-Donohue T (2005) Mechanisms of disease: the role of intestinal barrier function in the pathogenesis of gastrointestinal autoimmune diseases. Nat Clin Pract Gastroenterol Hepatol 2(9):416–422

Friberg IM, Taylor JD, Jackson JA (2019) Diet in the driving seat: natural diet-immunity-microbiome interactions in wild fish. Front Immunol 10:243. https://doi.org/10.3389/fimmu.2019.00243

Gagliardi A, Totino V, Cacciotti F, Iebba V, Neroni B, Bonfiglio G, Trancassini M, Passariello C, Pantanella F, Schippa S (2018) Rebuilding the gut microbiota ecosystem. Int J Environ Res Public Health 15(8):1679. https://doi.org/10.3390/ijerph15081679

Galindo-Villegas J, Garcia-Moreno D, de Oliveira S, Meseguer J, Mulero V (2012) Regulation of immunity and disease resistance by commensal microbes and chromatin modifications during zebrafish development. Proc Natl Acad Sci U S A 109:E2605–E2614. https://doi.org/10.1073/pnas.1209920109

Gallagher ML, Luczkovich JJ, Stellwag EJ (2001) Fish gut microbiota Characterization of the ultrastructure of the gastrointestinal tract mucosa, stomach contents and liver enzyme activity of the pinfish during development. J Fish Biol 58(6):1704–1713. https://doi.org/10.1111/j.1095-8649.2001.tb02324.x

Geurden I, Mennigen J, Plagnes-Juan E, Veron V, Cerezo T (2014) High or low dietary carbohydrate: protein ratios during first-feeding affect glucose metabolism and intestinal microbiota in juvenile rainbow trout. J Exp Biol 217(19):3396–3406. https://doi.org/10.1242/jeb.106062

Gibson GR, Probert HM, Loo JV, Rastall RA, Roberfroid MB (2010) Dietary modulation of the human colonic microbiota: updating the concept of prebiotics. Nutr Res Rev 17:259–275
Gomez GD, Balcaazar JL (2008) A review on the interactions between gut microbiota and innate immunity of fish. FEMS Immunol Med Microbiol 52(2):145–154
Gomez D, Sunyer JO, Salinas I (2013) The mucosal immune system of fish: the evolution of tolerating commensals while fighting pathogens. Fish Shellfish Immunol 35(6):1729–1739. https://doi.org/10.1016/j.fsi.2013.09.032. PMID: 24099804
González-Félix ML, Gatlin DM III, Urquidez-Bejarano P, de la Reé-Rodríguez C, DuarteRodríguez L, Sánchez F, Casas-Reyes A, Yamamoto FY, Ochoa-Leyva A, Perez-Velazquez M (2018) Effects of commercial dietary prebiotic and probiotic supplements on growth, innate immune responses, and intestinal microbiota and histology of Totoaba macdonaldi. Aquaculture 491:239–251
Green TJ, Smullen R, Barnes AC (2013) Dietary soybean protein concentrate-induced intestinal disorder in marine farmed Atlantic salmon, Salmo salar is associated with alterations in gut microbiota. Vet Microbiol 166:286–292
Hagi T, Tanaka D, Iwamura Y, Hoshino T (2004) Diversity and seasonal changes in lactic acid bacteria in the intestinal tract of cultured freshwater fish. Aquaculture 234(1-4):335–346. https://doi.org/10.1016/j.aquaculture.2004.01.018
Hanning I, Diaz-Sanchez S (2015) The functionality of the gastrointestinal microbiome in non-human animals. Microbiome 3:51
Hansen GH, Olafsen JA (1999) Bacterial interactions in early life stages of marine cold water fish. Microb Ecol 38(1):1–26. https://doi.org/10.1007/s002489900158
Harrison FC (1929) The discoloration of halibut. Can J Res 1:214–239
Hassaan M, Soltan M, Jarmołowicz S, Abdo H (2018) Combined effects of dietary malic acid and Bacillus subtilis on growth, gut microbiota and blood parameters of Nile tilapia (Oreochromis niloticus). Aquac Nutr 24:83–93
He S, Ran C, Qin C, Li S, Zhang H, De Vos WM, Ringø E, Zhou Z (2017) Anti-infective effect of adhesive probiotic lactobacillus in fish is correlated with their spatial distribution in the intestinal tissue. Sci Rep 7:13195. https://doi.org/10.1038/s41598-017-13466-1
Holben WE, Williams P, Gilbert MA, Saarinen M, Sarkilahti LK, Apajalahti JH (2002) Phylogenetic analysis of intestinal microflora indicates a novel Mycoplasma phylotype in farmed and wild salmon. Microb Ecol 44:175–185
Holt CC, Bass D, Stentiford GD, van der Giezen M (2020) Understanding the role of the shrimp gut microbiome in health and disease. J Invertebr Pathol 2020:107387. https://doi.org/10.1016/j.jip.2020.107387
Honda K, Littman DR (2016) The microbiota in adaptive immune homeostasis and disease. Nature 535:75–84
Hooper LV, Littman DR, Macpherson AJ (2012) Interactions between the microbiota and the immune system. Science 336(6086):1268–1273. https://doi.org/10.1126/science.1223490
Hoseinifar SH, Esteban MÁ, Cuesta A, Sun Y-Z (2015a) Prebiotics and fish immune response: a review of current knowledge and future perspectives. Rev Fish Sci Aquac 23:315–328
Hoseinifar SH, Mirvaghefi A, Amoozegar MA, Sharifian M, Esteban MÁ (2015b) Modulation of innate immune response, mucosal parameters and disease resistance in rainbow trout (Oncorhynchus mykiss) upon synbiotic feeding. Fish Shellfish Immunol 45:27–32
Hoseinifar SH, Roosta Z, Hajimoradloo A, Vakili F (2015c) The effects of Lactobacillus acidophilus as feed supplement on skin mucosal immune parameters, intestinal microbiota, stress resistance and growth performance of black swordtail (Xiphophorus helleri). Fish Shellfish Immunol 42:533–538
Hoseinifar S, Mirvaghefi A, Amoozegar M, Merrifield D, Ringø E (2017a) In vitro selection of a synbiotic and in vivo evaluation on intestinal microbiota, performance and physiological response of rainbow trout (Oncorhynchus mykiss) fingerlings. Aquac Nutr 23:111–118

Hoseinifar SH, Dadar M, Ringø E (2017b) Modulation of nutrient digestibility and digestive enzyme activities in aquatic animals: the functional feed additives scenario. Aquac Res 48: 3987–4000

Hoseinifar SH, Safari R, Dadar M (2017c) Dietary sodium propionate affects mucosal immune parameters, growth and appetite related genes expression: insights from zebrafish model. Gen Comp Endocrinol 243:78–83

Hoseinifar SH, Sun Y, Zhou Z (2017d) Prebiotics and synbiotics. In: Austin B, Newaj-Fyzul A (eds) Diagnosis and control of diseases of fish and shellfish. Wiley, Oxford

Hoseinifar SH, Doan HV, Dadar M, Ringø E, Harikrishnan R (2019) Feed additives, gut microbiota, and health in finfish aquaculture. In: Microbial communities in aquaculture ecosystems. Springer, Cham, pp 121–142

Hosseini M, Miandare HK, Hoseinifar SH, Yarahmadi P (2016) Dietary Lactobacillus acidophilus modulated skin mucus protein profile, immune and appetite genes expression in gold fish (Carassius auratus gibelio). Fish Shellfish Immunol 59:149–154

Ichinohe T, Pang IK, Kumamoto Y, Peaper DR, Ho JH, Murray TS, Iwasaki A (2011) Microbiota regulates immune defense against respiratory tract influenza A virus infection. Proc Natl Acad Sci U S A 108:5354–5359

Jami M, Ghanbari M, Kneifel W, Domig KJ (2015) Phylogenetic diversity and biological activity of culturable Actinobacteria isolated from freshwater fish gut microbiota. Microbiol Res 175:6–15. https://doi.org/10.1016/j.micres.2015.01.009

Jandhyala SM, Talukdar R, Subramanyam C, Vuyyuru H, Sasikala M, Reddy D (2015) Role of the normal gut microbiota. World J Gastroenterol 21:8787–8803

Jangi S, Gandhi R, Cox LM, Li N, von Glehn F, Yan R, Patel B, Mazzola MA, Liu S, Glanz BL, Cook S, Tankou S, Stuart F, Melo K, Nejad P, Smith K, Topçuolu BD, Holden J, Kivisäkk P, Chitnis T, De Jager PL, Quintana FJ, Gerber GK, Bry L, Weiner HL (2016) Alterations of the human gut microbiome in multiple sclerosis. Nat Commun 7:12015

Johnson CN, Barnes S, Ogle J, Grimes DJ, Chang Y-J, Peacock AD, Kline L (2008) Microbial community analysis of water, foregut, and hindgut during growth of pacific white shrimp, Litopenaeus vannamei, in closed-system aquaculture. J World Aquacult Soc 39:440. https://doi.org/10.1111/j.1749-7345.2008.00155.x

Kamada N, Seo SU, Chen GY, Núñez G (2013) Role of the gut microbiota in immunity and inflammatory disease. Nat Rev Immunol 13:321–335

Kanther M, Rawls JF (2010) Host-microbe interactions in the developing zebrafish. Curr Opin Immunol 22:10–19. https://doi.org/10.1016/j.coi.2010.01.006

Khan A, Ghosh K (2012) Characterization and identification of gut-associated phytase-producing bacteria in some freshwater fish cultured in ponds. Acta Ichthyol Piscat 42(1):37–45

Khan A, Mandal S, Samanta D, Chaterjee S, Ghosh K (2011) Phytase-producing Rhodococcus sp. (MTCC 9508) from fish gut: a preliminary study. Proc Zool Soc 64(1):29–34

Khosravi S, Rahimnejad S, Herault M, Fournier V, Lee CR, Dio Bui HT (2015) Effects of protein hydrolysates supplementation in low fish meal diets on growth performance, innate immunity and disease resistance of red sea bream Pagrus major. Fish Shellfish Immunol 45:858–868. https://doi.org/10.1016/j.fsi.2015.05.039

Kim DH, Brunt J, Austin B (2007) Microbial diversity of intestinal contents and mucus in rainbow trout (Oncorhynchus mykiss). J Appl Microbiol 102(6):1654–1664. https://doi.org/10.1111/j.1365-2672.2006.03185.x

Kiron V (2012) Fish immune system and its nutritional modulation for preventive health care. Anim Feed Sci Technol 173:111–133. https://doi.org/10.1016/j.anifeedsci.2011.12.015

Korsnes K, Nicolaisen O, Skar CK, Nerland AH, Bergh Ø (2006) Bacteria in the gut of juvenile cod Gadus morhua fed live feed enriched with four different commercial diets. ICES J Mar Sci 63(2):296–301

Kotzamanis YP, Gisbert E, Gatesoupe FJ, Infante JZ, Cahu C (2007) Effects of different dietary levels of fish protein hydrolysates on growth, digestive enzymes, gut microbiota, and resistance to Vibrio anguillarum in European sea bass (Dicentrarchus labrax) larvae. Comp Biochem Physiol A Mol Integr Physiol 147(1):205–214. https://doi.org/10.1016/j.cbpa.2006.12.037

Kubinak JL, Petersen C, Stephens WZ, Soto R, Bake E, O'Connell RM, Round JL (2015) MyD88 signaling in T cells directs IgA-mediated control of the microbiota to promote health. Cell Host Microbe 17:153–163

Lai Y, Cogen AL, Radek KA, Park HJ, Macleod DT, Leichtle A, Ryan AF, Di Nardo A, Gallo RL (2010) Activation of TLR2 by a small molecule produced by Staphylococcus epidermidis increases antimicrobial defense against bacterial skin infections. J Invest Dermatol 130:2211–2221

Lai HC, Ng TH, Ando M, Lee CT, Chen IT, Chuang JC, Mavichak R, Chang SH, Yeh MD, Chiang YA, Takeyama H, Hamaguchi HO, Lo CF, Aoki T, Wang HC (2015) Pathogenesis of acute hepatopancreatic necrosis disease (AHPND) in shrimp. Fish Shellfish Immunol 47:1006–1014. https://doi.org/10.1016/j.fsi.2015.11.008

Larsen AM, Mohammed HH, Arias CR (2014) Characterization of the gut microbiota of three commercially valuable warmwater fish species. J Appl Microbiol 116(6):1396–1404. https://doi.org/10.1111/jam.12475

Lauzon HL, Gudmundsdottir S, Steinersson A, Oddgeirsson M, Petursdottir SK, Reynisson E, Bjornsdottir R, Gudmundsdottir BK (2010) Effects of bacterial treatment at early stages of Atlantic cod (Gadus morhua L.) on larval survival and development. J Appl Microbiol 108(2):624–632

Legrand TPRA, James WW, Laura SW, Andrew PAO (2019) A microbial sea of possibilities: current knowledge and prospects for an improved understanding of the fish microbiome. Rev Aquac 12(2):1101–1134. https://doi.org/10.1111/raq.12375

Li T, Long M, Li H, Gatesoupe FJ, Zhang X (2017) Multi-omics analysis reveals a correlation between the host phylogeny, gut microbiota and metabolite profiles in cyprinid fishes. Front Microbiol 8:454. https://doi.org/10.3389/fmicb.2017.00454

Liu Y, Yao Y, Li H, Qiao F, Wu J, Du ZY, Zhang M (2016) Influence of endogenous and exogenous estrogenic endocrine on intestinal microbiota in zebrafish. PLoS One 11(10):e0163895

Liu H, Guo X, Gooneratne R, Lai R, Zeng C (2018) The gut microbiome and degradation enzyme activity of wild freshwater fishes influenced by their trophic levels. Sci Rep 6:24340. https://doi.org/10.1038/srep24340

Llewellyn MS, Boutin S, Hoseinifar SH, Derome N (2014) Teleost microbiomes: the state of the art in their characterization, manipulation and importance in aquaculture and fisheries. Front Microbiol 5:207–224. https://doi.org/10.3389/fmicb.2014.00207

Llewellyn MS, McGinnity P, Dionne M (2016) The biogeography of the Atlantic salmon (Salmo salar) gut microbiome. ISME J 10:1280–1284

Lødemel JB, Mayhew TM, Myklebust R, Olsen RE, Espelid S (2001) Effect of three dietary oils on disease susceptibility in Arctic charr (Salvelinus alpinus L.) during cohabitant challenge with Aeromonas salmonicida ssp. salmonicida. Aquac Res 32(12):935–945. https://doi.org/10.1046/j.1365-2109.2001.00621.x

Macfarlane GT, Englyst HN (1986) Starch utilization by the human large intestinal microflora. J Appl Bacteriol 60(3):195–201. https://doi.org/10.1111/j.1365-2672.1986.tb01073.x

Mandal S, Ghosh K (2013) Isolation of tannaseproducing microbiota from the gastrointestinal tracts of some freshwater fish. J Appl Ichthyol 29(1):145–153. https://doi.org/10.1111/j.1439-0426.2012.02054.x

Mcintosh D, Ji B, Forward BS, Puvanendran V, Boyce D, Ritchie R (2008) Culture-independent characterization of the bacterial populations associated with cod (Gadus morhua L.) and live feed at an experimental hatchery facility using denaturing gradient gel electrophoresis. Aquaculture 275(1-4):42–50

Merrifield DL, Rodiles A (2015) The fish microbiome and its interactions with mucosal tissues. In: Benjamin H, Peatman E (eds) Mucosal health in aquaculture. Academic, Cambridge, pp 273–295. https://doi.org/10.1016/B978-0-12-417186-2.00010-8

Mohapatra S, Chakraborty T, Kumar V, DeBoeck G, Mohanta KN (2012) Aquaculture and stress management: a review of probiotic intervention. J Anim Physiol Anim Nutr 97:405–430
Montalban-Argues A, De Schryver P, Bossier P, Gorkiewicz G, Mulero V, Gatlin DM, Galindo-Villegas J (2015) Selective manipulation of the gut microbiota improves immune status in vertebrates. Front Immunol 6:512. https://doi.org/10.3389/fimmu.2015.00512
Montero D, Mathlouthi F, Tort L, Afonso JM, Torrecillas S, Fernández-Vaquero A, Negrin D, Izquierdo MS (2010) Replacement of dietary fish oil by vegetable oils affects humoral immunity and expression of pro-inflammatory cytokines genes in gilthead sea bream Sparus aurata. Fish Shellfish Immunol 29:1073–1081. https://doi.org/10.1016/j.fsi.2010.08.024
Moya A, Ferrer M (2016) Functional redundancy-induced stability of gut microbiota subjected to disturbance. Trends Microbiol 24(5):402–513. https://doi.org/10.1016/j.tim.2016.02.002
Mueller K, Ash C, Pennisi E, Smith O (2012) The gut microbiome. Science 336:1245
Nakajima A, Vogelzang A, Maruya M, Miyajima M, Murata M, Son A, Kuwahara T, Tsuruyama T, Yamada S, Matsuura M, Nakase H, Peterson DA, Fagarasan S, Suzuki K (2018) IgA regulates the composition and metabolic function of gut microbiota by promoting symbiosis between bacteria. J Exp Med 215:2019–2034
Nandi A, Dan SK, Banerjee G, Ghosh P, Ghosh K, Ringo E, Ray AK (2017) Probiotic potential of autochthonous bacteria isolated from the gastrointestinal tract of four freshwater teleosts. Probiot Antimicrob Proteins 9(1):12–21
Navarrete P, Mardones P, Opazo R, Espejo R, Romero J (2008) Oxytetracycline treatment reduces bacterial diversity of intestinal microbiota of Atlantic salmon. J Aquat Anim Health 20:177–183. https://doi.org/10.1577/H07-043.1
Nayak SK (2010) Role of gastro intestinal microbiota in fish. Aquac Res 41:1553–1573. https://doi.org/10.1111/j.1365-2109.2010.02546.x
Nelson JS (2006) Fishes of the world, 4th edn. Wiley, Hoboken, p 624
Nelson JA, Wubah DA, Whitmer ME, Johnson EA, Stewart DJ (1999) Wood-eating catfishes of the genus Panaque: gut microflora and cellulolytic enzyme activities. J Fish Biol 54(5):1069–1082
Nelson TM, Rogers TL, Carlini AR, Brown MV (2013) Diet and phylogeny shape the gut microbiota of Antarctic seals: a comparison of wild and captive animals. Environ Microbiol 15:1132–1145. https://doi.org/10.1111/1462-2920.12022
Nie L, Zhou QJ, Qiao Y, Chen J (2017) Interplay between the gut microbiota and immune responses of ayu (Plecoglossus altivelis) during Vibrio anguillarum infection. Fish Shellfish Immunol 68:479–487. https://doi.org/10.1016/j.fsi.2017.07.054
Nikouli E, Meziti A, Smeti E, Antonopoulou E, Mente E (2020) Gut microbiota of five sympatrically farmed marine fish species in the Aegean Sea. Microb Ecol. https://doi.org/10.1007/s00248-020-01580-z
Okumura R, Takeda K (2018) Maintenance of intestinal homeostasis by mucosal barriers. Inflamm Regen 38:5
Olivier G, Lallier R, Lariviere S (1981) A toxigenic profile of Aeromonas hydrophila and Aeromonas sobria isolated from fish. Can J Microbiol 27:330–333. https://doi.org/10.1139/m81-050
Oyarbide U, Iturria I, Rainieri S, Pardo MA (2015) Use of gnotobiotic zebrafish to study Vibrio anguillarum pathogenicity. Zebrafish 12(1):71–80
Perez T, Balcazar J, Ruiz-Zarzuela I, Halaihel N, Vendrell D, De Blas I (2010) Host-microbiota interactions within the fish intestinal ecosystem. Mucosal Immunol. https://doi.org/10.1038/mi.2010.1229
Perry WB, Lindsay E, Payne CJ, Brodie C, Kazlauskaite R (2020) The role of the gut microbiome in sustainable teleost aquaculture. Proc R Soc B Biol Sci 287:20200184
Popovic NT (2017) The effects of diet supplemented with Lactobacillus rhamnosus on tissue parameters of rainbow trout, Oncorhynchus mykiss (Walbaum). Aquac Res 48:2388–2401. https://doi.org/10.1111/are.13074
Rai AK, Jini R, Swapna HC, Sachindra NM, Bhaskar N (2011) Application of native lactic acid bacteria (LAB) for fermentative recovery of lipids and proteins from fish processing wastes:

bioactivities of fermentation products. J Aquat Food Prod Technol 20(1):32–44. https://doi.org/10.1080/10498850.2010.528174

Raulo A (2018) Social behaviour and gut microbiota in red-bellied lemurs (Eulemur rubriventer): in search of the role of immunity in the evolution of sociality. J Anim Ecol 87:388–399. https://doi.org/10.1111/1365-2656.12781

Rawls JF, Samuel BS, Gordon JI (2004) From the cover: gnotobiotic zebrafish reveal evolutionarily conserved responses to the gut microbiota. Proc Natl Acad Sci U S A 101:4596–4601. https://doi.org/10.1073/pnas.0400706101

Ray AK, Roy T, Mondal S, Ringo E (2010) Identification of gut-associated amylase, cellulase and protease-producing bacteria in three species of Indian major carps. Aquac Res 41:1462–1469

Reid HI, Treasurer JW, Adam B, Birbeck TH (2009) Analysis of bacterial populations in the gut of developing cod larvae and identification of Vibrio logei, Vibrio anguillarum and Vibrio splendidus as pathogens of cod larvae. Aquaculture 288(1-2):36–43

Reikvam DH, Erofeev A, Sandvik A, Grcic V, Jahnsen FL, Gaustad P, McCoy KD, Macpherson AJ, Meza-Zepeda LA, Johansen FE (2011) Depletion of murine intestinal microbiota: effects on gut mucosa and epithelial gene expression. PLoS One 6:e17996

Ringø E, Birkbeck TH (1999) Intestinal microflora of fish and fry: a review. Aquac Res 30(2):73–93

Ringø E, Myklebust R, Mayhew T, Olsen R (2007) Bacterial translocation and pathogenesis in the digestive tract of larvae and fry. Aquaculture 268(1–4):251–264. https://doi.org/10.1016/j.aquaculture.2007.04.047

Ringø E, Sperstad S, Kraugerud OF, Krogdahl A (2008) Use of 16S rRNA gene sequencing analysis to characterize culturable intestinal bacteria in Atlantic salmon (Salmo salar) fed diets with cellulose or non-starch polysaccharides from soy. Aquac Res 39:1087–1100. https://doi.org/10.1111/j.1365-2109.2008.01972.x

Ringø E, Zhou Z, He S, Olsen RE (2014) Effect of stress on intestinal microbiota of Arctic charr, Atlantic salmon, rainbow trout and Atlantic cod: a review. Afr J Microbiol Res 8:609–618

Ringø E, Zhou Z, Vecino JLG, Wadsworth S, Romero J, Kroghdal Å, Olsen RE, Dimitoglou A, Foey A, Davies S, Owen M, Lauzon HL, Martisen LL, De Schryver P, Bossier P, Sperstad S, Merrifield DL (2015) Effect of dietary components on the gut microbiome of aquatic animals: the never-ending story? Aquac Nutr 22:219–282

Roeselers G, Mittge EK, Stephens WZ, Parichy DM, Cavanaugh CM (2011) Evidence for a core gut microbiota in the zebrafish. ISME J 5(10):1595–1608. https://doi.org/10.1038/ismej.2011.38

Romero J, Navarrete P (2006) 16S rDNA-based analysis of dominant bacterial populations associated with early life stages of coho salmon (Oncorhynchus kisutch). Microb Ecol 51(4):422–430. https://doi.org/10.1007/s00248-006-9037-9

Romero J, Ringø E, Merrifield DL (2014) The gut microbiota of fish. In: Merrifield D, Ringø E (eds) Aquaculture nutrition: gut health, probiotics and prebiotics. Wiley, Hoboken, pp 75–100

Roy T, Mondal S, Ray AK (2009) Phytase-producing bacteria in the digestive tracts of some freshwater fish. Aquac Res 40(3):344–353

Rungrassamee W, Klanchui A, Maibunkaew S, Karoonuthaisiri N (2016) Bacterial dynamics in intestines of the black tiger shrimp and the Pacific white shrimp during Vibrio harveyi exposure. J Invertebr Pathol 133:12–19. https://doi.org/10.1016/j.jip.2015.11.004

Safari R, Hoseinifar SH, Dadar M, Khalili M (2018) Dietary white bottom mushroom powder improved immunomodulatory and health promoting effects of Lactobacillus casei in zebrafish (Danio rerio). Int J Med Mushrooms 20:695–704

Sakata T, Sugita H, Mitsuoka T, Kakimoto D, Kadota H (1980) Isolation and distribution of obligate anaerobic bacteria from the intestines of freshwater fish. Bull Jpn Soc Sci Fish 46(10):1249–1255. https://doi.org/10.2331/suisan.46.1249

Salinas I, Zhang YA, Sunyer JO (2015) Mucosal immunoglobulins and B cells of teleost fish. Dev Comp Immunol 35:1346–1365

Salzman NH (2010) Paneth cell defensins and the regulation of the microbiome: detente at mucosal surfaces. Gut Microbes 1:401–406

Scher JU, Abramson SB (2011) The microbiome and rheumatoid arthritis. Nat Rev Rheumatol 7: 569–578

Schroeder BO (2019) Fight them or feed them: how the intestinal mucus layer manages the gut microbiota. Gastroenterol Rep 7:3–12

Seetharam AS, Emily K, Zhi QD, Max FR, Andrew JS (2015) Microbiome analyses of pacific white shrimp (Litopenaeus vannamei) collected from disparate geographical locations. Genomics Data 6:67–69. https://doi.org/10.1016/j.gdata.2015.08.009

Sila A, Nedjar-Arroume N, Hedhili K, Chataigné G, Balti R (2014) Antibacterial peptides from barbel muscle protein hydrolysates: activity against some pathogenic bacteria. LWT-Food Sci Technol 55(1):183–188. https://doi.org/10.1016/j.lwt.2013.07.021

Siriyappagouder P, Galindo-Villegas J, Lokesh J, Mulero V, Fernandes JMO (2018) Exposure to yeast shapes the intestinal bacterial community assembly in zebrafish larvae. Front Microbiol 9: 1868. https://doi.org/10.3389/fmicb.2018.01868

Soltan MA, El-Laithy S (2008) Effect of probiotics and some spices as feed additives on the performance and behaviour of the Nile tilapia, Oreochromis niloticus. Egypt J Aquat Biol Fish 12:63–80

Sommer F, Backhed F (2013) The gut microbiota-masters of host development and physiology. Nat Rev Microbiol 11(4):227–238. https://doi.org/10.1038/nrmicro2974

Stagaman K, Burns AR, Guillemin K, Bohannan BJ (2017) The role of adaptive immunity as an ecological filter on the gut microbiota in zebrafish. ISME J 11:1630–1639. https://doi.org/10.1038/ismej.2017.28

Stephens WZ, Burns AR, Stagaman K, Wong S, Rawls JF, Guillemin K, Bohannan BJM (2016) The composition of the zebrafish intestinal microbial community varies across development. ISME J 10:644–654. https://doi.org/10.1038/ismej.2015.140

Subramanian S, Huq S, Yatsunenko T, Haque R, Mahfuz M, Alam MA, Benezra A, DeStefano J, Meier MF, Muegge BD, Barratt MJ, VanArendonk LG, Zhang Q, Province MA, Petri WA Jr, Ahmed T, Gordon JI (2014) Persistent gut microbiota immaturity in malnourished Bangladeshi children. Nature 510:417–421. https://doi.org/10.1038/nature13421

Sugita H, Miyajima C, Deguchi Y (1991) The vitamin B12-producing ability of the intestinal microflora of freshwater fish. Aquaculture 92:267–276. https://doi.org/10.1016/0044-8486(91)90028-6

Sugita H, Kawasaki J, Deguchi Y (1997) Production of amylase by the intestinal microflora in cultured freshwater fish. Lett Appl Microbiol 24(2):105–108. https://doi.org/10.1046/j.1472-765X.1997.00360.x

Sullam KE, Essinger SD, Lozupone CA, O'connor MP, Rosen GL, Knight R, Kilham SS, Russell JA (2012) Environmental and ecological factors that shape the gut bacterial communities of fish: a meta-analysis. Mol Ecol 21:3363–3378. https://doi.org/10.1111/j.1365-294X.2012.05552.x

Sumathi C, Mohanapriya D, Mandal AB, Sekaran G (2012) Production of different proteases from fish gut microflora utilizing tannery fleshing. Eng Life Sci 12(2):223–237

Talwar C, Nagar S, Lal R, Negi RK (2018) Fish gut microbiome: current approaches and future perspectives. Indian J Microbiol 58(4):397–414. https://doi.org/10.1007/s12088-018-0760-y

Tarnecki AM, Burgos FA, Ray CL, Arias CR (2017) Fish intestinal microbiome: diversity and symbiosis unravelled by metagenomics. J Appl Microbiol 123(1):2–17

Tran NT, Zhang J, Xiong F, Wang GT, Li WX (2018) Altered gut microbiota associated with intestinal disease in grass carp (Ctenopharyngodon idellus). World J Microbiol Biotechnol 34(6):71. https://doi.org/10.1007/s11274-018-2447-2

Tsuchiya C, Sakata T, Sugita H (2008) Novel ecological niche of Cetobacterium somerae, an anaerobic bacterium in the intestinal tracts of freshwater fish. Lett Appl Microbiol 46(1):43–48. https://doi.org/10.1111/j.1472-765X.2007.02258.x

Turner TR, James EK, Poole PS (2013) The plant microbiome. Genome Biol 14:209. https://doi.org/10.1186/gb-2013-14-6-209

Uchii K, Matsui K, Yonekura R, Tani K, Kenzaka T (2006) Genetic and physiological characterization of the intestinal bacterial microbiota of bluegill (Lepomis macrochirus) with three different feeding habits. Microb Ecol 51(3):277–284. https://doi.org/10.1007/s00248-006-9018-z

Uddin N, Al-harbi AH (2012) Bacterial flora of polycultured common carp (Cyprinus carpio) and African catfish (Clarias gariepinus). Int Aquat Res 4(1):10

Vijaybaskar P, Somasundaram ST (2008) Isolation of bacteriocin producing lactic acid bacteria from fish gut and probiotic activity against common fresh water fish pathogen Aeromonas hydrophila. Biotechnology 7(1):124–128

Villamil SI, Roger H, Dean RJ (2020) Microbiome diversity and dysbiosis in aquaculture. Rev Aquac 13(2):1077–1096. https://doi.org/10.1111/raq.12513

Wang J, Tao Q, Wang Z, Mai K, Xu W, Zhang Y, Ai Q (2017) Effects of fish meal replacement by soybean meal with supplementation of functional compound additives on intestinal morphology and microbiome of Japanese seabass (Lateolabrax japonicus). Aquac Res 48:2186–2197. https://doi.org/10.1111/are.13055

Wanke I, Steffen H, Christ C, Krismer B, Gotz F, Peschel A, Schaller M, Schittek B (2011) Skin commensals amplify the innate immune response to pathogens by activation of distinct signaling pathways. J Invest Dermatol 131:382–390

Xia JH, Lin G, Fu GH, Wan ZY, Lee M (2014) The intestinal microbiome of fish under starvation. BMC Genomics 15:266. https://doi.org/10.1186/1471-2164-15-266

Xiong J, Zhu J, Zhang D (2014) The application of bacterial indicator phylotypes to predict shrimp health status. Appl Microbiol Biotechnol 98:8291–8299. https://doi.org/10.1007/s00253-014-5941-y

Xiong J, Zhu J, Wenfang D, Chunming D, Qiongfen Q, Li C (2017) Integrating gut microbiota immaturity and disease discriminatory taxa to diagnose the initiation and severity of shrimp disease. Environ Microbiol 19(4):1490–1501. https://doi.org/10.1111/1462-2920.13701

Yoshimizu M, Takizawa H, Kamei Y, Kim UT (1986) Interaction between fish pathogenic viruses and microorganisms in fish rearing water: survival and inactivation of infectious pancreatic necrosis virus, infectious hematopoietic necrosis virus and virus in rearing water. Fish Pathol 21(4):223–231

Yukgehnaish K, Praveen K, Parimannan S, Marimuthu K, Aziz A, Bilal AP, Jesu A (2020) Gut microbiota metagenomics in aquaculture: factors influencing gut microbiome and its physiological role in fish. Rev Aquac 1:1–25. https://doi.org/10.1111/raq.12416

Zhang YA, Salinas I, Li J, Parra D, Bjork S, Xu Z, Lapatra SE, Bartholomew J, Sunyer JO (2010) IgT, a primitive immunoglobulin class specialized in mucosal immunity. Nat Immunol 11:827–835. https://doi.org/10.1038/ni.1913

Zhang Z, Li D, Refaey MM, Xu W, Tang R, Li L (2018) Host age affects the development of southern catfish gut bacterial community divergent from that in the food and rearing water. Front Microbiol 9:495. https://doi.org/10.3389/fmicb.2018.00495

Zhao Q, Elson CO (2018) Adaptive immune education by gut microbiota antigens. Immunology 154:28–37

Zheng D, Liwinski T, Elinav E (2020) Interaction between microbiota and immunity in health and disease. Cell Res 30:492–506

Zhu J, Dai W, Qiu Q, Dong C, Zhang J, Xiong J (2016) Contrasting ecological processes and functional compositions between intestinal bacterial community in healthy and diseased shrimp. Microb Ecol 72:975–985. https://doi.org/10.1007/s00248-016-0831-8

Fish Microbiome and Metagenomics

4

Akhila Dharnappa Sannejal, Mithoor Divyashree,
Deekshit Vijaya Kumar, M. S. Nithin, and Praveen Rai

Abstract

The microbiome is the group of microorganisms, i.e. bacteria, fungi, archaea, and viruses, that reside in a particular environment and consist of both useful and potentially harmful microbes. Useful microbiomes play an integral role in host health by activating the immune system, aiding in nutrient absorption, and withstanding the effect of opportunistic pathogens. In the case of fish, such microbiomes also help in determining feeding habits in different ecological settings. These microbiomes can be effectively studied using a metagenomics approach that will help in understanding host-microbiome symbiosis and identify the core set of microbiomes that are absolute requirements to carry out essential functions for the betterment of fish health. This chapter aims to summarize the role of the gut microbiome in fish health, thereby aiding aquaculture productivity.

A. D. Sannejal
Department of Food Safety and Nutrition, Nitte University Centre for Science Education and Research (NUCSER), Nitte (Deemed to be University), Deralakatte, Mangaluru, Karnataka, India

M. Divyashree
Department of Environmental Health and Toxicology, Nitte University Centre for Science Education and Research (NUCSER), Nitte (Deemed to be University), Deralakatte, Mangaluru, Karnataka, India

D. V. Kumar · P. Rai (✉)
Department of Infectious Diseases and Microbial Genomics, Nitte University Centre for Science Education and Research (NUCSER), Nitte (Deemed to be University), Deralakatte, Mangaluru, Karnataka, India
e-mail: raiprav@nitte.edu.in

M. S. Nithin
College of Fisheries, Karnataka Veterinary, Animal and Fisheries Sciences University, Mangaluru, Karnataka, India

A. Diwan et al. (eds.), *Microbiome of Finfish and Shellfish*,
https://doi.org/10.1007/978-981-99-0852-3_4

Keywords

Fish · Microbiome · Metagenomics · Gut · Immunity · Aquaculture

4.1 Introduction

It is a well-accepted fact that the vertebrate gut microbiome plays an integral role in host health by activating the immune system, aiding in nutrient acquisition, and withstanding the effect of opportunistic pathogens (Tarnecki et al. 2017; Mueller et al. 2012; Llewellyn et al. 2014). The useful and harmful effects of microbiomes can be studied by metagenomics which is one of the important technologies in next-generation sequencing. Metagenomics is a rapidly developing field in next-generation sequencing technologies. It has been widely utilized to understand the host microbes association within the gut. Metagenomics helps to evaluate microbial communities' gene function and identify the various genetic and physiological traits of microorganisms in the host (Diwan et al. 2022). Functional metagenomics can further help in the discovery of new molecules in the form of enzymes or bioactive compounds that can be effectively used in therapeutics (Yukgehnaish et al. 2020). However, the majority of the metagenomics study on mammals comprises less than 10% of total vertebrate diversity (Sullam et al. 2012). Further, very few studies have made an attempt to understand the association of microbes in the fish gut. For instance, a study by Llewellyn et al. (2014) analysed and characterized the microbiome of many fish species, such as Atlantic salmon, carp, and tilapia. Similarly, Ringø et al. (2016) reviewed the effect of dietary compositions on the protective role of gut microbiota in fish. Studies have also highlighted the role of genetic predisposition, gut microbiome, and its contribution to disabling immune systems against infection (Prabhakara et al. 2022). There might be a core microbiota in the fish gut that co-evolves with the host's genetic background to maintain some essential functions (Ghanbari et al. 2015). At the beginning, attempts were made to analyse the fish gut microbiota using culture-based methods. However, it underestimated the actual diversity of microbes as only <10% could be isolated and characterized from these methods (Amann et al. 1995). More recently, culture-independent methods such as fluorescence in situ hybridization (FISH) that uses confocal or fluorescence microscopy with minimal processing and the use of target-specific probes (Amann et al. 1995), DNA fingerprinting methods that work based on the principle of PCR amplification of variable regions within the rRNA operon of bacterial strains (Fjellheim et al. 2012; Green et al. 2013; Le Nguyen et al. 2008; Zhou et al. 2009; Larsen et al. 2014), and lastly the 16S rRNA gene sequencing can be accurately used for the identification of gut microflora (Rosello-Mora 2005; Prabhakara et al. 2022). Further, cost-effective next-generation sequencing (NGS) technologies have made tremendous progress in understanding the microbial communities in the gastrointestinal tract of fish (Tarnecki et al. 2017). Metagenomics would provide a suitable framework for understanding the complex interactions between microbes, aquatic animals, and their surrounding environments. This chapter aims to understand the use of NGS technology in studying fish gut microbiomes for aquaculture productivity.

4.2 Gut Microbiota of Marine Fishes

To better understand the symbiotic relationship between indigenous microbes and the host, it is important to study the gut microbial diversity of non-mammalian vertebrates, that is fish. To know the co-evolution of vertebrates and gut microbes, a wide range of fish are to be studied. Approximately 33,000 species have been reported in fish and exhibit maximum species diversity among groups of vertebrates (Froese and Pauly 2019). Gut microbiome analysis has been limited to a few model fishes, including guppy, zebrafish, rainbow trout, Atlantic cod, Atlantic salmon, carp, and sturgeon (Kim et al. 2021). Available literature studies are insufficient to largely understand the patterns of co-evolution among the fishes and their gut microbial composition.

Culture-independent and traditional culture-dependent techniques are currently used to analyse fish gut microbiomes. The culture-dependent techniques generally use sequencing of *16S rRNA* gene to categorize bacteria. To study the uncultured microbiota, amplification and sequencing of partial or whole regions of the *16S rRNA* gene are widely employed. The similarity among the sequences was grouped and compared with existing databases (Talwar et al. 2018). Available nutrients, pH, and redox potential determine the distribution of microbes in different gut parts. *E. coli* could survive in extreme conditions pH of 2.5 or less. *Burkholderia cepacia, Chromobacterium violaceum, Aeromonas hydrophila, Aeromonas veronii, Citrobacter freundii, E. coli, Flavimonas oryzihabitans,* and *Plesiomonas shigelloides* are the most frequently isolated bacterial microflora from the gastrointestinal tract of Nile tilapia. *Plesiomonas shigelloides* was predominant in the posterior gut when compared to the anterior gut and stomach in tilapia (Molinari et al. 2003). Interaction and bacterial colonization in the fish begin as soon as the eggs are laid. Once the eggs hatch, the larval is colonized by environmental microbes where they are originally present. The early digestive colonization begins once the larva commences feeding. Bacterial taxa strongly bear a resemblance to those associated with the source of food. Microbes develop and mature in the region leading to the build-up of a stable, diversified and resilient endogenous microbial community (Llewellyn et al. 2014). Majority of the fish's gut microbiota is dominated by aerobes or facultative anaerobes. It has been reported that 90% of fish intestinal microbiota studied to date are composed of *Proteobacteria, Bacteroidetes,* and *Firmicutes* (Ghanbari et al. 2015).

4.3 Factors Influencing the Diversity and Function of the Gut Microbiome of Fish

According to studies, various factors outline the composition of the gut in various animals, including fish (Campbell 2012). The research highlights that both environmental and host-related factors significantly contribute to the microbial composition of the fish gut. However, the degree to which these aspects contribute to variation in the gut microbiome load is unclear.

4.4 Host Habitat or Environment

Host environment is among the most important key determinant that determines the gut microbiome composition. Several factors may cause the correlation between the environment and composition of the gut; for example, the availability of nutrients in freshwater or marine water, biotic and abiotic factors, geographical location, and water quality status (Yukgehnaish et al. 2020). Environmentally attributed factors, such as pH, total suspended and dissolved solids, conductivity, dissolved oxygen, ions NO_3-N and PO_3-P and are also significantly correlated with fish gut microbiota composition (Giatsis et al. 2016). Habitat contaminated with heavy metals, industrial effluents, and oil pollution also disturbs the gut microbiota. Gut microbiome analysis of herbivores, carnivores, and omnivores, and filter feeders revealed the dissimilar composition in their gut. It is suggested that the tropical level of the fish has an impact on their gut composition (Liu et al. 2016). Filter-feeding fishes show higher diversity because they filter huge volumes of water and cover large areas by swimming (Li et al. 2017). *Aeromonas, Acinetobacter, Lactococcus, Pseudomonas, Flavobacterium, obligate anaerobes Bacteroides, Clostridium*, and *Fusobacterium* and members of the family *Enterobacteriaceae* are found to be dominant species in the gut of freshwater species (Gómez and Balcázar 2008). In order to study the host dietary gradient in the microbial community, it is imperative to investigate the trophic level of individual fish species. The ecological position of the fish in the food chain is significantly associated with the gut microbial communities (Froese and Pauly 2019).

4.5 Salinity of the Sampling Site

The salinity of the water where the fish live principally determines the microbial composition in the fish gut. It was found that estuarine fish have similar gut microbiota as that of freshwater fish, but the intestinal microbiota of fish from mixed salinity environments was found to be more similar to the intestinal microbiota of salt water fish (Zhang et al. 2016). Hence it is quite tough to evaluate the exact effect of salinity on forming the gut microbial composition.

4.6 Fish Diet

The fish diet influences the composition of gut microbiota at different stages, like planktonic and mature (Ingerslev et al. 2014). The gut microbiome changes according to the change in the diet; as it changes from carnivorous to omnivorous or herbivorous the diversity increases (Ye et al. 2014). The source of the dietary nutrients, either plant or animal-based or protein-rich foods, has a potential role in microbiome composition change. Lactic acid-producing bacteria (LAB) were abundantly observed among the fish fed with fishmeal-based diets (Catalan et al. 2018). In aquaculture industries, prebiotics and probiotics are commonly used as environmental friendly feed additives to increase productivity. Administration of probiotic feed to fish for a short period significantly affects the gut microbiota composition at later stages of life in axenic cultured fish *Oreochromis niloticus*. Among these axenic cultured fish, dominant phyla were found to be *Actinobacteria, Firmicutes,* and *Proteobacteria* (Giatsis et al. 2016). Probiotic species of *Lactobacillus* and *Bacillus* were proven to have a beneficial effect on host immunity which alternatively prevents diseases or infection in the aquaculture field (Perry et al. 2020). Hence, the possibility of controlling the fish health status by modulating the gut microbiome through the use of prebiotics and probiotics is gaining interest in aquaculture. In the *Silurus meridionalis* fish gut, the level of *Firmicutes* was increased and *Fusobacteria* decreased after feeding at different intervals (Zhang et al. 2017).

4.7 Host Immunity Status

Studies have revealed that gut microbial communities have a solid link to immunity. Fish are in continuous contact with water, which is the source of various microorganisms, including pathogens (Perry et al. 2020). In aquaculture, the fishes having diverse gut microbiomes can prevent unfavourable microbial colonization. The mechanism of the process is not completely understood. It was reported that probiotic bacteria *Bacillus* and *Lactobacillus*, used in the aquaculture field, were able to stimulate the expression of inflammatory cytokines in the fish gut (He et al. 2017).

The host's immunological genotype could affect the composition of the gut microbiota. Change in the gut composition is also dependent upon the host's genetic background, life history or age of the host, health status, and other individual differences of the host (Talwar et al. 2018). Several pathogenic bacteria such as *Aeromonas, Vibrios, Edwardsiella, Flavobacterium, Streptococcus, Pseudomonas,* and *Yersinia* cause infection or diseases to freshwater fish (Austin 2011). Distinct alterations in the gut microbial diversity among the diseased healthy crucian carp (*Carassius auratus*) were observed (Li et al. 2017). Population effects were greater than habitat effects in explaining microbial variation and far more important than the variation due to host sex. The beneficial microorganisms present in the gut have certain immunological benefits on host fish by interacting with host fish neutrophils, NK cells, and monocytes, meanwhile modulating the host innate immune system (Yukgehnaish et al. 2020).

Overuse of antibiotics in disease control has led to severe ecological and biological problems, especially due to antibiotic-resistant strains. The presence of such strains and antibiotic residues in the host habitat can alter the microbial diversity in the fish gut (Meek et al. 2015). Oxytetracycline antibiotic is commonly used in aquaculture and has been seen to reduce gut microbial diversity in Atlantic salmon. Increased use has led to multiple tetracycline resistance genes enriching possible opportunistic pathogens belonging to the genus *Aeromonas* (Perry et al. 2020).

4.8 Role of the Fish Gut Microbiome

Fish gut microbiome confers various effects on the host such as fish metabolism, its feeding behaviour, overall size, and immune response. Studies revealed that gut microbiomes are considered to be a biomarker for stress. Stress can impact the microbiome in different ways with consequences for hormonal, physiological, and cellular function. During normal conditions, commensals and pathogens maintain the 'homeostasis' by controlling the constitutive molecules and receptors of the host innate immune system. 'Dysbiosis' could happen due to the disturbance of commensals, which leads to susceptibility to diseases. For example, WSSV (white spot syndrome virus), the pathogenic virus of shrimp, affects shrimp, which express only when there is stress (Sanchez-Paz 2010). Gut microbial dysbiosis could be a useful predictor of stress-related syndromes in fish and also acts as a behavioural indicator, which could ultimately lead to mortality (Webster et al. 2020). Animal experiments reveal that gut microbe influence behaviour, serotonin, brain-derived neurotrophic factor (BDNF) levels, and metabolism. Selected gut microbes could also be used to modulate endogenous neuroactive molecules in the zebrafish model system (Borrelli et al. 2016).

The gut microbiome provides information on disease diagnosis. Potential utilization of gut microbiome as a biomarker for monitoring health status in farmed fish, especially in salmon, was studied by Bozzi et al. (2021). High fish weight was positively correlated with *Mycoplasma*-dominated gut microbiota and was negatively correlated with *Aliivibrio* spp. (Bozzi et al. 2021). In parrot fish, *Bacillus subtilis* E20 improves the growth performance and disease resistance against *Vibrio alginolyticus* (Liu et al. 2010a, b). Fish feed enriched with probiotics encouraged enzymes related to weight gain, digestive enzymes, and overall good health (Anguiano et al. 2013). The gut microbes that have been shown to have a positive effect on the health of the fish can be used as probiotic candidates in the future.

4.9 Microbiome and Aquaculture Productivity

The significant role of the gut microbiome in maintaining the optimal health and well-being of fish has been thoroughly studied by several researchers. The gut microbiome composition primarily depends on the type of ecosystem or environment fish inhabit (Eichmiller et al. 2016). The diversity of the gut microbiome can help in protecting the host against diseases, modulate the immune system, increase the absorption of essential nutrients, regulate the metabolic processes, and synthesize vitamins (Cornejo-Granados et al. 2017; Gómez and Balcázar 2008; Romero et al. 2014; Sullam et al. 2012; Diwan et al. 2022). The possibility of enhancing aquaculture production by studying functional aspects of the fish gut microbiome was emphasized by Talwar et al. (2018). Later on, there have been many studies highlighting the importance of the gut microbiome in aquaculture productivity (Legrand et al. 2019; Perry et al. 2020; Villamil et al. 2020). The effect of dietary lipids, proteins, functional glycomic ingredients, nutraceuticals, antibiotics, dietary iron, and chromic oxide on fish gut microbiota has been extensively reviewed by Ringø et al. (2016). The manipulation of dietary components such as high and low dietary lipids in the feed resulted in the modulation of species composition of microbes in the gut of Arctic charr (Ringø and Olsen 1999). It was found that *Carnobacteria* spp. inhibited the growth of *A. salmonicida* in fishes fed with soya bean oil and linseed oil. This suggests that both soyabean oil and linseed oil enhanced the growth of *Carnobacteria,* thereby inhibiting the growth of *A. salmonicida* after challenge (Ringø et al. 2002). An important bacterial species *Cetobacterium somerae,* which usually occupies the gut of freshwater fish, is known to produce large quantities of vitamin B12 that can inhibit the growth of pathogens (Etyemez and Balcazar 2015). There is a complex interaction between the fish gut microbiome and the surrounding aquatic environment in the aquaculture systems (Ding et al. 2022; Chen et al. 2022a). A gnotobiotic model suggested by Villamil et al. (2020) highlights the close relationship between aquaculture productivity and microbial composition in the gut. According to this model, changes in the microbial composition reflect the fish's health status (Duan et al. 2020). However, no further investigations have been done to confirm this model in the aquaculture systems (Zhang et al. 2020).

Microbiome studies have also been carried out in shellfishes, particularly in shrimps, wherein both culture-based and molecular methods, including the metagenomics approach, have been employed to determine the diversity of microbes in different organs of shrimp such as the intestine and hepatopancreatic tissues (Rungrassamee et al. 2014; Huang et al. 2019; Cheung et al. 2015). Changes in the microbial composition of shrimps affected by Acute hepatopancreatic necrosis disease/ Early Mortality Syndrome (AHPND/EMS) was observed during its early developmental stages. The bacterial species such as *Aeromonas taiwanensis*, *Simiduia agarivorans,* and *Photobacterium angustum* were dominant in diseased shrimps while healthy shrimps mostly harboured *Faecalibacterium prausnitzii* and *Pantoea agglomeran* (Cornejo-Granados et al. 2017). This might be due to the reduced ability of affected shrimps to select resident flora in the gut, thereby making environmental factors more important in deciding gut microbiota. Similar studies were carried out by Chen et al. (2017) wherein the *16S rRNA* gene with Illumina sequencing technology was employed to understand the microbial diversity of the shrimp (*Littopaeneus vannamei*) and pond affected with AHPND. The study found diverse microbial communities dominated by *Vibrio* and *Candidatus Bacilloplasma* in infected shrimps. These studies further highlight that the outbreaks of AHPND can be effectively controlled by understanding the microbial dynamics that further contribute to aquaculture productivity (Chen et al. 2017). However, some useful microorganisms are difficult to grow in laboratory conditions and warrant further future research.

4.10 Metagenomics

Metagenomics, a branch of science aged only a few years, makes it feasible to study microbes in the diverse communities where they typically reside. It resulted in changes in ecology, biotechnology, medicine, and biology that could be as significant as what was first brought on by the development of the microscope. In addition to being a subject of study, metagenomics is also a collection of research methods that includes a variety of connected approaches and techniques. Meta means 'transcendent' in Greek. Metagenomics avoids the microorganism's unculturability and genomic diversity through its methodology and techniques. Every aspect of human life—indeed, all life on earth—depends on microbes. The seemingly infinite ability of bacteria to change their environment impacts every ecological process. Microbes are reliant upon the chemical processes that transform the essential components (oxygen, carbon, sulphur, and nitrogen) into biologically usable forms. All animals and plants have closely related microbial communities that provide their hosts with essential minerals, vitamins, and nutrients. The microorganisms associated with the vertebrates in the mouth and colon allow us to absorb energy from food that we could not digest without them and shield us against pathogens.

Microbes were first brought to attention by the development of microscopes in the late eighteenth century. Few microbes live as an individual colony through the development of laboratory cultivation techniques in the mid-1800s, and the genomics revolutions and molecular biology of the latter half of the twentieth century brought this physiological knowledge together with a detailed understanding of its fundamental genetic basis. The last 10 years have seen the emergence of metagenomics, which aims to explore the genomes found within a microbe. Metagenomics is derived from the Greek word meaning 'beyond', which refers to the study of things beyond the simple genome (Nielsen et al. 2014). Genomic DNA can provide information about the diversity of microorganisms. It can disclose information about microbes' biological and functional roles instead of culturing the microorganism in the laboratory (Martínez-Porchas and Vargas-Albores 2017).

The developing field of metagenomics, in association with next-generation sequencing, can revolutionize pathogen identification in laboratories which aids the simultaneous analysis of microbes in a clinical or environmental sample without prior identification. Using a metagenomics analysis, it is possible to identify novel and uncommon pathogens and learn how dysbiotic microbiomes contribute to both infectious and chronic diseases. Recent research has demonstrated that metagenomics can discover the whole-genome sequences of pathogens, permitting conclusions regarding antibiotic resistance, pathogenicity, evolution, and transmission. This is made possible by developments in sequencing technologies and bioinformatics tools. As opposed to conventional laboratory techniques, metagenomics-based tools can be used to identify new infectious illnesses. The aquaculture and livestock industries are under pressure to incorporate metagenomics methods into their diagnostic toolkits.

Metagenomics research in public health can take one of two forms, analogous to metagenomics studies in environmental and human microbiology (Miller et al. 2013; Jackson et al. 2016; Ko et al. 2022). Deep amplicon sequencing (DAS) uses a Polymerase Chain Reaction (PCR) amplification step before sequencing to specifically amplify a taxonomic identification like the rRNA gene. The second strategy is referred to as shotgun metagenomics and is more thorough. In this method, the complete nucleic acid content of a sample is sequenced either right away or after applying an enrichment step, which might be a subtraction or capture-based strategy before sequencing. The two types of study that makeup metagenomics are environmental single-gene surveys and random shotgun analyses of all available genes in any environment. Studying particular genes amplified by PCR, analysing (orthologs/ paralogs), and sequencing are the core objective of metagenomics based on single-gene surveys. Contrarily, full theoretical coverage of all genes is achieved by randomly sequencing all the DNA recovered from a particular sample. But in reality, this is not the case because technology is still in its infancy and has yet to sequence the whole genomes of the thousands of distinct bacteria that might be present in a randomly selected environmental sample (Gilbert and Dupont 2011). Therefore, research is restricted to the gene sequences identified through amplification and sequencing.

Using shotgun metagenomics (single-molecule real-time sequencing) can sequence a nucleic acid strand without amplification (DeCastro et al. 2016; Tanveer et al. 2022).

4.11 Significance of Metagenomics in Aquaculture

With genomic methods, it is easier to demonstrate microorganisms' diversity and potential functions. Genomics, in combination with Chemical ecology, functional genomics and metagenomics, could resolve these issues (Riesenfeld et al. 2004). Studies on microorganisms in aquaculture are primarily concerned with understanding the symbiotic and antagonistic associations that microbes have with aquatic organisms. In this context, metagenomics can offer a greater understanding of these connections by performing associations between the data provided by the collected DNA and ecosystems or specific hosts (Suttle 2007; Gianoulis et al. 2009). Even it has been challenging to study intracellular pathogenic bacteria since few among them are obligatory intracellular microbes that can only be cultivated in cell culture and/or semi-aqueous conditions (Avila-Villa et al. 2011). In addition to exploring the diversity of internal bacteria, new sequencing and bioinformatics methods enable the elucidation of relevant genetic data from such communities.

To enhance water quality, waste treatment, and disease prevention, biofloc technology (BFT) is one of the most promising technologies in the aquaculture sector. It has been demonstrated that BFT improves immunological response due to the immunostimulant and probiotic effects of the microorganisms in the biofloc. BFT can also increase an organism's growth performance, elevate digestive enzyme activity, and boost antioxidant status (Yu et al. 2020). In a biofloc-based aquaculture system, metagenomic sequencing can be used to examine the structure and functions of bacterial communities in aquatic systems. Further it can be used to study the development of antibiotic resistance in biofloc-based aquaculture systems (Chen et al. 2022b).

Metagenomics can be used to study the occurrence of microbial community imbalances, pathogen outbreak scenarios, and probiotics' roles and interactions.

According to Aguilera-Rivera et al. (2014), numerous studies have concentrated on studying microorganisms as probiotics, sentinels, pathogens, and food sources. Moreover, knowledge about their interactions, unculturable microbial diversity, and genes still needs to be enhanced. Krishnani et al. (2010a, b) illustrate the role of sulphate-oxidizing bacteria in a coastal aquaculture green water system. They revealed the presence of *Pseudoxanthomonas* sp., which in turn maintains the sulphide concentration in a safe level.

Therefore, understanding the capability and biochemical processes a specific microbial community uses to recycle nutrients depends on the genes and operons that encode the proteins responsible for detecting, absorbing, and using particular metabolites. Use of culture-based systems to understand the microbial community is laborious and time-consuming. Several studies have used quantitative PCR and microarray technologies to identify functional genomics (Perreten et al. 2005). In comparison to other techniques like PCR and microarrays, metagenomics has been shown to perform better in terms of precision and efficiency in the identification of the genome. Viral metagenomics does not require expensive (and frequently ineffective) antibody laboratory tests, culture media, or preliminary genomic information about the pathogen. Additionally, in terms of efficiency and coverage, viral metagenomics performs better than any other random identification technique.

4.12 Role of the Microbiome in Fish Health

The fish is protected from dangerous bacteria by the gut microbial population, which also regulates endocrine organs, metabolic processes, and immunity. Earlier culture techniques, however, have few applications for detecting microorganisms. Development has been achieved in the study of the fish gut microbiome thanks to the use of molecular biologic technology, particularly next-generation sequencing technology which aided in metagenomic sequencing. Studying the variety and dysbiosis of the intestinal microbiome and how it relates to health and disease can be done using metagenomics. Although there are still certain restrictions, functional metagenomics can also find new functional genes, microbial pathways, genes for antibiotic resistance, functional dysbiosis of the gut microbiome, and interactions and co-evolution between the microbiota and host. The fields of metabolomics, metaproteomics, and metatranscriptomics are major advances in our comprehension of the fish gut microbiome (Wang et al. 2015; Xiang et al. 2022).

The microbiota is a diverse community of microorganisms that inhabits an organism's exposed surfaces and epithelia. In addition to boosting the immune system and avoiding disease, these microorganisms also aid in food absorption, serve as a barrier against possible pathogens, and regulate the expression of genes involved in epithelial growth. The gastrointestinal tract (GIT) has drawn the greatest attention to their function. These reasons make the microbiota a reliable source of probiotic potentials. Probiotics are defined as live microbiological dietary supplements that benefit the animal host, provide a defense (antagonism) against pathogens, promote the development of the immune system, and provide nutritional benefits. Probiotics were first utilized as a preventative measure against diseases in the aquaculture industry at the end of the 1980s, largely because of their capacity to trigger the immune system (innate immune response), which is notable by its nonspecific mode of action against diverse bacteria. According to Fig. 4.1, which shows the fish microbiota (and, by extension, the microbiome) as a collection of dynamic microbial communities, the disruption of one of these microbial communities directly affects the host's health and well-being. The GIT, which represents low and high microbial complexity communities, respectively, has been thoroughly defined as a result of current high-tech advancements and is currently the most researched fish micro-ecosystems. Despite the fact that the GIT has been primarily associated with dysbiosis, it can occur on any unprotected and uncovered surface or mucous membrane present in the skin of the fish or the gill system, which are the primary barriers. In fact, such microbial population fluctuation has a significant effect on fish health (Tello et al. 2019; Xiang et al. 2022).

Additionally, marine microbes may be employed as probiotics. These produce additional biodegradable polymer polyhydroxy butyrate (PHB), which has the power to slay sicknesses in crustacean models (Artemia), fish, and/or shrimps, which can be stored by these bacteria. In the stomach of the target species, PHB is broken down into monomers of short-chain fatty acids (SCFAs), which alter pH, boost bacterial diversity, or strengthen immune protection while also supplying energy to cells. This breakdown appears to be the cause of the probiotic action. Baruah et al. (2011) demonstrated that a commercial PHB source increased the survival of Artemia challenged with pathogens by increasing the production of the heat shock protein Hsp 70, which is connected to protective innate immune responses. The European Sea bass (*Dicentrarchus labrax*), in an experiment where the feed of juveniles was supplemented with 2% and 5% PHB (w/w), demonstrated yet another positive PHB effect. Having a high and diverse microbial richness in the gut was associated with juveniles performing better in terms of growth (Abt et al. 2012; Chu and Mazmanian 2013; Tello et al. 2019).

According to studies, the immune system and microbiota interact chemically. This interaction includes (a) the microbiota's specific provocation of the host's receptors by molecules from microbes (archaea, bacteria, fungi, viruses, and yeast); (b) the microbiota's metabolism's secretion of bioactive molecules; and (c) the stimulation of epigenetic mechanisms that regulate the gene expression of immune genes (Tello et al. 2019).

4.13 Specific Host Receptor Activation by Chemicals from Microorganisms

The pattern of various structural compounds from microbes, whether commensal or pathogenic, can be recognized by the immune system. A family of receptors termed pattern recognition receptors (PRR) found in immune cells, specifically a subtype of them known as toll-like receptors (TLR), recognize these pathogen-associated molecular patterns (PAMP), which are specific to each species of bacterium and

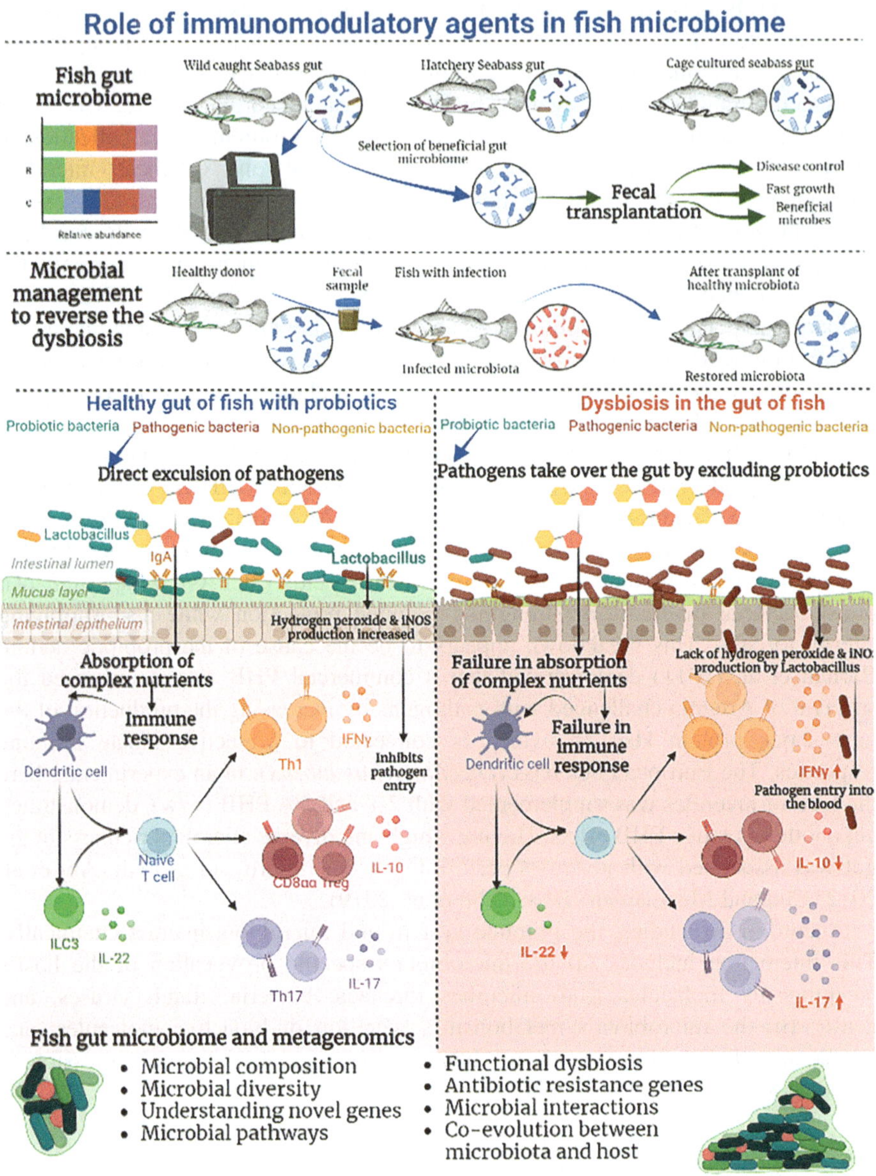

Fig. 4.1 Using metagenomics to study the fish gut microbiota. Metagenomic analysis has the ability to give adequate data in the following research fields when examining the fish gut microbiome: determination of relationships and co-evolution between the host's microbiota and itself, as well as the identification of new genes, microbial pathways, functional dysbiosis, antibiotic resistance genes, and microbial composition and diversity

were originally noticed in infections. This interaction activates NF-k, interferon-response factors, and the inflammasome, which in turn causes a variety of inflammatory cytokines (e.g. interferon, IL-1, and IL-22) to be produced. These cytokines can have systemic or local effects, depending on the cell type, environment, and microorganism involved. On the other side, the commensal microbiota controls the immune response by releasing or exposing substances that encourage an anti-inflammatory response. The polysaccharide A of *Bacteroides fragilis* can induce an anti-inflammatory response by increasing IL-10 production, the number of regulatory T cells, and decreasing the inflammatory response mediated by Th-1 and Th-17 (Martín et al. 2014; Ost and Round 2018; Pietretti and Wiegertjes 2014; Tello et al. 2019).

It has been discovered that the fish microbiota, especially lactobacilli, exerts its protective effect against pathogens through two main mechanisms: (1) exclusion, fuelled by competition for epithelial cell receptors, and (2) growth inhibition, brought on by the production of antimicrobial compounds (Fig. 4.1). A minimum of three bacteriocins, including lactocine, salivaricin CRL 1328, and L23, have been found in different *Lactobacillus* strains. In fact, there has been a correlation shown between the predominance of H_2O_2-producing strains and a decline in bacterial infections in fish guts. The most well-studied microbial short-chain fatty acid is butyrate, which promotes the development of regulatory T lymphocytes outside of the thymus, the maturation and function of microglia, the activity of PMN lymphocytes, and the reduction of proinflammatory cytokines (IFN, INF-, IL-1, and TNF) in macrophages. Additionally, butyrate enhances the synthesis of anti-inflammatory cytokines in dendritic cells (IL-C3, IL-10, IL-17, IL-22, and IL-23), decreases the exposure of MHC II, and inhibits the proliferation and death of T CD4 lymphocytes. In *Sparus aurata* it increases intestinal microvilli and nutrient absorption and in *Cyprinus carpio* it increases the expression of shock protein-70 (HSP70), proinflammatory cytokines (IL-1β and TNF-α), and anti-inflammatory cytokines (transforming growth factor-β) (Fig. 4.1) (Kawashima et al. 2013; Martín et al. 2014; Ost and Round 2018; Pietretti and Wiegertjes 2014; Rauta et al. 2014; Salazar et al. 2016; Tello et al. 2019; Wang et al. 2015; Wlodarska et al. 2014).

Understanding the interactions between the fish, microbiota, and ambient microbes could be a vital element in the development of logical ways to increase production by lowering the usage of antibiotics and mitigating their harmful effects on the environment. Metagenomics has limited applications for analysing fish-microbes-environment interactions in the farming of salmons, which runs counter to the lucrative nature of the sector. However, it has been useful in evaluating the gut and skin microbiota of farmed salmonid exposed, (1) to different nutritional and growth settings, (2) the role of the microbiota as an antagonist against infections, and (3) the possible role in nutrition and environmental adaptation.

References

Abt MC, Osborne LC, Monticelli LA, Doering TA, Alenghat T, Sonnenberg GF et al (2012) Commensal bacteria calibrate the activation threshold of innate antiviral immunity. Immunity 37(1):158–170

Aguilera-Rivera D, Prieto-Davó A, Escalante K, Chávez C, Cuzon G, Gaxiola G (2014) Probiotic effect of FLOC on Vibrios in the pacific white shrimp Litopenaeus vannamei. Aquaculture 424: 215–219

Amann RI, Ludwig W, Schleifer KH (1995) Phylogenetic identification and in situ detection of individual microbial cells without cultivation. Microbiol Rev 59:143–169

Anguiano M et al (2013) The effects of prebiotics on the digestive enzymes and gut histomorphology of red drum (Sciaenops ocellatus) and hybrid striped bass (Morone chrysops × M. saxatilis). Br J Nutr 109(4):623–629

Austin B (2011) Taxonomy of bacterial fish pathogens. Vet Res 42:1–3

Avila-Villa LA, Martínez-Porchas M, Gollas-Galván T, López-Elías JA, Mercado L, Murguia-López Á, Mendoza-Cano F, Hernández-López J (2011) Evaluation of different microalgae species and Artemia (Artemia franciscana) as possible vectors of necrotizing hepatopancreatitis bacteria. Aquaculture 318(3-4):273–276

Baruah K, Ranjan J, Sorgeloos P, MacRae TH, Bossier P (2011) Priming the prophenoloxidase system of Artemia franciscana by heat shock proteins protects against Vibrio campbellii challenge. Fish Shellfish Immunol 31(1):134–141

Borrelli L et al (2016) Probiotic modulation of the microbiota-gut-brain axis and behaviour in zebrafish. Sci Rep 6(1):1–9

Bozzi D et al (2021) Salmon gut microbiota correlates with disease infection status: potential for monitoring health in farmed animals. Anim Microbiome 3(1):30

Campbell JH et al (2012) Host genetic and environmental effects on mouse intestinal microbiota. ISME J 6(11):2033–2044

Catalan N, Villasante A, Wacyk J, Ramírez C, Romero J (2018) Fermented soybean meal increases lactic acid bacteria in gut microbiota of Atlantic salmon (Salmo salar). Probiotics Antimicrob Proteins 10:566–576

Chen CY, Chen PC, Weng FC, Shaw GT, Wang D (2017) Habitat and indigenous gut microbes contribute to the plasticity of gut microbiome in oriental river prawn during rapid environmental change. PLoS One 12:e0181427

Chen CZ, Li P, Liu L, Li ZH (2022a) Exploring the interactions between the gut microbiome and the shifting surrounding aquatic environment in fisheries and aquaculture: a review. Environ Res 214:114202

Chen X, He Z, Zhao J, Liao M, Xue Y, Zhou J, Hoare R, Monaghan SJ, Wang N, Pang H, Sun C (2022b) Metagenomic analysis of bacterial communities and antibiotic resistance genes in Penaeus monodon biofloc-based aquaculture environments. Front Mar Sci 8

Cheung MK, Yip HY, Nong W, Law PTW, Chu KH, Kwan HS, Hui JHL (2015) Rapid change of microbiota diversity in the gut but not the hepatopancreas during gonadal development of the new shrimp model Neocaridina denticulata. Mar Biotechnol 17:811–819

Chu H, Mazmanian SK (2013) Innate immune recognition of the microbiota promotes host-microbial symbiosis. Nat Immunol 14(7):668–675. https://doi.org/10.1038/ni.2635

Cornejo-Granados F, Lopez-Zavala AA, Gallardo-Becerra L, Mendoza-Vargas A, Sánchez F, Vichido R, Brieba LG, Viana MT, Sotelo-Mundo RR, Ochoa-Leyva A (2017) Microbiome of Pacific white leg shrimp reveals differential bacterial community composition between wild, aquacultured and AHPND/EMS outbreak conditions. Sci Rep 7:11783

DeCastro ME, Rodríguez-Belmonte E, González-Siso MI (2016) Metagenomics of thermophiles with a focus on discovery of novel thermozymes. Front Microbiol 7:1521

Ding X, Jin F, Xu J, Zhang S, Chen D, Hu B, Hong Y (2022) The impact of aquaculture system on the microbiome and gut metabolome of juvenile Chinese softshell turtle (*Pelodiscus sinensis*). iMeta 1(2):e17

Diwan AD, Harke SN, Panche AN (2022) Aquaculture industry prospective from gut microbiome of fish and shellfish: an overview. J Anim Physiol Anim Nutr 106(2):441–469
Duan Y, Huang J, Wang Y, Zhang J (2020) Characterisation of bacterial community in intestinal and rearing water of Penaeus monodon differing growth performances in outdoor and indoor ponds. Aquac Res 51:4279–4289
Eichmiller JJ, Hamilton MJ, Staley C, Sadowsky MJ, Sorensen PW (2016) Environment shapes the fecal microbiome of invasive carp species. Microbiome 4:44
Etyemez M, Balcazar JL (2015) Bacterial community structure in the intestinal ecosystem of rainbow trout (Oncorhynchus mykiss) as revealed by pyro sequencing-based analysis of 16S rRNA genes. Res Vet Sci 100:8–11
Fjellheim AJ, Playfoot KJ, Skjermo J, Vadstein O (2012) Inter-individual variation in the dominant intestinal microbiome of reared Atlantic cod (Gadus morhua L.) larvae. Aquac Res 43:1499–1508
Froese R, Pauly D (2019) Fish Base: species list: world Wide Web electronic publication
Ghanbari M, Kneifel W, Domig KJ (2015) A new view of the fish gut microbiome: advances from next-generation sequencing. Aquaculture 448:464–475
Gianoulis TA, Raes J, Patel PV, Bjornson R, Korbel JO, Letunic I, Yamada T, Paccanaro A, Jensen LJ, Snyder M, Bork P (2009) Quantifying environmental adaptation of metabolic pathways in metagenomics. Proc Natl Acad Sci 106(5):1374–1379
Giatsis C et al (2016) Probiotic legacy effects on gut microbial assembly in tilapia larvae. Sci Rep 6 (1):1–11
Gilbert JA, Dupont CL (2011) Microbial metagenomics: beyond the genome. Annu Rev Mar Sci 3: 347–371
Gómez GD, Balcázar JL (2008) A review on the interactions between gut microbiota and innate immunity of fish. FEMS Immunol Med Microbiol 52:145–154
Green TJ, Smullen R, Barnes AC (2013) Dietary soybean protein concentrate-induced intestinal disorder in marine farmed Atlantic salmon, Salmo salar is associated with alterations in gut microbiota. Vet Microbiol 166:286–292
He Y, Mao C, Wen H, Chen Z, Lai T, Li L, Lu W, Wu H (2017) Influence of ad libitum feeding of piglets with Bacillus subtilis fermented liquid feed on gut flora, luminal contents and health. Sci Reports 7(1):44553
Huang X, Fan X, Ying J (2019) Emerging trends and research foci in gastrointestinal microbiome. J Transl Med 17:67
Ingerslev HC et al (2014) Diet type dictates the gut microbiota and the immune response against Yersinia ruckeri in rainbow trout (Oncorhynchus mykiss). Fish Shellfish Immunol 40(2):624–633
Jackson KA, Stroika S, Katz LS, Beal J, Brandt E, Nadon C, Reimer A, Major B, Conrad A, Tarr C, Jackson BR (2016) Use of whole genome sequencing and patient interviews to link a case of sporadic listeriosis to consumption of prepackaged lettuce. J Food Prot 79(5):806–809
Kawashima T, Kosaka A, Yan H, Guo Z, Uchiyama R, Fukui R et al (2013) Double-stranded RNA of intestinal commensal but not pathogenic bacteria triggers production of protective interferon-β. Immunity 38(6):1187–1197. http://www.ncbi.nlm.nih.gov/pubmed/23791646
Kim PS et al (2021) Host habitat is the major determinant of the gut microbiome of fish. Microbiome 9(1):1–16
Ko KK, Chng KR, Nagarajan N (2022) Metagenomics-enabled microbial surveillance. Nat Microbiol 7(4):486–496
Krishnani KK, Gopikrishna G, Pillai SM, Gupta BP (2010a) Abundance of sulphur-oxidising bacteria in coastal aquaculture using soxB gene analyses. Aquac Res 41(9):1290–1301
Krishnani KK, Kathiravan V, Natarajan M, Kailasam M, Pillai SM (2010b) Diversity of sulfur-oxidising bacteria in greenwater system of coastal aquaculture. Appl Biochem Biotechnol 162(5):1225–1237
Larsen AM, Mohammed HH, Arias CR (2014) Comparison of DNA extraction protocols for the analysis of gut microbiota in fishes. FEMS Microbiol Lett 362:fnu031

Le Nguyen DD, Ngoc HH, Dijoux D, Loiseau G, Montet D (2008) Determination of fish origin by using 16S rDNA fingerprinting of bacterial communities by PCR-DGGE: an application on Pangasius fish from Viet Nam. Food Control 19:454–460

Legrand TPRA, James WW, Laura SW, Andrew PAO (2019) A microbial sea of possibilities: current knowledge and prospects for an improved understanding of the fish microbiome. Rev Aquac 12(2):1101–1134

Liu H et al (2016) The gut microbiome and degradation enzyme activity of wild freshwater fishes influenced by their trophic levels. Sci Rep 6(1):1–12

Liu KF et al (2010a) Effects of the probiotic, Bacillus subtilis E20, on the survival, development, stress tolerance, and immune status of white shrimp, Litopenaeus vannamei larvae. Fish Shellfish immunol 6:837–844

Liu KF et al (2010b) Effects of the probiotic, Bacillus subtilis E20, on the survival, development, stress tolerance, and immune status of white shrimp, Litopenaeus vannamei larvae. Fish & Shellfish Immunol 28(5–6):837–844

Llewellyn MS, Boutin S, Hoseinifar SH, Derome N (2014) Teleost microbiotas: the state of the art in their characterisation, manipulation, and importance in aquaculture and fisheries. Front Microbiol 5:207

Martín R, Miquel S, Langella P, Bermúdez-Humarán LG (2014) The role of metagenomics in understanding the human microbiome in health and disease. Virulence 5(3):413–423. https://doi.org/10.4161/viru.27864. Epub 2014 Feb 11. PMID: 24429972; PMCID: PMC3979869

Martínez-Porchas M, Vargas-Albores F (2017) Microbial metagenomics in aquaculture: a potential tool for a deeper insight into the activity. Rev Aquac 9(1):42–56

Meek RW, Vyas H, Piddock LJ (2015) Nonmedical uses of antibiotics: time to restrict their use? PLoS Biol 13(10):e1002266

Miller RR, Montoya V, Gardy JL, Patrick DM, Tang P (2013) Metagenomics for pathogen detection in public health. Genome Med 5(9):1–4

Molinari LM et al (2003) Bacterial microflora in the gastrointestinal tract of Nile tilapia, Oreochromis niloticus, cultured in a semi-intensive system. Acta Sci Biol Sci 25(2):267–271

Mueller K, Ash C, Pennisi E, Smith O (2012) The gut microbiota. Science 336:1245

Nielsen NH, Backes G, Stougaard J, Andersen SU, Jahoor A (2014) Genetic diversity and population structure analysis of European hexaploid bread wheat (Triticum aestivum L.) varieties. PLoS One. 9(4):e94000

Ost KS, Round JL (2018) Communication between the microbiota and mammalian immunity. Annu Rev Microbiol 72(1):399–422

Perreten V, Vorlet-Fawer L, Slickers P, Ehricht R, Kuhnert P, Frey J (2005) Microarray-based detection of 90 antibiotic resistance genes of gram-positive bacteria. J Clin Microbiol 43(5): 2291–2302

Perry WB, Lindsay E, Payne CJ, Brodie C, Kazlauskaite R (2020) The role of the gut microbiome in sustainable teleost aquaculture. Proc R Soc B Biol Sci 287:20200184

Pietretti D, Wiegertjes GF (2014) Ligand specificities of toll-like receptors in fish: indications from infection studies. Dev Comp Immunol 43(2):205–222. http://www.ncbi.nlm.nih.gov/pubmed/23981328

Prabhakara AM, Nanjappa DP, Kotian A, Kalladka K, Chakraborty G, Vittal R, Raj JR, Deekshit VK, Chakraborty A (2022) Wild-type and cancer-prone zebrafish exhibit distinct gut microbial diversity and differential anti-inflammatory response upon infection. J Biosci 47(4):65

Rauta PR, Samanta M, Dash HR, Nayak B, Das S (2014) Toll-like receptors (TLRs) in aquatic animals: signaling pathways, expressions and immune responses. Immunol Lett 158(1–2): 14–24. http://www.ncbi.nlm.nih.gov/pubmed/24291116

Riesenfeld CS, Schloss PD, Handelsman J (2004) Metagenomics: genomic analysis of microbial communities. Annu Rev Genet 38(1):525–552

Ringø E, Olsen RE (1999) The effect of diet on aerobic bacterial flora associated with intestine of Arctic charr (Salvelinus alpinus L.). J Appl Microbiol 86:22–28

Ringø E, Lødemel JB, Myklebust R, Jensen L, Lund V, Mayhew TM, Olsen RE (2002) Effect of soybean, linseed and marine oils on aerobic gut microbiota of Arctic charr Salvelinus alpinus L. before and after challenge with Aeromonas salmonicida ssp. salmonicida. Aquac Res 33: 591–606

Ringø E, Zhou Z, Vecino JLG, Wadsworth S, Romero J, Kroghdal Å, Olsen RE, Dimitoglou A et al (2016) Effect of dietary components on the gut microbiota of aquatic animals: the never-ending story? Aquac Nutr 22:219–282

Romero J, Ringø E, Merrifield DL (2014) The gut microbiota of fish. In: Merrifield D, Ringø E (eds) Aquaculture nutrition: gut health, probiotics and prebiotics. Wiley Blackwell Publishing, pp 75–100

Rosello-Mora R (2005) Updating prokaryotic taxonomy. J Bacteriol 187:6225–6257

Rungrassamee W, Klanchui A, Maibunkaew S, Chaiyapechara S, Jiravanichpaisal P, Karoonuthaisiri N (2014) Characterisation of intestinal bacteria in wild and domesticated adult black tiger shrimp (Penaeus monodon). PLoS One 9(3):e91853

Salazar C, Haussmann D, Kausel G, Figueroa J (2016) Molecular cloning of Salmo salar toll-like receptors (TLR1, TLR22, TLR5M and TLR5S) and expression analysis in SHK-1 cells during Piscirickettsia salmonis infection. J Fish Dis 39(2):239–248. http://www.ncbi.nlm.nih.gov/pubmed/25903926

Sanchez-Paz A (2010) White spot syndrome virus: an overview on an emergent concern. Vet Res 41(6):43

Sullam KE, Essinger SD, Lozupone CA, O'Connor MP, Rosen GL, Knight R, Kilham SS, Russell JA (2012) Environmental and ecological factors that shape the gut bacterial communities of fish: a meta-analysis. Mol Ecol 21:3363–3378

Suttle CA (2007) Marine viruses—major players in the global ecosystem. Nat Rev Microbiol 5(10): 801–812

Talwar C, Nagar S, Lal R, Negi RK (2018) Fish gut microbiome: current approaches and future perspectives. Indian J Microbiol 58(4):397–414

Tanveer A, Dwivedi S, Gupta S, Gaur RK, Yadav D (2022) Deciphering soil microbiota using metagenomic approach for sustainable agriculture: an overview. Bioinform Agric:439–454

Tarnecki AM, Burgos FA, Ray CL, Arias CR (2017) Fish intestinal microbiome: diversity and symbiosis unravelled by metagenomics. J Appl Microbiol 123(1):2–17

Tello M, Valdes N, Vargas R, Rojas J, Parra M, Gajardo G, Gonzalez A (2019) Application of metagenomics to Chilean aquaculture. In: Metagenomics—basics, methods and applications. IntechOpen. https://doi.org/10.5772/intechopen.86302

Villamil SI, Roger H, Dean RJ (2020) Microbiome diversity and dysbiosis in aquaculture. Rev Aquac 13(2):1077–1096

Wang WL, Xu SY, Ren ZG, Tao L, Jiang JW, Zheng SS (2015) Application of metagenomics in the human gut microbiome. World J Gastroenterol 21(3):803–814. https://doi.org/10.3748/wjg.v21.i3.803. PMID: 25624713; PMCID: PMC4299332

Webster U, Tamsyn M et al (2020) Cortisol-related signatures of stress in the fish microbiome. Front Microbiol 11:1621

Wlodarska M, Thaiss CA, Nowarski R, Henao-Mejia J, Zhang J-P, Brown EM et al (2014) NLRP6 inflammasome orchestrates the colonic host-microbial interface by regulating goblet cell mucus secretion. Cell 156(5):1045–1059

Xiang X, Wang R, Chen L et al (2022) Immunomodulatory activity of a water-soluble polysaccharide extracted from mussel on cyclophosphamide-induced immunosuppressive mice models. NPJ Sci Food 6:26. https://doi.org/10.1038/s41538-022-00140-8

Ye L et al (2014) Fish gut microbiota analysis differentiates physiology and behavior of invasive Asian carp and indigenous American fish. ISME J 8(3):541–551

Yu Z, Li L, Zhu R, Li M, Duan J, Wang JY, Liu YH, Wu LF (2020) Monitoring of growth, digestive enzyme activity, immune response and water quality parameters of Golden crucian carp (Carassius auratus) in zero-water exchange tanks of biofloc systems. Aquac Rep 16:100283

Yukgehnaish K, Praveen K, Parimannan S, Marimuthu K, Aziz A, Bilal AP, Jesu A (2020) Gut microbiota metagenomics in aquaculture: factors influencing gut microbiome and its physiological role in fish. Rev Aquac 1:1–25

Zhang C et al (2016) Ecological robustness of the gut microbiota in response to ingestion of transient food-borne microbes. ISME J 10(9):2235–2245

Zhang Z et al (2017) High spatial and temporal variations of microbial community along the southern catfish gastrointestinal tract: insights into dynamic food digestion. Front Microbiol 8:1531

Zhang M, Shan C, Tan F, Limbu SM, Chen L, Du ZY (2020) Gnotobiotic models: powerful tools for deeply understanding intestinal microbiota-host interactions in aquaculture. Aquaculture 517:734800

Zhou ZK, Shi P, He S, Liu Y, Huang GH, Yao B, Ringø E (2009) Identification of adherent microbiota in the stomach and intestine of emperor red snapper (Lutjanus sebae Cuvier) using 16S rRNA-DGGE. Aquac Res 40:1213–1218

Ready Reckoner for Using Bioinformatics Tools in Fish and Shellfish Microbiome Analysis

5

Pavan Gollapalli, Kattapuni Suresh Prithvisagar, and Ballamoole Krishna Kumar

Abstract

The aquaculture industry is experiencing significant growth and offers sustainable solutions for ensuring global food security. Past research has attempted to elucidate the intricate relationship between fish/shellfish hosts and the microbial population that surrounds/associated with them, as well as how it impacts on their health, growth, reproduction, metabolism, and nutritional absorption. Despite obvious advantages for the bio/pharma industry, the use of gut microbiomes, particularly those from fish and shellfish, has not advanced greatly. Lack of knowledge of microbial species diversity or functional traits may be one of the reasons. Priority is currently being given to modifying the microbiota in the digestive system to enhance the health and production of cultivable organisms due to the enormous growth of the aquaculture sector and the growing threat of emerging infections as a result of culture intensification. Recent advancements in next-generation sequencing (NGS) technologies have made it possible for researchers to conduct more comprehensive and affordable analyses of the microbiome community. The application of bioinformatics expertise to the progress of fish and shellfish studies has not gotten much attention up until now. In this chapter, we present a best-practice approach to speed up the execution of such studies. Assessments of the target gene and shotgun metagenomic sequencing

P. Gollapalli
Center for Bioinformatics and Biostatistics, Nitte (Deemed to Be University), Mangaluru, Karnataka, India

K. S. Prithvisagar · B. K. Kumar (✉)
Department of Infectious Diseases and Microbial Genomics, Nitte University Centre for Science Education and Research, Nitte (Deemed to be University), Deralakatte, Mangaluru, Karnataka, India
e-mail: krishnakumarb@nitte.edu.org

A. Diwan et al. (eds.), *Microbiome of Finfish and Shellfish*,
https://doi.org/10.1007/978-981-99-0852-3_5

data will become more reliable and repeatable with the use of these standardized processes.

Keywords

Aquaculture · Host–pathogen interaction · Microbiome · NGS technology · Metagenomics · Systems biology

5.1 Introduction: Microbiome Analysis in Fish and Shellfish

Aquaculture is the rapidly growing industry with sustainable means of achieving food security in the world. According to the FAO statistics (2019), the aquaculture production stands at more than 85 million tons valued at 275 billion US dollars globally (Ottinger et al. 2021). The "Pathobiome" describes the interaction among the host, its associated and surrounding microbes, and their beneficial and/or harmful effects on host (Bass et al. 2019). The richness of the microbiome plays a key role in the host's health, nutrition, physiological activities, and processes. Due to significant growth of the aquaculture industry and the growing threat of the emerging superbugs due to the extensive use of antibiotics, priority is now being given to manipulating the microbiota in the digestive tract to improve the health and production of cultivable organisms (Diwan et al. 2022). The traditional techniques used for microbiome analysis relied on the phenotypic, biochemical, and genomic characteristics and were less expensive, semi-quantitative, yet they have a number of disadvantages such as labor-intensive, time-consuming, low resolution, technical bias, and no phylogenetic approach (Rezasoltani et al. 2020). With the cutting-edge next-generation sequencing (NGS) technologies, researchers have recently been able to analyze the microbiome community affordably at greater depth (Tarnecki et al. 2017). The 16s rRNA sequencing has since been accepted as the gold standard for isolate identification and characterization of the whole bacterial community (Diwan et al. 2022).

5.1.1 Factors Influencing the Microbiome in Fish and Shellfish

The host-associated microbiome in aquaculture candidates are influenced by several factors which could be classified as host dependent/endogenous factors (host genotype, host developmental stages, physiological conditions of the host, population dynamics, and starvation of the animals) and the host independent attributes/exogenous factors (environmental stress due to water chemistry, temperature, type of the diet or nutrition given into the host and impact of infectious diseases) (Derome and Filteau 2020; Rajeev et al. 2021). In addition, the use of xenobiotics has shown adverse effect on native microflora and their beneficial effects on the host (Yukgehnaish et al. 2020) (Fig. 5.1).

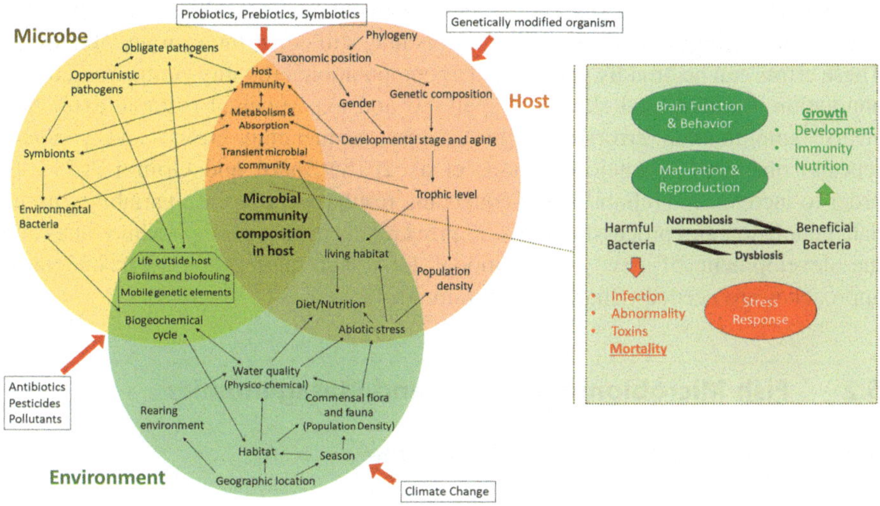

Fig. 5.1 Illustration of host, pathogen, and environmental factors influencing the microbiome of fish and shellfish

5.1.2 Ontogeny of the Microbiome in Fish and Shellfish

There are two basic ways that germs might spread among aquatic animals. In the case of oviparous species, the eggs are exposed directly to horizontal transmission of microbes from the surrounding environment, but in viviparous species, the larvae initially exposed to vertical transmission of microbes from the maternal environment (Derome and Filteau 2020). These microbial communities begin to develop in the early stages of yolk sac larvae and continue to change as the larvae progress through developmental stages (Gómez and Balcázar 2008). These initial microorganisms are essential as they influence the epigenetic patterns, which may have long-term effects on the host's general fitness (Deng et al. 2021). The host begins active feeding upon larval stage begins, exposing the gastrointestinal tract and gills to the rearing environment and food, which may affect additively or negatively on the pre-existing microbiota (Olafsen 2001). A variety of biotic and abiotic factors have an impact on the ontogeny of the microbiota in aquatic species (Derome and Filteau 2020).

5.1.3 Challenges and Future Perspectives in Metagenomics

Numerous studies have come up to describe the microbiome of commercially significant fish and shellfish which sheds light on the impact of the associated microbiome on immune function, growth, reproduction, metabolism, and nutrient absorption from food (Diwan et al. 2022). Despite recent improvements in computational analysis tools and sequencing technologies, when contrasting the microbiota findings of various studies, errors may still occur for a variety of reasons. These errors can be broadly categorized as computational (quality control, assembly, annotation, binning, and statistical analysis) and experimental (sample extraction errors, amplification errors/bias, primer bias, sequencing errors, and genomic repeats) obstacles (Cornejo-Granados et al. 2018; Bharti and Grimm 2021). In order to increase the reliability of sequencing results and decrease variability in the data generation and processing steps, improvements in sequencing technologies and the development of new computational tools and algorithms with integrated approaches are needed (Bharti and Grimm 2021).

5.2 Fish Microbiome Diversity and Metagenomics

5.2.1 Introduction to Fish Microbiome

Fish make up nearly half of all vertebrate species and have a large diversity of more than 28,000 species, showing a wide range of physiologies, ecologies, and natural histories and due to their lifestyle, are constantly colonized by a variety of commensals, opportunistic, and obligate pathogens on its skin, gills, gut, luminescent organs, and other mucosal surfaces histories (Wong and Rawls 2012; Austin 2006; Salinas and Magadán 2017). These fish gut microbes can be divided into two groups: temporary microbes that are ingested to the fish through feed and permanent microbes that reside in the host's intestine and are intimately involved in symbiosis (Zhang et al. 2016).

5.2.2 Significance of Fish Microbiome

Microbes play an important role in the development of the digestive system, the production of enzymes and vitamins, the host's ability to absorb nutrients in addition to being involved in host defense (Romero et al. 2014). The mucus layer of the skin and gills constantly interacting with the aquatic environment, enabling for the simple exchange of bacteria with the environment, also promotes the growth and homeostasis of the host immunity, the recycling and reusing of waste products, and the resistance against colonizing pathogens by competing for nutrients and space and affects vulnerability to various infections and probably also to environmental stress (Legrand et al. 2018; Gomez and Primm 2021).

5.2.3 Early Research in Fish Microbiome Analysis

The study of fish's intestinal and "slime flora" dates back to the first half of the twentieth century, with the earliest publications on the subject appearing in the late 1920s to 1930s (Reed and Spence 1929; Gibbons 1933). In 1953, a study on effect of fasting on gut microflora of fishes was conducted by culture-based method (Margolis 1953). Understanding the gut-associated microbes and their translocation in the gastrointestinal tracts of various fish species has been aided by the use of electron microscopy as a tool (Ringø et al. 2003). 16S rDNA-based methodology was used to identify the bacterial populations in early life stages of salmon (Oncorhynchus kisutch) (Romero and Navarrete 2006) and combined 16S ribosomal DNA (rDNA) amplicons subjected to temporal temperature gradient gel electrophoresis (TTGE) was used to analyze digestive tract of juvenile Atlantic salmon (*Salmo salar* L.) (Navarrete et al. 2009). Reveco et al. (2014) used PCR-DGGE that allowed screening of changes in the intestinal bacterial community of healthy and inflamed intestines of Atlantic salmon (*Salmo salar* L.). These conventional studies showed a very low percentage of important microorganisms of the overall microbiome associated with respective host environments (Talwar et al. 2018).

5.2.4 Metagenomics in Fish Microbiome

Drawbacks of traditional techniques (Amann et al. 1995) led to the development of most advanced culture-independent methods which has made it possible to conduct the detailed analysis of the dense composition and functional potential of the associated microflora (Rodrigue et al. 2010; de Bruijn et al. 2018). Such studies have largely used 454/Roche pyrosequencing and Illumina technologies since they offer a large number of reads in a single run with the quick, inexpensive acquisition of accurate data, and allow for the detection of both dominant and limited abundance microbial community members (Ghanbari et al. 2015). In order to identify the microbiota and comprehend their mechanisms involved in host immunity, the effects of environmental, host-related, dietary, and genetic factors on the composition of fish gut microbiota, a number of studies have analyzed and compared the microbiomes from eggs, skin, and gut of healthy and diseased individuals which are in the same developmental stages and grown in the same conditions (Ghanbari et al. 2015; de Bruijn et al. 2018).

5.3 Shellfish Microbiome Diversity and Metagenomics

5.3.1 Introduction to Shellfish Microbiome

Shellfish are one of the primary culture candidates in the aquaculture industry (Azra et al. 2021). It has grown significantly over the past two decades; however, this development has occasionally led to low productivity, the emergence of diseases, and environmental deterioration (Ringø 2020). The majority of infections are brought on by ubiquitous opportunistic pathogens that are part of the host's native or habitat microflora and cause disease upon favorable conditions (Diwan et al. 2022). The shellfish gut's resident microbes interact with their host and support a number of vital physiological functions, such as immunity and digestion (Holt et al. 2021).

5.3.2 Importance of Shellfish Microbiome

The water and sediment in the area have a significant impact on the microbiota composition of aquatic animals like crustaceans (Cornejo-Granados et al. 2018). The studies have shown that in shellfishes, depending on the host's health, growth stage, diet, culture capacity, altered environment and water quality control of the culture environment contribute to changes in the composition of the gut microbiome (Omont et al. 2020). A healthy microbiome is composed primarily of a functional core that is resistant, resilient, and healthy which is made up of a variety of specific microbial gene families, metabolic modules, and regulatory pathways that can withstand any agitation caused by internal or external factors and then return to a healthy normal state to maintain a stable and balanced host-associated ecology (Rajeev et al. 2021).

5.3.3 Early Research in Shellfish Microbiome Analysis

In 1961, the first study of its kind on the shellfish microbiome was conducted on shrimp, using conventional methods to examine the gut bacterial community (Tysset et al. 1961). Furthermore, similar works were done on microbiome study in digestive tract of Penaeus japonicus Bate (Yasuda and Kitao 1980), aerobic heterotrophic bacterial flora in hatchery-reared eggs, larvae, and post-larvae of Penaeus indicus (Hameed 1993), and microbiome of Penaeus indicus (Milne Edwards) reared in ponds (Singh et al. 1998). Denaturing gradient gel electrophoresis, RFLP, and clone libraries were established for microbiome analysis with the advancement of polymerase chain reaction and other molecular techniques (Romero et al. 2002; Cornejo-Granados et al. 2018; Diwan et al. 2022). Conventional methods were also used to study microbiome in bivalves and their antimicrobial properties against known pathogens (Cavallo et al. 2009; Desriac et al. 2014; Pierce and Ward 2018).

5.3.4 Metagenomics in Shellfish Microbiome

Recently, NGS-based microbiome analyses of aquatic animals have emphasized the dysbiosis of the gut microbiota in response to host diseases (Sha et al. 2022). A next-generation sequencing (NGS) analysis of the gut microbiome of Mytilus chilensis using the V6-V8 16S rDNA gene regions as the target (Santibáñez et al. 2022), bacterial community microbiome in the water of a commercial shellfish hatchery of Pacific geoduck clam larvae (Timmins-Schiffman et al. 2021), the total microbiota in the hemolymph of three aquatic invertebrates—Scylla paramamosain (mud crab), Litopenaeus vannamei (white leg shrimp), and Crassostrea angulata (Portuguese oysters) by a combined filtration method, fluorescence microscopy, and high-throughput sequencing technique (Zhang et al. 2018) have all shown significant bacterial diversity and the influence of habitat on the bacterial community structure, identified the potential versus actual metabolic activity of bacterial taxa, which affects water quality and intertrophic relationships and the metabolic processes of non-cultivable hemolymph microorganisms.

5.4 Bioinformatics Skills in Fish and Shellfish Microbiome Research

Through various studies on the health of humans and livestock, where these kinds of studies have been going on for many years, microbiome research has gained a lot of relevance. The study of fish and shellfish microbiomes, in contrast, is still very young. Utilizing bioinformatics expertise for the advancement of research on fish and shellfish has received very little attention in this context up to now. Understanding the composition of a resident microbiota, as well as the organization and interactions between them and their hosts, can shed light on the health and dysfunction of the host organism. Nearly half of all species of vertebrates, or over 28,000 species, are fish, with a diverse range of physiologies, ecologies, and natural histories. Fish, therefore, are an important group for comprehending the variety and nature of symbioses in vertebrate gut microbial ecosystems (Ghanbari et al. 2015). Sanger sequencing is expensive, but technological challenges and poor coverage make it impossible to analyze many samples simultaneously. However, quick and affordable methods for next-generation sequencing (NGS) technologies have recently been developed to examine the genetic potential and makeup of densely populated microbial communities, such as the gut microbiota (Foster et al. 2012). The composition and functional characteristics of fish microbial communities have been examined using these techniques over the past few years. The Illumina and 454/Roche pyrosequencing technologies have been used most frequently in fish microbiota research (for example, the Roche 454 FLX Titanium and FLX+) (e.g., the Illumina MiSeq and HiSeq 2000). These NGS platforms offer a greater number of reads in a single run than the Sanger sequencing method, allowing for the quick and economical capture of detailed and accurate sequence data and identifying both dominant and low abundance (rare) microbial community members (Wong et al. 2013). In this section, we discuss the advances in NGS techniques for studying complex microbial ecosystems and the growing appreciation of the importance of the indigenous microbiota of fish.

5.4.1 Computational Protocol for Microbiome Acquisition

We demonstrate a best-practice workflow to streamline the process of carrying out such research. With these standardized procedures, assessments of the target gene and shotgun metagenomic sequencing data will become more robust and reproducible.

5.4.1.1 Target Gene Approach

In the section below, OTU-based data processing and analysis are performed using raw reads that have undergone quality filtering and have been processed using either OTU-based (mothur) (Schloss et al. 2009) or ASV-based (DADA2) (Callahan et al. 2016) techniques,

Taxonomy

The microbial abundance contained in the sequencing data is profiled using the OTU table created from the processed reads. Multiple pan genome alignments utilizing specialized databases, like SILVA, Greengenes, and RDP of different gene families, are used to identify microbial communities.

Functional Annotation

Tax4Fun (Douglas et al. 2018) or PICRUSt (Aßhauer et al. 2015) can be used to predict the functional profile of detected microbial communities based on the output from mothur or DADA2. While PICRUSt is a bioinformatics pipeline that clusters protein sequences based on KEGG or COG gene families and 16S rRNA gene copy counts, Tax4Fun is an R-based approach that uses SILVA as a reference. In a given dataset, both offer approximations of functional profiles.

Data Visualization

The R Bioconductor tool phyloseq (McMurdie and Holmes 2013) may then examine differences between and within samples using the resultant OTUs and the dissimilarity indices.

5.4.1.2 Shotgun Metagenomics Approach

With this method, every gene from a wide range of microbial genomes found in a sample is exhaustively catalogued. The two main types of sequencing platforms used today are long-read (500–4000 bp; PacBio and Oxford Nanopore) technologies that provide superior contig assembly or short-read (250–300 bp; Illumina) technologies that enable higher sequencing depths. Ultimately, both of them still depend on ideal sequence processing, including correct assembly (Fig. 5.2):

Quality Filtering

Compared to the target gene technique, the reads from metagenomic sequencing contain significantly more variability. The majority of the time, low-quality or auxiliary adapter/hairpin sequences that are eliminated during the quality filtering processes are the cause of heterogeneity in metagenomics reads. Trimmomatic, Sickle, and BBTools can be used to trim paired-end and single-end reads for short-read sequencing platforms like Illumina and Solexa. Long-read sequences can also be used, in addition to modules specifically designed for shortening short-

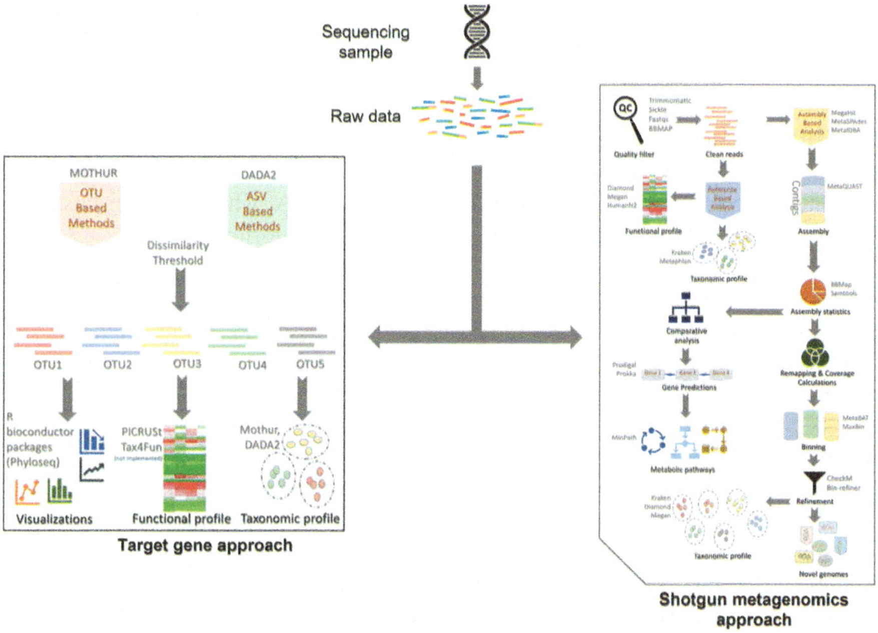

Fig. 5.2 Targeted amplicon and shotgun metagenomics data acquisition and analysis best practices, from sequencing to functional annotation. This diagram illustrates the fundamental order of experimental stages, which are followed by preprocessing and analysis steps. The optional tools used during the procedure are also displayed at each phase

read sequences. The processed reads that have undergone quality filtering are next subjected to assembly- and reference-based analysis.

Reference-Based Analysis

To create a taxonomy catalogue of the communities, this study aligns with databases (NCBI or a customized repository like SILVA):

Taxonomy

The most effective method for compositional profiling of communities from metagenomic sequencing data is to use particular clade-specific marker genes discovered from 3000 reference genomes (MetaPhlAn) or to use exact k-mer alignments in conjunction with a classification algorithm (Kraken).

Functional Annotation

The HUMAnN2 or Megan pipelines are the best options for performing the functional profiling of metagenomic communities. A biphasic alignment screen using MetaPhlAn is implemented by HUMAnN2, the functionally annotated pangenomes of the discovered species are presented. Megan does annotations using COG/NOG and KEGG orthology seed classifications. Pairwise and frame shift alignments for lengthy readings can be done using Megan alone or in conjunction with the DIAMOND sequence aligner.

Assembly-Based Analysis

De novo assemblers are used in this more thorough study of the metagenomic sequencing data. MegaHit, MetaSPAdes, and MetaIDBA, which were discussed in the parts above, are the three most effective assembly algorithms.

Contig Assembly

The contiguously grouped assembled reads are assessed using MetaQUAST (Mikheenko et al. 2016), which compares them with metagenome assemblies using alignments to nearby references.

Assembly Statistics

Remapping/coverage calculations and comparison analysis are prerequisites to this stage. Sorting and indexing are efficiently carried out along with alignment construction by SAM (Sequence Alignment Map) tools. Similar to this, for very big genomes, BBMap, which can accommodate both short- and long-read sequences from Illumina, PacBio, or MinION, might be preferred.

Comparative Analysis

Algorithm-based gene predictions and metabolic pathway identification are included in comparative analysis. Prokka uses Prodigal to predict genes, which it then uses to do functional annotation on the genes (Seemann 2014). On the output of prodigally translated proteins, Prokka employs the CDD, PFAM, and TIGRFAM databases for homology searches. The MinPath algorithm could also be used to reconstruct biological pathways based on the predictions of protein families (Ye and Doak 2009).

Binning

After remapping and coverage calculations, the resulting contigs are binned or grouped before going through further downstream processing. For metagenomic contig binning, one might use either MaxBin, which uses an EM technique, or MetaBAT, which has an adaptive binning algorithm.

Refinement

Taxonomic profiles are generated and any novel genomes found in the samples are annotated using post-binning remapping and refining processes. The best method for determining genome completeness and contamination uses both CheckM (Parks et al. 2015) and bin-refiner (Song and Thomas 2017). The above-described Kraken (Wood and Salzberg 2014) and Diamond (Buchfink et al. 2015) algorithms can be used with or without the Megan pipeline (Huson et al. 2016) to achieve the best results for taxonomic profiles and novel genome identification.

5.4.2 Taxonomic Profile Based on 16S Amplicon Data

The 16S rRNA gene was proposed to be a marker for taxonomic identification by Woese and Fox in 1977 (Woese and Fox 1977; Fox et al. 1977). Identifying gut microbiota that are extremely resistant to culture-based assays was made possible by profiling microbial communities in a culture-independent way. This technique uses highly conserved sequences within the hypervariable sections of the 16S rRNA gene (like V3-V5) to selectively bind universal primer pairs, allowing amplification and sequencing of a region that includes enough phylogenetic information for taxonomic categorization. Using this data, operational taxonomic units (OTUs) with 97% sequence identity to a bacterial taxon are identified.

5.4.2.1 Picking Operational Taxonomic Units

Analysis of the taxonomic composition of microbial communities is a crucial step. Several preliminary processing procedures, including raw data filtering, chimaera detection, and the elimination of erroneous non-biological sequences, are necessary for these analyses to be robust (Schloss et al. 2011). Grouping sequences into operational taxonomic units is a key idea utilized in microbial community analysis (OTUs). Sokal (1963) employed this concept for the first time in botanical study, but when molecular technologies advanced, microbiologists started to apply it as well (McCaig et al. 1999). A random degree of sequence identity (for example, 97% identity representing genus and 95% identity roughly indicating family) is used to group several DNA sequences into a single OUT (Schloss and Handelsman 2005). Once the number of sequences is reduced by selecting a representative sequence from a pool of sequences in an OTU, the main benefit of grouping sequences into OTUs is the decrease in computational requirements. Although this idea is extensively used and accepted by the scientific community, its practical applicability is in doubt since various biological entities have varying levels of identity and because the similarity cutoffs applied to incomplete 16S gene sequences have no biological significance. However, its continued usage is justified by the absence of a better strategy to address this problem.

The method of selecting OTUs has been used in microbial community analysis since its inception, and it can be applied with three different methods: closed reference-based (BLAST) (Altschul et al. 1990), open reference-based (UCLUST, USEARCH) (Edgar 2010), and de novo (CD-HIT (Fu et al. 2012), Mothur (Schloss et al. 2009), prefix/suffix, trie, UCLUST, USEARCH) (Navas-Molina et al. 2013) (Fig. 5.1). A reference database (such as Greengenes DeSantis et al. 2006) and amplicon sequences are compared for identity in the closed-reference technique (Fig. 5.3a). Although the de novo approach is used to cluster sequences that do not cluster with the reference, the open-reference strategy is also dependent on alignment against a reference database (Fig. 5.3b). Amplicon sequences are clustered by pairwise comparison using the de novo method, which does not require a reference database (Fig. 5.3c). Numerous benchmarking studies have examined these algorithms, which are used in various software programs (Lemos et al. 2017).

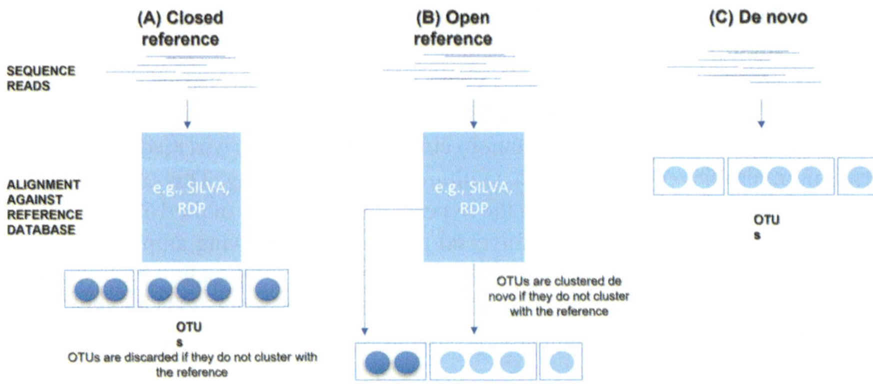

Fig. 5.3 Data analysis flowchart showing many methods for determining operational taxonomic units. (**a**) Closed reference-based, (**b**) open reference-based, and (**c**) de novo are the three options

5.4.2.2 Assigning Taxonomy

Based on 16S rRNA gene partial sequences, several techniques have been developed to predict microbial taxonomy. The naive Bayesian classifier used by the Ribosomal Database Project (RDP) is the most popular (Wang et al. 2007). This technique can classify 400 bp long sequences at the genus level and employs bootstrap confidence scores to back up the taxonomic assignment (Wang et al. 2007). QIIME (Caporaso et al. 2010) and Mothur (Schloss et al. 2009) both use additional approaches that are available. QIIME's default categorization relies solely on sequence similarity to determine taxonomy (Kuczynski et al. 2012). Like the RDP technique, Mothur employs k-mer counting and the Wang naive Bayesian classifier.

5.4.2.3 Measuring Alpha and Beta Diversity

An ecological community's alpha and beta diversity can be measured using a variety of approaches. These include statistical software programs (such as Vegan Oksanen et al. 2016) used in common pipelines like QIIME and Mothur. Beta diversity is the variety among distinct samples, whereas alpha diversity is the local diversity of a single sample (Lemos et al. 2011). There are specific ways to identify each variety form (alpha or beta). The species richness and evenness of the community structure are gaged using alpha diversity indices, such as the Shannon diversity index (Shannon 1948) and the Simpson diversity index (Simpson 1949).

5.4.3 Application of NGS Technologies in Recent Studies of Fish Gut-Associated Microbiota

The term "next-generation sequencing" (NGS) refers to a group of sequencing techniques that produce vast amounts of sequence data quickly and affordably. The Illumina system, Roche 454 system, and Ion Torrent PGM (Personal Genome

Machine) (second-generation sequencing technologies) are the main NGS platforms used in fish gut microbiome research. Recent developments in sequencing technology have given rise to third-generation sequencing technologies like PacBio (Pacific Biosciences) sequencing and Oxford Nanopore, which offer larger read lengths. However, compared to older platforms, they have a larger error rate, which might be reduced with a considerable increase in sequencing depth (Johny et al. 2022). Despite the fact that they have not yet been applied to studies of the fish gut microbiome, they should be mentioned nonetheless because they are expected to fundamentally alter this field of study in the near future.

5.4.3.1 Illumina Sequencing Platforms

Fish gut metagenomics and metabarcoding applications almost exclusively use the Illumina platform, which employs the sequencing-by-synthesis methodology as illustrated by (Johny et al. 2022). To create the local colonies known as clusters, the template DNA coupled to adaptors is amplified and affixed to a glass slide or flow cell. Sequencing primers, DNA polymerase, and fluorescently labeled, reversible terminator nucleotides are added to begin DNA synthesis. After inclusion, a fluorescence scanner records the nucleotide's position (Metzker 2010). MiSeq, NextSeq, and NovaSeq are the three Illumina systems that are most frequently used. The longest Illumina read lengths are produced by MiSeq, however, NovaSeq and NextSeq have higher throughput (Besser et al. 2018). The ultra-high throughput NovaSeq and NextSeq platforms may be chosen for large sequencing projects, whereas MiSeq may be favored for amplicon sequencing projects.

5.4.3.2 Roche Platforms

Pyrosequencing chemistry is used on Roche 454 platforms, which similarly employ the sequencing-by-synthesis principle. Deoxynucleotides are incorporated into the DNA chain during sequencing, and pyrophosphate is released. This reaction is performed by the enzyme ATP sulfurylase, and it acts as the substrate for the synthesis of ATP. A luciferase reaction is then ·catalyzed by ATP, converting luciferin to oxyluciferin and causing light to be released. The photons are then recorded by a camera, and the information about the DNA sequence is translated from the amount of light that is emitted (Siqueira et al. 2012).

5.4.3.3 Ion Torrent PGM

Sequencing is carried out on semiconductor chips in Ion Torrent PGM. The emission of a proton that alters the pH of the surrounding area allows for the detection of the integration of a base into DNA. Micro-detectors attached to the chip are used to find the pH change. As a result, the voltage created by the flooding of nucleotides one after another is translated into the DNA sequence (Flusberg et al. 2010).

5.4.3.4 PacBio Sequencers

Like other sequencing techniques, PacBio sequencers use fluorescently tagged nucleotides and operate on the principle of sequencing by synthesis. The primary distinction is that it makes real-time signal detection possible after adding nucleotides. Hairpin adaptors are joined to both ends of a target double-stranded DNA to create the template DNA, also known as SMRTbellTM, which is a closed, circular single-stranded DNA. When a template is loaded into a chip (SMRT cell), the SMRTbellTM diffuses into the sequencing unit, which is a single immobilized DNA polymerase that can bind to either of the SMRTbellTM's hairpin adaptors. When various fluorescently tagged nucleotides are added, DNA polymerase incorporates the nucleotides and emits light pulses that are unique to each label. These pulses are analyzed to determine the DNA base sequence (Eid et al. 2009; Rhoads and Au 2015).

5.4.3.5 Oxford Nanopore Technologies

One base at a time, a DNA molecule passes through a nanopore as measured by Oxford Nanopore Technologies' (MinIONTM, GridIONTM, and PromethIONTM) instruments. A tiny potential gradient is created across the protein pore, which comprises a protein and conductive electrolytic solution. According to the characteristics of each nucleotide base, the passage of each nucleotide obstructs the nanopore to a different degree. The DNA base sequence is inferred from the variations in current that are observed (Mikheyev and Tin 2014; Jain et al. 2016).

5.4.3.6 Other NGS Platforms

The SOLiD platform employs the sequencing-by-ligation technique, starting with an emulsion PCR step (Metzker 2010). The SOLiD platform can generate a lot of sequences in a run (up to three billion reads). Still, its application for microbiomics projects has been severely constrained by its chronically small read length (maximum single read 75 bp) (Hodkinson and Grice 2015). Pacific BioSciences' single-molecule, real-time sequencing is more suited to addressing questions in genome, transcriptome, and epigenetics research since it has longer read lengths than second-generation sequencing (SGS) technology. PacBio sequencing's extremely contiguous de novo assemblies can fill in gaps in the existing reference assemblies and describe structural variation (SV) in individual genomes. We can sequence through lengthy repetitive areas with longer reads and find mutations, many of which are linked to diseases. Due to its capability to sequence full-length transcripts or significant-length segments, PacBio transcriptome sequencing is also helpful for identifying gene isoforms. It helps discover novel genes and isoforms of recognized genes (Rhoads and Au 2015).

5.4.4 Functional Profile Based on Metagenomic and Metatranscriptomic Data

5.4.4.1 Gene Prediction and Functional Gene Annotation

The first step in predicting protein function is identifying the genes encoded in metagenomic DNA sequences. Bioinformatics has a significant barrier here because the prediction must be made using brief fragmented readings (incomplete genes). Orphelia (Hoff et al. 2009), FragGeneScan (Rho et al. 2010), MetaGeneMark (Zhu et al. 2010), and Glimmer-MG (Kelley et al. 2012) are just a few of the programs that have been created to annotate short metagenomic reads. Examples include Orphelia (Hoff et al. 2009), which use fragment length-specific models for gene prediction, and FragGeneScan (Rho et al. 2010), which also incorporates probabilistic modeling with information on codon usage and sequencing error. The prediction of coding sequences is now more accurate because of this information. BLASTx (Altschul et al. 1990) and DIAMOND (Buchfink et al. 2015) are two methods that exclusively employ homology and do not involve ab-initio gene prediction. Protein databases, such as the non-redundant database (nr/National Center for Biology Information [NCBI]), can be searched for comparable sequences using homology-based approaches (NCBI Resource Coordinators 2017). Since the sequences must be translated into the six reading frames, the similarity search takes longer than a direct comparison of sequences that were predicted from scratch. Due to its efficiency in quickly annotating millions of sequences, DIAMOND is now a substitute for annotating metagenomic data.

5.4.4.2 Assigning Taxonomy

Short metagenomic reads can be classified in two ways: (1) by methods based on comparisons with every section of the genome, including conserved housekeeping genes and highly variable genes; or (2) by methods based only on similarity to conserved housekeeping genes. Most tools, including homology-based techniques like MEGAN (Huson et al. 2007), adopt the first choice. For taxonomy prediction from the least common ancestor, MEGAN uses an output BLAST score (best hits) search. All metagenomic reads in this instance are aligned, for example, against a protein database of all microbial genomes stored at the NCBI. The slow BLAST search, which requires millions of reads, is the algorithm's limitation. However, other programs, such as DIAMOND (Buchfink et al. 2015), Kraken (Wood and Salzberg 2014), and Centrifuge (Kim et al. 2016), have been developed to align metagenomic readings to databases. There are additional taxonomic prediction techniques that compare each read to a database of gene markers specific to a given clade, like that found in MetaPhlAn (Segata et al. 2012).

5.4.4.3 Genome Assembly from Metagenomic Data

Gene-centric and genome-centric methods are now used to investigate metagenomes. Individual unassembled genomes serve as the foundation for gene-centric techniques, and individual genes are predicted from brief fragmented reads. On the other hand, whole metagenome assembly-based techniques consider specific microbial communities (Lemos et al. 2017). The length of short metagenomic reads restricts gene-centric analytical strategies. Even though there is specialized software for gene prediction based on short sequences, contigs are a more effective way to combine small reads. Short reads are currently divided into graphs based on k-mer lengths by software like MetaVelvet (Namiki et al. 2012) and metaSPAdes (Bankevich et al. 2012) (De Brujin graphs). Since De Brujin techniques are the most widely used metagenomic assemblers, we solely focus on them here when it comes to assembling short reads. Each sub-graph in the graph created by MetaVelvet reflects a different genome.

Improved functional annotation and gene prediction may be made possible by post-assembly analysis. Contigs can be used to reconstruct close, partial, or entire microbial genomes of uncultivated bacteria because they are longer than typical short reads (Albertsen et al. 2013). The method is referred to as "binning." Binding's main principle is the grouping of assembled contigs into distinct populations based on the compositional characteristics of the sequences, such as the guanine-cytosine (GC) content, tetranucleotide frequency, and sequence coverage. MaxBin (Wu et al. 2014), GroopM (Imelfort et al. 2014), and MetaBAT (Kang et al. 2015) are a few software programs for binning and rebuilding individual microbial genomes from metagenomic data.

5.4.5 Hypothesis-Driven Functional Analyses

It is usually believed to be impossible to fully analyze every functional feature of a microbiome dataset and to query every potential longitudinal and cross-sectional aspect. Multiple testing problems severely reduce the analysis power even when it is computationally viable. Determining the pertinent functional characteristics is a key step in overcoming these constraints, even while methods like the elimination of collinear variables and the confirmation of putative correlations in independent datasets can partially solve these problems (Falony et al. 2016). Such a method involves using a particular database to investigate a specific hypothesis, like in the case of certain enzyme classes or a collection of enzymatic pathways. Investigating the functional and metabolic functions of microbiome member species, as well as their alterations under various conditions/treatments, depends on the functional classification of metagenomics data. Tools used for whole genome studies and tools for functional classification both have a number of traits in common. These methods and tools are divided into four main groups: context-based, motif- or pattern-based, homology-based, and other functional predictions.

5.4.5.1 Homology-Based Tools

Predicted protein sequences are compared to reference protein sequences using one of the early methods. These reference protein sequence databases include NCBI RefSeq (O'Leary et al. 2016), UniProt (UniProt Consortium 2008), and SMART (Schultz et al. 2000). The query matching functionality of the MG-RAST (Keegan et al. 2016) and IMG/M (Chen et al. 2016) servers extend to additional databases, such as COGs (Tatusov et al. 2000), Pfam (Finn et al. 2014) and TIGRFAM (Haft et al. 2003). Due to database propagation, this strategy has significant drawbacks, including lengthy computation times and high error rates (~15%).

5.4.5.2 Motif- or Pattern-Based Tools

Short reads and complex samples that could not be matched using homology-based techniques are good candidates for this method. To find frequent patterns in metagenomic sequences, one can use databases like PROSITE (De Castro et al. 2006), PRINTS (Attwood 2002), or InterPro. IMG/M performs the majority of motif- and pattern-based annotations, albeit with low statistical significance and substantial false-positive rates.

5.4.5.3 Context-Based Tools

A context-based annotation is used to process novel metagenomic sequences that do not share any homology, pattern, or motif with the two prior techniques. For the purpose of screening metagenomic sequences, this method primarily uses genomic neighborhoods. Common context-based mapping algorithms are IMG/M and SmashCommunity (Arumugam et al. 2010).

5.4.5.4 Other Functional Predictions

Utilizing specific tools for predicting functional properties, such as carbohydrate-active enzymes (CAZy), protein localizations (PSORT, CELLO), lipoproteins (DOLLOP, Lipo, SignalP), insertion sequences (ISsaga), and virulence factors (VFDB, MvirDB), putative annotations of novel metagenomic sequences are typically carried out (Nakai 1999; Lee et al. 2006; Zhou et al. 2007; Lombard et al. 2014).

5.4.6 Assembly-Independent Strain-Level Characterization

The 16S rRNA gene sequences of probiotic species like *Bifidobacterium longum* subsp. *longum* and *Bifidobacterium longum* subsp. *infantis*, which exhibit two different phenotypes with functional consequences for infant nutrition (Underwood et al. 2015), differ only marginally from one another (Lawley et al. 2017). Traditional operational taxonomic unit (OTU) clustering-based taxonomic analyses do not account for these discrepancies. By organizing reads into amplicon sequence variations, novel techniques as UNOISE2 (Edgar 2016) and DADA2 (Callahan et al. 2016) avoid clumping and enable single-nucleotide sequence differentiation (ASVs). The phylogenetic depth at which microbiome investigations can be interpreted could greatly benefit from this. Insights into oral (Mukherjee et al. 2018) and vaginal microbiomes were gained owing to the notable uses of these novel algorithms (Callahan et al. 2017). Algorithms like PanPhlAn (Scholz et al. 2016) and StrainPhlAn (Truong et al. 2017) allow strain-level analyses using shotgun metagenome datasets without the necessity for metagenome assembly in situations where many strains of a species of interest have identical 16S rRNA sequences. These techniques enable a routine compositional analysis to confirm the presence of desirable strains or locate possible infections in final products.

5.4.7 Long-Read Sequencing and Other Advances

Long-read sequencing systems such as Pacific Biosciences (PacBio) and Oxford Nanopore Technologies (ONT) have interesting possibilities for the aforementioned industrial applications, despite their relatively uncommon use in microbiome studies. For example, the circular consensus sequencing application by PacBio, which enables the bioinformatic fusion of multiple reads derived from a circularized amplicon molecule into a high-quality, full-length (16S) sequence (Callahan et al. 2019), offers the required phylogenetic resolution for applications like fermentation studies, which is not feasible with short-read amplicon sequencing. While the high error rate makes reliable, strain-level detection difficult, the on-demand sequencing feature of ONT appears suited for quality control applications for detecting different diseases.

5.4.8 Machine Learning and Data Science

Microbiome studies' dataset sizes and the amount of sequencing done on each sample have increased due to falling sequencing prices. This resulted in studies with greater statistical power. As a result, OTU tables and functional profiles were transformed from final deliverables into raw data for later analysis, such as those carried out by machine learning (ML) tools (Pasolli et al. 2016). Many people have effectively employed techniques like random forests (RF) in the context of diseases, for example, for precisely predicting irritable bowel syndrome (Saulnier et al. 2011) and bacterial vaginosis (Beck and Foster 2014) based on taxonomic profiles.

In addition to taxonomic profiling and metagenomics-based prediction, there are few software tools that include statistical methodologies and machine learning modules to extract microbiome-phenotype connections. For instance, MetAML (Pasolli et al. 2016) was created to perform prediction tasks based on metagenomics and to quantitatively analyze the strength of putative connections between the microbiome and phenotype. To predict the phenotype of a genomic sample based on its microbial taxonomic abundance profile, Reiman et al. (2017) investigated convolutional neural networks. A k-mer frequency-based machine learning model from metagenome assemblies was also built to identify virus contigs in VirFinder (Ren et al. 2017). They all diverge from our research's objective, which is to evaluate two widely used methodological approaches—read-based and assembly-based—for metagenomics research using a variety of machine learning techniques with an emphasis on very big data sets.

5.4.9 Systems Biology Approaches to Understand the Host–Microbiome Interactions

The host may experience positive or negative consequences from gut-associated microorganisms. A condition known as "necrobiosis" exists when the bulk of the microbiome is made up of helpful microorganisms. Any changes to its regular makeup cause "dysbiosis," a situation in which harmful microorganisms prevail and cause sickness. The gut microbiome is shaped by both host intrinsic traits and the environment, but the microbiota also has an impact on the biology of the host. Thus, a three-way interaction between the host, microbes, and environment keeps the fish gut's homeostasis balanced (Talwar et al. 2018).

To understand how the microbiota affects host cellular responses in mice, the gut was colonized with various microorganisms. This colonization revealed that these microorganisms modulate the expression of several cellular genes that are important for the host's metabolism, nutrient absorption, immune response production, and intestinal maturation (Hooper et al. 2001). Similar conclusions were reached using zebrafish models (Kanther and Rawls 2010) and in other fish species where microorganisms were shown to control metabolism (Li et al. 2013b). Studies on transgenic fast-growing common carp, *Cyprinus carpio* L., reveal a crucial function for gut flora in fish growth and that Firmicutes confer fast growth on fish over Bacteroidetes (Li et al. 2013a, b). According to many studies (Wu et al. 2012), microbial components help herbivorous fishes digest cellulose. Additionally, they affect fish's innate immune responses and train the host immune system to be more effective in defending against pathogenic intruders (Talwar et al. 2018). The interactions between residing microorganisms and their host fish are seen in the light of these discoveries.

5.5 Microbiome Research and Development of Probiotics, Aquaculture Sector and Bioindustry

Consumer health could be at risk due to antibiotic-resistant microbes and an imbalance in the gut microbiota, both of which impact the health of fish and residue build up in the fish muscle. Probiotics have been proposed as a substitute method by several professionals to prevent infections in cultivable species in the aquaculture sector, and some others have suggested its use not only to manage the disease but also to promote growth and reproduction (Newaj-Fyzul et al. 2014; Hai 2015; Dawood and Koshio 2016). Additionally, Hoseinifar et al. (2015) stressed the significant effect of dietary prebiotics (mannan oligosaccharides, fructooligosaccharides, galacto-oligosaccharides, and yeast by-products) in boosting fish immune response in aquaculture systems. Besides serving as nutrient sources, probiotics can also modulate the immune system, improve digestion, and boost the immunological response to harmful microorganisms (Adel et al. 2017a, b). The most popular probiotics employed in aquaculture include yeast, Saccharomyces cerevisiae, and lactic acid bacteria such Lactobacillus sp., Bacillus sp., and Enterococcus sp. (Adel et al. 2017a, b). The idea behind the use of probiotics is to feed an adequate amount of healthy microbes, which replace the harmful microbes in the gut flora of fish and shellfish, with the result that these exogenous bacteria compete with pathogens, preventing their adhesion to the intestinal wall and secreting antibacterial substances. Increased growth results from the growth of beneficial microorganisms in the gut, which also aids in better food digestion and absorption (Ringø et al. 2010). Utilizing bioinformatics knowledge and expertise for the growth of aquaculture farming and the bioindustry has not received much attention to date. When bacterial and viral disease outbreaks caused losses for the aquaculture business, various steps were taken and put into place to lessen the impact. In recent years, bioinformatics knowledge has been used to address some of the disease management issues in the aquaculture industry to understand disease dynamics and its interaction with cultivable organisms as well as the interaction of pathogens present in aquatic bodies and their relationship with aquatic animals.

For instance, one of the most devastating diseases to shrimp aquaculture in commercial shrimp farms is the white spot syndrome virus (WSSV) (Liu et al. 2009). In the past 10 years, a growing number of studies on shrimp have been conducted to suppress this virus. With the help of a DNA microarray approach, Lan et al. (2006) reported finding the transcriptional profile of the WSSV genes in shrimp. Numerous researchers have discovered numerous host genes and proteins responding to WSSV infection (Liu et al. 2011; Li et al. 2013a). According to these investigations, many host genes and proteins are discovered to be up- or down-regulated during WSSV infection. To properly understand the pathophysiology of WSSV in shrimp, information on the direct interaction between WSSV and the host proteins is still insufficient. Based on the public protein–protein interaction (PPI) databases, bioinformatics analysis will offer a highly effective method for identifying genes and proteins implicated in the WSSV/shrimp interaction in such a setting (Sun et al. 2014). When predicting WSSV-host protein–protein interactions in the shrimp *Fenneropenaeus chinensis*, Sun et al. (2014) noted that the WSSV genome data (van Hulten et al. 2001; Yang et al. 2001) and the shrimp transcriptome data make it possible to identify WSSV/shrimp interacting proteins with bioinformatics techniques. The Search Tool for the Retrieval of Interacting Genes/Proteins (STRING), the Database of Interacting Proteins (DIP), and Reactome are the PPI databases that are utilized the most frequently. A database of known and projected protein interactions called STRING was created using data from high-throughput experiments, co-expression, genetic context, and prior research (Szklarczyk et al. 2021). The use of DIP databases of interacting proteins as a bioinformatics search tool for examining cellular networks of protein interaction was reported by Xenarios et al. (2002). To determine the biological pathways connected to PPI information, Joshi-Tope et al. (2005) stated another database tool named "Reactome."

Despite clear benefits to the bio/pharma business, there has not been much advancement in the usage of gut microbiomes, notably those from fish and shellfish. One of the primary explanations given in this particular subject may be a lack of understanding of metagenomics or the functional characteristics of microbial species. The study of functional features of microbial species using molecular methods has, therefore, received a lot of attention in recent years. The metagenomics part of this publication describes the efforts undertaken so far to study the microbial species in the fish gut system through metagenomics. Few biopharma companies have recently taken the initiative to manufacture cutting-edge products from various microbiomes found in fish and shellfish under various brand names for use in the aquaculture industry. Probiotics or feed are two possible product forms. These products are very helpful in maintaining the quality of the soil, water, and cultivable organisms in farming systems. There is an urgent need for rigorous research to be done on fish and shellfish microbiomes, particularly in the areas of genomics and metagenomics, to support the future possibilities of the bioindustry.

References

Adel M, El-Sayed AF, Yeganeh S, Dadar M, Giri SS (2017a) Effect of potential probiotic Lactococcus lactis subsp. lactis on growth performance, intestinal microbiota, digestive enzyme activities, and disease resistance of Litopenaeus vannamei. Probiot Antimicrob Prot 9(2): 150–156

Adel M, Yeganeh S, Dawood MA, Safari R, Radhakrishnan S (2017b) Effects of Pediococcus pentosaceus supplementation on growth performance, intestinal microflora and disease resistance of white shrimp, Litopenaeus vannamei. Aquac Nutr 23(6):1401–1409

Albertsen M, Hugenholtz P, Skarshewski A, Nielsen KL, Tyson GW, Nielsen PH (2013) Genome sequences of rare, uncultured bacteria obtained by differential coverage binning of multiple metagenomes. Nat Biotechnol 31(6):533–538

Altschul SF, Gish W, Miller W, Myers EW, Lipman DJ (1990) Basic local alignment search tool. J Mol Biol 215(3):403–410

Amann RI, Ludwig W, Schleifer KH (1995) Phylogenetic identification and in situ detection of individual microbial cells without cultivation. Microbiol Rev 59(1):143–169

Arumugam M, Harrington ED, Foerstner KU, Raes J, Bork P (2010) SmashCommunity: a metagenomic annotation and analysis tool. Bioinformatics 26(23):2977–2978

Aßhauer KP, Wemheuer B, Daniel R, Meinicke P (2015) Tax4Fun: predicting functional profiles from metagenomic 16S rRNA data. Bioinformatics 31(17):2882–2884

Attwood TK (2002) The PRINTS database: a resource for identification of protein families. Brief Bioinform 3(3):252–263

Austin B (2006) The bacterial microflora of fish, revised. Sci World J 6:931–945

Azra MN, Okomoda VT, Tabatabaei M, Hassan M, Ikhwanuddin M (2021) The contributions of shellfish aquaculture to global food security: assessing its characteristics from a future food perspective. Front Mar Sci 8:654897

Bankevich A, Nurk S, Antipov D, Gurevich AA, Dvorkin M, Kulikov AS, Lesin VM, Nikolenko SI, Pham S, Prjibelski AD, Pyshkin AV (2012) SPAdes: a new genome assembly algorithm and its applications to single-cell sequencing. J Comput Biol 19(5):455–477

Bass D, Stentiford GD, Wang HC, Koskella B, Tyler CR (2019) The pathobiome in animal and plant diseases. Trends Ecol Evol 34(11):996–1008

Beck D, Foster JA (2014) Machine learning techniques accurately classify microbial communities by bacterial vaginosis characteristics. PloS one 9(2):e87830

Besser J, Carleton HA, Gerner-Smidt P, Lindsey RL, Trees E (2018) Next-generation sequencing technologies and their application to the study and control of bacterial infections. Clin Microbiol Infect 24(4):335–341

Bharti R, Grimm DG (2021) Current challenges and best-practice protocols for microbiome analysis. Brief Bioinform 22(1):178–193

de Bruijn I, Liu Y, Wiegertjes GF, Raaijmakers JM (2018) Exploring fish microbial communities to mitigate emerging diseases in aquaculture. FEMS Microbiol Ecol 94(1):fix161

Buchfink B, Xie C, Huson DH (2015) Fast and sensitive protein alignment using DIAMOND. Nat Methods 12(1):59–60

Callahan BJ, McMurdie PJ, Rosen MJ, Han AW, Johnson AJ, Holmes SP (2016) DADA2: high-resolution sample inference from Illumina amplicon data. Nat Methods 13(7):581–583

Callahan BJ, DiGiulio DB, Goltsman DS, Sun CL, Costello EK, Jeganathan P, Biggio JR, Wong RJ, Druzin ML, Shaw GM, Stevenson DK (2017) Replication and refinement of a vaginal microbial signature of preterm birth in two racially distinct cohorts of US women. Proc Natl Acad Sci 114(37):9966–9971

Callahan BJ, Wong J, Heiner C, Oh S, Theriot CM, Gulati AS, McGill SK, Dougherty MK (2019) High-throughput amplicon sequencing of the full-length 16S rRNA gene with single-nucleotide resolution. Nucleic Acids Res 47(18):e103

Caporaso JG, Kuczynski J, Stombaugh J, Bittinger K, Bushman FD, Costello EK, Fierer N, Peña AG, Goodrich JK, Gordon JI, Huttley GA (2010) QIIME allows analysis of high-throughput community sequencing data. Nat Methods 7(5):335–336

Cavallo RA, Acquaviva MI, Stabili L (2009) Culturable heterotrophic bacteria in seawater and Mytilus galloprovincialis from a Mediterranean area (Northern Ionian Sea–Italy). Environ Monit Assess 149(1):465–475

Chen IM, Markowitz VM, Chu K, Palaniappan K, Szeto E, Pillay M, Ratner A, Huang J, Andersen E, Huntemann M, Varghese N (2016) IMG/M: integrated genome and metagenome comparative data analysis system. Nucleic Acids Res:gkw929

Cornejo-Granados F, Gallardo-Becerra L, Leonardo-Reza M, Ochoa-Romo JP, Ochoa-Leyva A (2018) A meta-analysis reveals the environmental and host factors shaping the structure and function of the shrimp microbiota. PeerJ 6:e5382

Dawood MA, Koshio S (2016) Recent advances in the role of probiotics and prebiotics in carp aquaculture: a review. Aquaculture 454:243–251

De Castro E, Sigrist CJ, Gattiker A, Bulliard V, Langendijk-Genevaux PS, Gasteiger E, Bairoch A, Hulo N (2006) ScanProsite: detection of PROSITE signature matches and ProRule-associated functional and structural residues in proteins. Nucleic Acids Res 34(Suppl 2):W362–W365

Deng Y, Kokou F, Eding EH, Verdegem MC (2021) Impact of early-life rearing history on gut microbiome succession and performance of Nile tilapia. Anim Microbiome 3(1):1–7

Derome N, Filteau M (2020) A continuously changing selective context on microbial communities associated with fish, from egg to fork. Evol Appl 13(6):1298–1319

DeSantis TZ, Hugenholtz P, Larsen N, Rojas M, Brodie EL, Keller K, Huber T, Dalevi D, Hu P, Andersen GL (2006) Greengenes, a chimera-checked 16S rRNA gene database and workbench compatible with ARB. Appl Environ Microbiol 72(7):5069–5072

Desriac F, Le Chevalier P, Brillet B, Leguerinel I, Thuillier B, Paillard C, Fleury Y (2014) Exploring the hologenome concept in marine Bivalvia: haemolymph microbiota as a pertinent source of probiotics for aquaculture. FEMS Microbiol Lett 350(1):107–116

Diwan AD, Harke SN, Panche AN (2022) Aquaculture industry prospective from gut microbiome of fish and shellfish: an overview. J Anim Physiol Anim Nutr 106(2):441–469

Douglas GM, Beiko RG, Langille MG (2018) Predicting the functional potential of the microbiome from marker genes using PICRUSt. In: Microbiome analysis. Humana Press, New York, NY, pp 169–177

Edgar RC (2010) Search and clustering orders of magnitude faster than BLAST. Bioinformatics 26(19):2460–2461

Edgar RC (2016) UNOISE2: improved error-correction for Illumina 16S and ITS amplicon sequencing. BioRxiv:081257

Eid J, Fehr A, Gray J, Luong K, Lyle J, Otto G, Peluso P, Rank D, Baybayan P, Bettman B, Bibillo A (2009) Real-time DNA sequencing from single polymerase molecules. Science 323(5910): 133–138

FAO Yearbook (2019) Fishery and aquaculture statistics. https://doi.org/10.4060/cb7874t

Falony G, Joossens M, Vieira-Silva S, Wang J, Darzi Y, Faust K, Kurilshikov A, Bonder MJ, Valles-Colomer M, Vandeputte D, Tito RY (2016) Population-level analysis of gut microbiome variation. Science 352(6285):560–564

Finn RD, Bateman A, Clements J, Coggill P, Eberhardt RY, Eddy SR, Heger A, Hetherington K, Holm L, Mistry J, Sonnhammer EL (2014) Pfam: the protein families database. Nucleic Acids Res 42(D1):D222–D230

Flusberg BA, Webster DR, Lee JH, Travers KJ, Olivares EC, Clark TA, Korlach J, Turner SW (2010) Direct detection of DNA methylation during single-molecule, real-time sequencing. Nat Methods 7(6):461–465

Foster JA, Bunge J, Gilbert JA, Moore JH (2012) Measuring the microbiome: perspectives on advances in DNA-based techniques for exploring microbial life. Brief Bioinform 13(4):420–429

Fox GE, Magrum LJ, Balch WE, Wolfe RS, Woese CR (1977) Classification of methanogenic bacteria by 16S ribosomal RNA characterization. Proc Natl Acad Sci 74(10):4537–4541

Fu L, Niu B, Zhu Z, Wu S, Li W (2012) CD-HIT: accelerated for clustering the next-generation sequencing data. Bioinformatics 28(23):3150–3152

Ghanbari M, Kneifel W, Domig KJ (2015) A new view of the fish gut microbiome: advances from next-generation sequencing. Aquaculture 448:464–475

Gibbons NE (1933) The slime and intestinal flora of some marine fishes. Contribut Can Biol Fish 8(1):275–290

Gómez GD, Balcázar JL (2008) A review on the interactions between gut microbiota and innate immunity of fish. FEMS Immunol Med Microbiol 52(2):145–154

Gomez JA, Primm TP (2021) A slimy business: the future of fish skin microbiome studies. Microb Ecol 82(2):275–287

Haft DH, Selengut JD, White O (2003) The TIGRFAMs database of protein families. Nucleic Acids Res 31(1):371–373

Hai NV (2015) The use of probiotics in aquaculture. J Appl Microbiol 119(4):917–935

Hameed AS (1993) A study of the aerobic heterotrophic bacterial flora of hatchery-reared eggs, larvae and post-larvae of Penaeus indicus. Aquaculture 117(3–4):195–204

Hodkinson BP, Grice EA (2015) Next-generation sequencing: a review of technologies and tools for wound microbiome research. Adv Wound Care 4(1):50–58

Hoff KJ, Lingner T, Meinicke P, Tech M (2009) Orphelia: predicting genes in metagenomic sequencing reads. Nucleic Acids Res 37(Suppl 2):W101–W105

Holt CC, Bass D, Stentiford GD, van der Giezen M (2021) Understanding the role of the shrimp gut microbiome in health and disease. J Invertebr Pathol 186:107387

Hooper LV, Midtvedt T, Gordon JI (2001) Commensal host-bacterial relationships in the gut. Science 292:1115–1118

Hoseinifar SH, Esteban MÁ, Cuesta A, Sun YZ (2015) Prebiotics and fish immune response: a review of current knowledge and future perspectives. Rev Fish Sci Aquacult 23(4):315–328

van Hulten MC, Witteveldt J, Peters S, Kloosterboer N, Tarchini R, Fiers M, Sandbrink H, Lankhorst RK, Vlak JM (2001) The white spot syndrome virus DNA genome sequence. Virology 286(1):7–22

Huson DH, Auch AF, Qi J, Schuster SC (2007) MEGAN analysis of metagenomic data. Genome Res 17(3):377–386

Huson DH, Beier S, Flade I, Górska A, El-Hadidi M, Mitra S, Ruscheweyh HJ, Tappu R (2016) MEGAN community edition-interactive exploration and analysis of large-scale microbiome sequencing data. PLoS Comput Biol 12(6):e1004957

Imelfort M, Parks D, Woodcroft BJ, Dennis P, Hugenholtz P, Tyson GW (2014) GroopM: an automated tool for the recovery of population genomes from related metagenomes. PeerJ 2:e603

Jain M, Olsen HE, Paten B, Akeson M (2016) The Oxford Nanopore MinION: delivery of nanopore sequencing to the genomics community. Genome Biol 17(1):1–1

Johny TK, Puthusseri RM, Bhat SG (2022) Metagenomic landscape of taxonomy, metabolic potential and resistome of Sardinella longiceps gut microbiome. Arch Microbiol 204(1):1

Joshi-Tope G, Gillespie M, Vastrik I, D'Eustachio P, Schmidt E, de Bono B, Jassal B, Gopinath GR, Wu GR, Matthews L, Lewis S (2005) Reactome: a knowledgebase of biological pathways. Nucleic Acids Res 33(Suppl 1):D428–D432

Kang DD, Froula J, Egan R, Wang Z (2015) MetaBAT, an efficient tool for accurately reconstructing single genomes from complex microbial communities. PeerJ 3:e1165

Kanther M, Rawls JF (2010) Host–microbe interactions in the developing zebrafish. Curr Opin Immunol 22(1):10–19

Keegan KP, Glass EM, Meyer F (2016) MG-RAST, a metagenomics service for analysis of microbial community structure and function. In: Microbial environmental genomics (MEG). Humana Press, New York, NY, pp 207–233

Kelley DR, Liu B, Delcher AL, Pop M, Salzberg SL (2012) Gene prediction with Glimmer for metagenomic sequences augmented by classification and clustering. Nucleic Acids Res 40(1):e9

Kim D, Song L, Breitwieser FP, Salzberg SL (2016) Centrifuge: rapid and sensitive classification of metagenomic sequences. Genome Res 26(12):1721–1729

Kuczynski J, Lauber CL, Walters WA, Parfrey LW, Clemente JC, Gevers D, Knight R (2012) Experimental and analytical tools for studying the human microbiome. Nat Rev Genet 13(1): 47–58

Lan YS, Xu X, Yang F, Zhang X (2006) Transcriptional profile of shrimp white spot syndrome virus (WSSV) genes with DNA microarray. Arch Virol 151(9):1723–1733

Lawley B, Munro K, Hughes A, Hodgkinson AJ, Prosser CG, Lowry D, Zhou SJ, Makrides M, Gibson RA, Lay C, Chew C (2017) Differentiation of Bifidobacterium longum subspecies longum and infantis by quantitative PCR using functional gene targets. PeerJ 5:e3375

Lee S, Lee BC, Kim D (2006) Prediction of protein secondary structure content using amino acid composition and evolutionary information. Prot Struct Funct Bioinformatics 62(4):1107–1114

Legrand TP, Catalano SR, Wos-Oxley ML, Stephens F, Landos M, Bansemer MS, Stone DA, Qin JG, Oxley AP (2018) The inner workings of the outer surface: skin and gill microbiota as indicators of changing gut health in yellowtail kingfish. Front Microbiol 8:2664

Lemos LN, Fulthorpe RR, Triplett EW, Roesch LF (2011) Rethinking microbial diversity analysis in the high throughput sequencing era. J Microbiol Methods 86(1):42–51

Lemos LN, Morais DK, Tsai SM, Roesch L, Pylro V (2017) Bioinformatics for microbiome research: concepts, strategies, and advances. In: The Brazilian microbiome. Springer, Cham, pp 111–123

Li S, Zhang X, Sun Z, Li F, Xiang J (2013a) Transcriptome analysis on Chinese shrimp Fenneropenaeus chinensis during WSSV acute infection. PLoS One 8(3):e58627

Li X, Yan Q, Xie S, Hu W, Yu Y, Hu Z (2013b) Gut microbiota contributes to the growth of fast-growing transgenic common carp (Cyprinus carpio L.). PLoS One 8(5):e64577

Liu H, Söderhäll K, Jiravanichpaisal P (2009) Antiviral immunity in crustaceans. Fish Shellfish Immunol 27(2):79–88

Liu HP, Chen RY, Zhang QX, Peng H, Wang KJ (2011) Differential gene expression profile from haematopoietic tissue stem cells of red claw crayfish, Cherax quadricarinatus, in response to WSSV infection. Dev Comp Immunol 35(7):716–724

Lombard V, Golaconda Ramulu H, Drula E, Coutinho PM, Henrissat B (2014) The carbohydrate-active enzymes database (CAZy) in 2013. Nucleic Acids Res 42(D1):D490–D495

Margolis L (1953) The effect of fasting on the bacterial flora of the intestine of fish. J Fish Board Can 10(2):62–63

McCaig AE, Glover LA, Prosser JI (1999) Molecular analysis of bacterial community structure and diversity in unimproved and improved upland grass pastures. Appl Environ Microbiol 65(4): 1721–1730

McMurdie PJ, Holmes S (2013) phyloseq: an R package for reproducible interactive analysis and graphics of microbiome census data. PLoS One 8(4):e61217

Metzker ML (2010) Sequencing technologies—the next generation. Nat Rev Genet 11(1):31–46

Mikheenko A, Saveliev V, Gurevich A (2016) MetaQUAST: evaluation of metagenome assemblies. Bioinformatics 32(7):1088–1090

Mikheyev AS, Tin MM (2014) A first look at the Oxford Nanopore MinION sequencer. Mol Ecol Resour 14(6):1097–1102

Mukherjee C, Beall CJ, Griffen AL, Leys EJ (2018) High-resolution ISR amplicon sequencing reveals personalized oral microbiome. Microbiome 6(1):1–5

Nakai K (1999) PSORT: a program for detecting the sorting signals of proteins and predicting their subcellular localization. Trends Biochem Sci 24(1):34–35

Namiki T, Hachiya T, Tanaka H, Sakakibara Y (2012) MetaVelvet: an extension of Velvet assembler to de novo metagenome assembly from short sequence reads. Nucleic Acids Res 40(20):e155

Navarrete P, Espejo RT, Romero J (2009) Molecular analysis of microbiota along the digestive tract of juvenile Atlantic salmon (Salmo salar L.). Microb Ecol 57(3):550–561

Navas-Molina JA, Peralta-Sánchez JM, González A, McMurdie PJ, Vázquez-Baeza Y, Xu Z, Ursell LK, Lauber C, Zhou H, Song SJ, Huntley J (2013) Advancing our understanding of the human microbiome using QIIME. In: Methods in enzymology, vol 531. Academic Press, London, pp 371–444

NCBI Resource Coordinators (2017) Database resources of the National Center for Biotechnology Information. Nucleic Acids Res 45(D1):D12–D17

Newaj-Fyzul A, Al-Harbi AH, Austin B (2014) Developments in the use of probiotics for disease control in aquaculture. Aquaculture 431:1–1

O'Leary NA, Wright MW, Brister JR, Ciufo S, Haddad D, McVeigh R, Rajput B, Robbertse B, Smith-White B, Ako-Adjei D, Astashyn A (2016) Reference sequence (RefSeq) database at NCBI: current status, taxonomic expansion, and functional annotation. Nucleic Acids Res 44(D1):D733–D745

Oksanen J, Blanchet FG, Kindt R, Legendre P, Minchin PR, O'Hara RB, Simpson GL, Solymos P, Stevens MH, Wagner H. (2016) Vegan: community ecology package. R package 2.0.3. https://cran.r-project.org/package=vegan

Olafsen JA (2001) Interactions between fish larvae and bacteria in marine aquaculture. Aquaculture 200(1–2):223–247

Omont A, Elizondo-González R, Quiroz-Guzmán E, Escobedo-Fregoso C, Hernández-Herrera R, Peña-Rodríguez A (2020) Digestive microbiota of shrimp Penaeus vannamei and oyster Crassostrea gigas co-cultured in integrated multi-trophic aquaculture system. Aquaculture 521:735059

Ottinger M, Bachofer F, Huth J, Kuenzer C (2021) Mapping aquaculture ponds for the Coastal Zone of Asia with Sentinel-1 and Sentinel-2 time series. Remote Sens 14(1):153

Parks DH, Imelfort M, Skennerton CT, Hugenholtz P, Tyson GW (2015) CheckM: assessing the quality of microbial genomes recovered from isolates, single cells, and metagenomes. Genome Res 25(7):1043–1055

Pasolli E, Truong DT, Malik F, Waldron L, Segata N (2016) Machine learning meta-analysis of large metagenomic datasets: tools and biological insights. PLoS Comput Biol 12(7):e1004977

Pierce ML, Ward JE (2018) Microbial ecology of the Bivalvia, with an emphasis on the family Ostreidae. J Shellfish Res 37(4):793–806

Rajeev R, Adithya KK, Kiran GS, Selvin J (2021) Healthy microbiome: a key to successful and sustainable shrimp aquaculture. Rev Aquac 13(1):238–258

Reed GB, Spence CM (1929) The intestinal and slime flora of the haddock: a preliminary report. Contribut Can Biol Fish 4(1):257–264

Reiman D, Metwally A, Dai Y (2017) Using convolutional neural networks to explore the microbiome. In: 2017 39th annual international conference of the IEEE engineering in medicine and biology society (EMBC). IEEE, IEEE, Washington, DC, pp 4269–4272

Ren J, Ahlgren NA, Lu YY, Fuhrman JA, Sun F (2017) VirFinder: a novel k-mer based tool for identifying viral sequences from assembled metagenomic data. Microbiome 5(1):1–20

Reveco FE, Øverland M, Romarheim OH, Mydland LT (2014) Intestinal bacterial community structure differs between healthy and inflamed intestines in Atlantic salmon (Salmo salar L.). Aquaculture 420:262–269

Rezasoltani S, Bashirzadeh DA, Mojarad EN, Aghdaei HA, Norouzinia M, Shahrokh S (2020) Signature of gut microbiome by conventional and advanced analysis techniques: advantages and disadvantages. Mid E J Digest Dis 12(1):5

Rho M, Tang H, Ye Y (2010) FragGeneScan: predicting genes in short and error-prone reads. Nucleic Acids Res 38(20):e191

Rhoads A, Au KF (2015) PacBio sequencing and its applications. Genom Proteom Bioinformatics 13(5):278–289
Ringø E (2020) Probiotics in shellfish aquaculture. Aquacult Fish 5(1):1–27
Ringø E, Olsen RE, Mayhew TM, Myklebust R (2003) Electron microscopy of the intestinal microflora of fish. Aquaculture 227(1–4):395–415
Ringø E, Olsen RE, Gifstad TØ, Dalmo RA, Amlund H, Hemre GI, Bakke AM (2010) Prebiotics in aquaculture: a review. Aquac Nutr 16(2):117–136
Rodrigue S, Materna AC, Timberlake SC, Blackburn MC, Malmstrom RR, Alm EJ, Chisholm SW (2010) Unlocking short read sequencing for metagenomics. PLoS One 5(7):e11840
Romero J, Navarrete P (2006) 16S rDNA-based analysis of dominant bacterial populations associated with early life stages of coho salmon (Oncorhynchus kisutch). Microb Ecol 51(4): 422–430
Romero J, García-Varela M, Laclette JP, Espejo RT (2002) Bacterial 16S rRNA gene analysis revealed that bacteria related to Arcobacter spp. constitute an abundant and common component of the oyster microbiota (Tiostrea chilensis). Microb Ecol 44(4):365–371
Romero J, Ringø E, Merrifield DL (2014) The gut microbiota of fish. In: Aquaculture nutrition: gut health, probiotics and prebiotics. John Wiley & Sons, Ltd, Chichester, pp 75–100
Salinas I, Magadán S (2017) Omics in fish mucosal immunity. Dev Comp Immunol 75:99–108
Santibáñez P, Romalde J, Maldonado J, Fuentes D, Figueroa J (2022) First characterization of the gut microbiome associated with Mytilus chilensis collected at a mussel farm and from a natural environment in Chile. Aquaculture 548:737644
Saulnier DM, Riehle K, Mistretta TA, Diaz MA, Mandal D, Raza S, Weidler EM, Qin X, Coarfa C, Milosavljevic A, Petrosino JF (2011) Gastrointestinal microbiome signatures of pediatric patients with irritable bowel syndrome. Gastroenterology 141(5):1782–1791
Schloss PD, Handelsman J (2005) Introducing DOTUR, a computer program for defining operational taxonomic units and estimating species richness. Appl Environ Microbiol 71(3): 1501–1506
Schloss PD, Westcott SL, Ryabin T, Hall JR, Hartmann M, Hollister EB, Lesniewski RA, Oakley BB, Parks DH, Robinson CJ, Sahl JW (2009) Introducing Mothur: open-source, platform-independent, community-supported software for describing and comparing microbial communities. Appl Environ Microbiol 75(23):7537–7541
Schloss PD, Gevers D, Westcott SL (2011) Reducing the effects of PCR amplification and sequencing artifacts on 16S rRNA-based studies. PLoS One 6(12):e27310
Scholz M, Ward DV, Pasolli E, Tolio T, Zolfo M, Asnicar F, Truong DT, Tett A, Morrow AL, Segata N (2016) Strain-level microbial epidemiology and population genomics from shotgun metagenomics. Nat Methods 13(5):435–438
Schultz J, Copley RR, Doerks T, Ponting CP, Bork P (2000) SMART: a web-based tool for the study of genetically mobile domains. Nucleic Acids Res 28(1):231–234
Seemann T (2014) Prokka: rapid prokaryotic genome annotation. Bioinformatics 30(14): 2068–2069
Segata N, Waldron L, Ballarini A, Narasimhan V, Jousson O, Huttenhower C (2012) Metagenomic microbial community profiling using unique clade-specific marker genes. Nat Methods 9(8): 811–814
Sha H, Lu J, Chen J, Xiong J (2022) A meta-analysis study of the robustness and universality of gut microbiota-shrimp diseases relationship. Environ Microbiol 24:3924
Shannon CE (1948) A mathematical theory of communication. Bell Syst Tech J 27(3):379–423
Simpson EH (1949) Measurement of diversity. Nature 163:688
Singh IB, Lakshmanaperumalsamy P, Chandramohan D (1998) Bacterial flora of pond reared Penaeus indicus (Milne Edwards). J Aquac Tropics 13:133–142
Siqueira JF, Fouad AF, Rôcas IN (2012) Pyrosequencing as a tool for better understanding of human microbiomes. J microbiol 4(1):10743
Sokal RR (1963) The principles and practice of numerical taxonomy. Taxon 12:190–199

Song WZ, Thomas T (2017) Binning_refiner: improving genome bins through the combination of different binning programs. Bioinformatics 33(12):1873–1875

Sun Z, Li S, Li F, Xiang J (2014) Bioinformatic prediction of WSSV-host protein-protein interaction. Biomed Res Int 2014:416543

Szklarczyk D, Gable AL, Nastou KC, Lyon D, Kirsch R, Pyysalo S, Doncheva NT, Legeay M, Fang T, Bork P, Jensen LJ (2021) The STRING database in 2021: customizable protein–protein networks, and functional characterization of user-uploaded gene/measurement sets. Nucleic Acids Res 49(D1):D605–D612

Talwar C, Nagar S, Lal R, Negi RK (2018) Fish gut microbiome: current approaches and future perspectives. Indian J Microbiol 58(4):397–414

Tarnecki AM, Burgos FA, Ray CL, Arias CR (2017) Fish intestinal microbiome: diversity and symbiosis unravelled by metagenomics. J Appl Microbiol 123(1):2–17

Tatusov RL, Galperin MY, Natale DA, Koonin EV (2000) The COG database: a tool for genome-scale analysis of protein functions and evolution. Nucleic Acids Res 28(1):33–36

Timmins-Schiffman E, White SJ, Thompson RE, Vadopalas B, Eudeline B, Nunn BL, Roberts SB (2021) Coupled microbiome analyses highlights relative functional roles of bacteria in a bivalve hatchery. Environ Microbiome 16(1):1–2

Truong DT, Tett A, Pasolli E, Huttenhower C, Segata N (2017) Microbial strain-level population structure and genetic diversity from metagenomes. Genome Res 27(4):626–638

Tysset C, Mailloux M, Brisou J (1961) La microflore commensale de la crevette rouge des côtes Algériennes. Archives Del Institut Pasteur DAlgerie 39:287–301

Underwood MA, German JB, Lebrilla CB, Mills DA (2015) Bifidobacterium longum subspecies infantis: champion colonizer of the infant gut. Pediatr Res 77(1):229–235

UniProt Consortium (2008) The universal protein resource (UniProt). Nucleic Acids Res 36 (Database issue):D190–D195

Wang Q, Garrity GM, Tiedje JM, Cole JR (2007) Naive Bayesian classifier for rapid assignment of rRNA sequences into the new bacterial taxonomy. Appl Environ Microbiol 73(16):5261–5267

Woese CR, Fox GE (1977) Phylogenetic structure of the prokaryotic domain: the primary kingdoms. Proc Natl Acad Sci U S A 74(11):5088–5090

Wong S, Rawls JF (2012) Intestinal microbiota composition in fishes is influenced by host ecology and environment. Mol Ecol 21:3100–3102

Wong S, Waldrop T, Summerfelt S, Davidson J, Barrows F, Kenney PB, Welch T, Wiens GD, Snekvik K, Rawls JF, Good C (2013) Aquacultured rainbow trout (Oncorhynchus mykiss) possess a large core intestinal microbiota that is resistant to variation in diet and rearing density. Appl Environ Microbiol 79(16):4974–4984

Wood DE, Salzberg SL (2014) Kraken: ultrafast metagenomic sequence classification using exact alignments. Genome Biol 15(3):1–2

Wu S, Wang G, Angert ER, Wang W, Li W, Zou H (2012) Composition, diversity, and origin of the bacterial community in grass carp intestine. PLoS One 7(2):e30440

Wu YW, Tang YH, Tringe SG, Simmons BA, Singer SW (2014) MaxBin: an automated binning method to recover individual genomes from metagenomes using an expectation-maximization algorithm. Microbiome 2(1):1–8

Xenarios I, Salwinski L, Duan XJ, Higney P, Kim SM, Eisenberg D (2002) DIP, the Database of Interacting Proteins: a research tool for studying cellular networks of protein interactions. Nucleic Acids Res 30(1):303–305

Yang F, He J, Lin X, Li Q, Pan D, Zhang X, Xu X (2001) Complete genome sequence of the shrimp white spot bacilliform virus. J Virol 75(23):11811–11820

Yasuda K, Kitao T (1980) Bacterial flora in the digestive tract of prawns, Penaeus japonicus Bate. Aquaculture 19(3):229–234

Ye Y, Doak TG (2009) A parsimony approach to biological pathway reconstruction/inference for genomes and metagenomes. PLoS Comput Biol 5(8):e1000465

Yukgehnaish K, Kumar P, Sivachandran P, Marimuthu K, Arshad A, Paray BA, Arockiaraj J (2020) Gut microbiota metagenomics in aquaculture: factors influencing gut microbiome and its physiological role in fish. Rev Aquac 12(3):1903–1927

Zhang H, Sun Z, Liu B, Xuan Y, Jiang M, Pan Y, Zhang Y, Gong Y, Lu X, Yu D, Kumar D (2016) Dynamic changes of microbial communities in Litopenaeus vannamei cultures and the effects of environmental factors. Aquaculture 455:97–108

Zhang X, Sun Z, Zhang X, Zhang M, Li S (2018) Hemolymph microbiomes of three aquatic invertebrates as revealed by a new cell extraction method. Appl Environ Microbiol 84(8): e02824–e02817

Zhou CE, Smith J, Lam M, Zemla A, Dyer MD, Slezak T (2007) MvirDB—a microbial database of protein toxins, virulence factors and antibiotic resistance genes for bio-defence applications. Nucleic Acids Res 35(Suppl 1):D391–D394

Zhu W, Lomsadze A, Borodovsky M (2010) Ab initio gene identification in metagenomic sequences. Nucleic Acids Res 38(12):e132

Microbiome Applications for the Growth and Development of the Aquaculture/Biopharma Industry

6

Arvind Diwan, Sanjay N. Harke, and Archana Panche

Abstract

After the ban on the use of antibiotics in aquaculture farming, efforts are being made to find out other alternative methods which can replace antibiotics and augment aquaculture production safely without compromising the quality of the yield. Some of the alternative methods devised are the use of bacteriophage and biofloc technology and also the microbiome in the form of probiotics. But there is not much information regarding the use of gut microbiome species of fish and shellfish for the development of aquaculture farming/industry. There is a lot of potential for development in this particular area, provided the knowledge pertaining to the functional aspects and the ingredient molecules that are present in these microbiome species is available. Therefore, extensive efforts are needed to develop R and D programs on the metagenomics aspects of the gut microbiome species. In order to promote the future prospects of the bio industry, there is an urgent need for intensive research to be undertaken on fish and shellfish microbiomes, particularly in the areas of genomics and metagenomics. The Functionality aspects of the gut microbiome are an important issue for future investigation and therefore, in-depth knowledge of functional metagenomics is very much essential. Metagenomics has emerged in the last decade as a promising technology that attempts to analyze the complex genomes contained within a microbial niche.

A. Diwan (✉) · S. N. Harke
Institute of Biosciences and Technology, MGM University, Aurangabad, Maharashtra, India

A. Panche
Centre for Biosustanability, Technical University of Denmark, Copenhagen, Denmark

A. Diwan et al. (eds.), *Microbiome of Finfish and Shellfish*,
https://doi.org/10.1007/978-981-99-0852-3_6

Keywords

Microbiome · Industry · Ingredients · Probiotics · Metagenomics · Bioinformatics skills

6.1 Introduction

According to the estimates of the United Nations Food and Agriculture Organization (FAO 2020), aquaculture farming is the fastest-growing industry in the food-animal sector. It is speculated that by 2030 the global population demand for fish may increase to the tune of 30 million tons of fish. However, for achieving this surplus target, the industry has to surpass many challenges of scaling. These challenges may be technological innovations in breeding animals, climate change, and pollution, constraints of several outbreaks of diseases leading to huge losses, and maintaining the quality and safety of fishery products. In order to overcome these challenges, particularly the prevention of disease and health management and the safety of cultivable organisms, several technological breakthroughs have been made and also implemented. But due to a lack of sufficient knowledge and information regarding the infectious and pathogenic microbiota, disease control and maintaining the quality of harvested yield have become a great challenge to the aquaculturist. In the recent past, the importance of the gut microbiome and its role in controlling various physiological activities in fish and shellfish has been recognized by several workers (Holt et al. 2020; Sharma and Thaiss 2020; Diwan et al. 2021; Li et al. 2022). The microbiome present in the gut tract, in particular, plays a vital role in improving the immune system and preventing an attack from pathogenic bacteria and thus maintaining the proper health status of an organism. The microbiota associated with other tissues like skin, gill, and mucosa also has a functionally important role in preventing the entry of harmful pathogens using different types of mechanisms (Balcázar et al. 2007; Pérez-Sánchez et al. 2011; Pérez et al. 2010; Guivier et al. 2019). In recent years a lot of work has been carried out on the manipulation of the gut microbiota using microbiota as a supplementary diet in the form of probiotics with the achievements of enhanced aquaculture production (Rawls et al. 2006; Balcázar et al. 2007; Egerton et al. 2018; Rajeev et al. 2021; Diwan et al. 2021). The holobiont concept has been accepted for almost all organisms for successful growth and production of cultivable organisms in captive conditions (McFall-Ngai et al. 2013; Foster et al. 2017; Bordenstein and Theis 2015). There are few reports regarding attempts on revealing the mechanisms involved in host–microbiome interaction and also the interaction between pathogenic bacteria and beneficial bacteria. Such attempts have been made at the molecular and gene expression levels also (Simon et al. 2019; Li et al. 2022; Redfern et al. 2021; Zhang et al. 2021; Piazzon et al. 2020; Zhang and Sun 2022; Holt et al. 2020). After the ban on the use of antibiotics in aquaculture farming, efforts are being made to find out other alternative methods which can replace antibiotics and augment aquaculture production safely without compromising the quality of the yield. Some of the alternative methods devised are the use of bacteriophage and biofloc technology and also the microbiome in the form of probiotics. But there is not much information regarding the use of gut microbiome species of fish and shellfish for the development of aquaculture farming/industry. There is a lot of potential for development in this particular area, provided the knowledge pertaining to the functional aspects and the ingredient molecules that are present in these microbiome species is available. Therefore, extensive efforts are needed to develop R and D programs on the metagenomics aspects of the gut microbiome species. In this chapter, we have made our sincere attempts to generate a comprehensive knowledge on the use of gut microbiomes for the advancement of aquaculture and other related bio-pharma industries.

6.2 Fish and Shellfish Gut Microbiome and Their Ingredients

A considerable amount of work has been carried out on gut microbiomes in different fish and shellfish inhabiting different geographical locations and environments. Information is available on the gut microbiome of fish and shellfish in relation to a number of factors like changes in the diet formulation, the impact of the use of probiotics, prebiotics, and synbiotics, and in recent years studies have also been made to investigate the effect of biofloc technology on the gut microbiome diversity. In order to accelerate aquaculture and other related biopharma industries, it is very much essential to know what type of biochemical ingredients are produced by different microbial species present in the gut tract of these organisms and their functional properties. This aspect of the research study is now gaining a lot of importance because the role of the gut microbiome in controlling various physiological activities of an organism has been a well-established fact and proven experimentally also. Several workers are now focusing their research goals aiming at metagenomics aspects of microbial species present in the gut tract (Diwan et al. 2021; Li et al. 2022). Sanchez et al. (2012), while working on the fish gut microbiome, mentioned that the microbes that are available in the intestinal tract are a vital source of a number of natural products having significant functional properties. They observed that microbial group like Actinobacteria which is predominantly present in the gut tract of six varieties of fish has been identified to contain a novel bioactive lipid component as sebastenoic acid. This novel lipid compound found in this particular group of bacteria has been reported to have antibacterial properties against certain bacterial strains such as *Staphylococcus aureus, Bacillus subtilis, Enterococcus faecium,* and *Vibrio mimicus*. In recent years there are several published reports on the production of many such bioactive compounds from gut microbiome communities. Diwan et al. (2021) in their review paper mentioned that more than 1000 new bioactive compounds have been investigated in the marine environment and it is speculated that all these compounds are from microbial species which are associated with marine invertebrates.

Trindade et al. (2015) while on the microbiome mentioned that the microbial species belonging to the group Actinomycetales can produce a number of antimicrobial compounds and further it is stated that perhaps the antibiotic drug discovery has originated from these microbiome species. A lot of work has been therefore carried out on the genome sequencing of these bacterial communities. From the genome sequencing of the microbiome species like *Streptomyces avermitilis* Trindade et al. (2015) discovered a number of gene clusters that code for the production of 37 secondary metabolites, out of which only 13 natural products have been reported from this organism. Furthermore, it is mentioned that they have isolated several Actinomycetales bacterial species from the fish intestine and also some unique strains of Firmicutes and Proteobacteria. All these bacterial species have been shown to contain active biological ingredients which have functional properties against both Gram-positive and Gram-negative pathogenic bacteria and are able to inhibit the growth of a number of commercially important fish pathogens. In the recent past number of aqua/bio-industries have taken a lot of interest in manufacturing new and innovative products from microbiome species of both fish and shellfish under different commercial names for their use in aquaculture farming. The companies have launched these products under different commercial names and also sometimes in the form of feed. These products are also quite useful in maintaining the quality of water and soil as well as the health of the cultivable organisms in aquaculture farming. For the promotion of future prospects of the bio industry, there is an urgent need for intensive research to be undertaken on fish and shellfish microbiomes, particularly in the areas of genomics and metagenomics. The functional aspects of the gut microbiome are an important issue for future investigation and therefore, in-depth knowledge of functional metagenomics is very much essential. For the development of any commercial product from the bacterial species and its applications in the fish farming system, the industry has to face several challenges. Table 6.1 summarizes fish and shellfish microbiomes producing bioactive compounds for their potential use in the development of aqua-bio industries.

6.3 Fish and Shellfish Gut Microbiomes and Their Functionality

Though several studies are now available covering different aspects of fish and shellfish gut microbiomes, viz., microbiome diversity, their identification and composition, and changes in the microbiome profile in response to changing environments, microbiome ingredients, etc., however, there is a lack of information on the functional aspects of microbial species at genomic levels. Therefore, compared to microbial composition and abundance, microbial functionality in the fish and shellfish gut is poorly understood. It is now well understood that the gut microbiome plays a very significant role in most of the physiological functions in the body such as digestion, metabolism, strengthening the immune system, protection from infectious pathogens, and maintaining proper growth and health of the animals. Several investigations carried out in earlier years reported that the gut microbiomes of fish and shellfish were capable of producing various digestive

Table 6.1 Microbes produce bioactive compounds for the potential development of products for their possible applications in the aquaculture industry

Microbiomes	Chemical ingredients	Possible applications	Reference
Actinobacteria	Sebastenoic acid	Antibacterial properties	Sanchez et al. (2012)
Actinomycetales	Antimicrobial compounds	Antibiotic drugs	Trindade et al. (2015)
Cetobacterium sp.	Vitamin B12-like compounds	Vitamin B12	Tsuchiya et al. (2008), Roeselers et al. (2011), Ye et al. (2014), Liu et al. (2016), Eichmiller et al. (2016)
Gut microbiomes	Digestive enzymes (b-glucanases, carbohydrate, cellulases, chitinases)	Cellulose digestion enzyme products	Ray et al. (2012)
Streptomyces sp. Firmicutes sp. Proteobacteria sp.	Secondary metabolites Bioactive compounds	Anti-gram-positive and gram-negative bacteria and other anti-fish pathogens	Trindade et al. (2015)
Clostridium group	Enzymes for collegen digestion	Enzyme production for collagen degradation	Baldo et al. (2015)
	Propionate compounds, short-chain fatty acids, butyrate, collagen-degrading enzymes, acid and alkaline phosphatase enzymes	Growth promoting stimulants Probiotics	Eichmiller et al. (2016), Ramirez and Dixon (2003)
Luminous and non-luminous photo-bacteria	Enzyme chitinases	Chitin digestive enzymes	Itoi et al. (2006), Ward et al. (2009), Smriga et al. (2010)
Proteobacteria	Degrading enzymes for cellulose, hemicellulose, chitin, pectin, starch, etc.	Enzymes in the form of probiotics supplements	Tyagi et al. (2019)
Vibrio sp.	Hydrolytic enzymes, amylases, lipases, cellulases, and chitinases	Digestive enzymes products	Itoi et al. (2006), Sugita and Ito (2006)
Lactobacillus, enterococcus, Bacillus, Aeromonas, Alteromonas, Arthrobacter, Bifidobacterium,	Growth and immunostimulants bioactive compounds	Probiotics	Ringø (2020)

(continued)

Table 6.1 (continued)

Microbiomes	Chemical ingredients	Possible applications	Reference
Clostridium, Microbacterium, Paenibacillus, Phaeobacter, Pseudoalteromonas, pseudomonas, Rhodosporidium, Roseobacter, Streptomyces, and vibrio			

enzymes. There are some enzymes (b-glucanases, carbohydrases, cellulases, and chitinases) that are not produced by the fish particularly, herbivorous and detritivorous species and in such animals probably the microbiome present in the gut system takes over the function of digestion of cellulose materials (Ray et al. 2012). Tsuchiya et al. (2008), while working on the fish gut microbiome, reported that microbiome species like *Cetobacterium somerae* present in the gut tract produce an abundant amount of vitamin B12. Furthermore, Romero et al. (2014) mentioned that this bacterium produces vitamin B12 in those fish like Nile tilapia and common carp *Cyprinus carpio*, which have no dietary vitamin B12 requirement, whereas species in which the bacteria is not commonly present, such as channel catfish *Ictalurus punctatus* and Japanese eel *Anguilla japonica*, have a requirement for dietary vitamin B12. Some other workers reported that the production of various short-chain fatty acids occurs when gut tract microbes are involved in the digestion of dietary fiber, particularly in herbivorous fish species (Clements and Choat 1995; Mountfort et al. 2002). The functional aspects of short-chain fatty acids in a number of fish have been worked out and it has been mentioned that these fatty acids provide energy, reduce intestinal pH which can make the environment less favorable to some potential pathogens, and also increase the solubility of minerals potentially making them more easily absorbed (Mountfort et al. 2002; Merrifield and Rodiles 2015). It has been observed that in the absence of certain gut microbiome species, the fishes are unable to degrade cellulose material, so also xenobiotic compounds. For example, the presence of Verrucomicrobia in the gut of the fish facilitates the digestion of cellulose converting b-glycans to b-glucose. The presence of these bacteria now suggests that it is important for the digestion of plant cellulose in the gut of fish. This feature has been demonstrated in carp and further strengthened by reduced cellulase activity in antibiotic-treated fish (van Kessel et al. 2011). Eichmiller et al. (2016) reported that the microbes belonging to the Clostridia group facilitate the synthesis of propionate, short fatty acid chains, and butyrate. These compounds are produced in the host gut system and supplied to the host for better growth. Baldo et al. (2015) observed that these bacteria also help in collagen digestion as they produce collagen-degrading enzymes. Several workers have reported that the Fusobacteria present in the gut of the fish facilitate the production of vitamin B12 which is vital for the fish

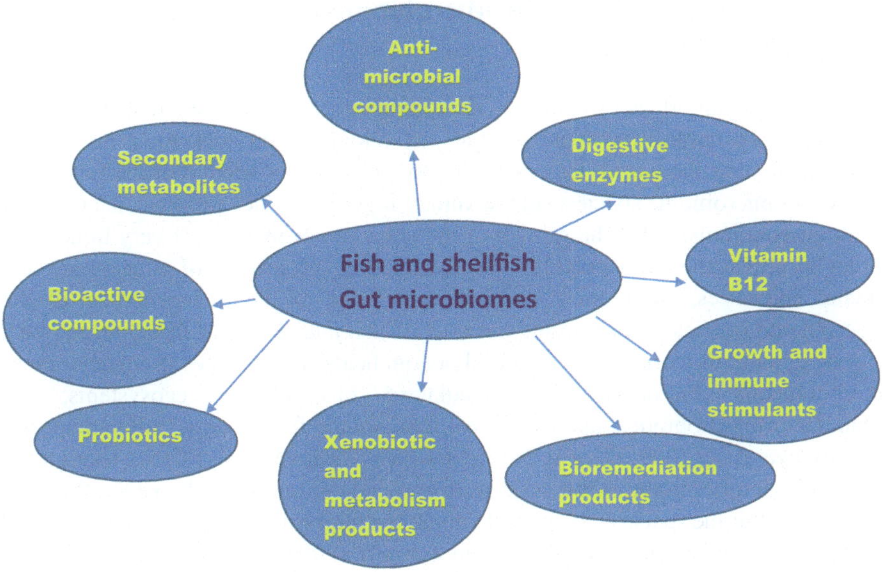

Fig. 6.1 Fish and shellfish microbiome products for enhancing the growth of the aquaculture and related bio-industries

development process (Roeselers et al. 2011; Ye et al. 2014; Liu et al. 2016; Eichmiller et al. 2016). Figure 6.1 describes fish and shellfish microbiomes and their ingredients for the growth of aquaculture and related bio-industries.

While discussing the functional aspects of the gut microbiome in humans it has been mentioned that although the bacterial composition varied greatly between individuals, the metabolic pathways varied to a lesser degree, being more stable across individuals. This indicates that the functional activity of the microbiome is very important than the phylogenetic composition (Human Microbiome Project Consortium, 2012). This concept of the comparison of the functional importance of the microbiome rather than phylogenetic composition may be also true in the case of fish and shellfish. Xing et al. (2013) while working on the gut microbiome metagenomics in turbot fish reported that the metabolic activities were dominant and the biofilm formation was more abundant in the gut metagenomes. Additionally, iron acquisition and metabolism were also more abundant in the turbot gut intestine tract metagenomes when compared with the striped bass. When a comparison was made with regard to functional metagenomics of the gut microbiome of freshwater and marine fish, it was reported that differences in metabolic potential may occur in marine and freshwater fish gut intestine microbiomes.

6.4 Metagenomics of the Microbiome and the Aquaculture Industry

The role and importance of the gut microbiomes in fish and shellfish have been recognized in controlling many of the physiological activities by a number of workers all over the world. However, due to a lack of knowledge of the functional aspects of microbiome species and the kind of ingredients they produce, the usage of such microorganisms for the growth of the aquaculture industry is very limited. In order to utilize their full potential for accelerating the growth of aquaculture and related industries, we require in-depth knowledge of the metagenomics of the microbiome species present in the gut tract of fish and shellfish. In the recent past, a good amount of work has been carried out on the use of microbes in aquaculture as environmental biomarkers, bioremediators of polluted ecosystems, for manufacturing probiotics, and as a direct food source for the aqua-cultured species (Ezemonye et al. 2009; Caruso 2013; Martınez-Cordova et al. 2015). Though the usage of microbes in the culture system is growing continuously, however, there are many microbiome species where we do not know their role and precise functions and also the mechanism involved. Streit and Schmitz (2004) reported that most of the bacterial species that are found in different environments are not readily culturable and therefore, for basic research purposes it becomes extremely difficult to do an investigation on such bacterial diversity. Nielsen et al. (2017) emphasized that in recent years metagenomics has emerged as a promising technology that attempts to analyze the complex genomes contained within a microbial niche. This technic involves the isolation of DNA which can provide information related to the microbial diversity of certain ecosystems and also the functions and biological roles of these bacterial species. It has been reported that there are two ways metagenomics studies can be carried out. One way is a single gene survey of the microbes present in the environment and the second one is random shotgun studies of all accessible genes in any environment. Compared to single gene protocol, shotgun studies cover sequencing all of the DNA isolated from a particular sample with full coverage of all genes; however, according to Gilbert and Dupont (2011), this does not occur in the practice because there is no single technology that sequences the entire genomes of the thousands of diverse bacteria. Now second-generation sequencing technology based on polymerase cluster of chain (PCR amplicon) derived from a single molecule has become a more popular technic for critical analysis of the bacterial genome (Porreca et al. 2006; Wooley et al. 2010).

In aquaculture farming, generally, fish culture is done in high densities in closed ponds that facilitate the transfer of microbes from the external environment as well as from the body. The microbiome community available in the gut system of fish and the culture environment is seen as an integral part to keep the whole ecosystem healthy and productive. Maintaining a healthy microbiome in fish is particularly important for closed recirculating culture systems, however, in the grow-out systems, there is the possibility of transfer of microbes between captive and wild populations. The link between the gut microbiome and immunity has been well-established in fish by several workers and the same has been substantiated by the scientific shreds of evidence. A diverse microbiome is beneficial to prevent infection by pathogens for several reasons. The use of probiotics in the form of *Lactobacillus* and *Bacillus* is common practice in aquaculture farming in recent times because this product facilitates the improvement of the fish gut barrier function and the germ-fighting abilities of the fish immune system. Many times antibiotics are used in aquaculture farming to control pathogens, however, the use of such products has been reported to create profound and long-lasting effects on the fish gut microbiome composition that leads to a decrease in productivity. Therefore, aquaculture microbiome studies are very much essential to understand the symbiotic and antagonistic interrelations between the microbes and aquaculture species. Several workers have pointed out that the knowledge of metagenomics can provide better insight with regard to the relationship between the microbiome species and the host inhabiting the different environments (Suttle 2007; Gianoulis et al. 2009). New sequencing tools and bio-informatics protocols enable us to investigate the diversity of intracellular microbial species as well as specific genomic information from such communities. In the field of metagenomics, researchers have focused their investigation on the diversity and quantity of different microbes or genes from different environmental patterns and have identified stronger associations between certain microbial communities and the host genotype (Monchy et al. 2011; Gilbert et al. 2012; Quince and Lundin 2013). The functional impact of the gut microbiome on the host fish is perceived through a number of interactions of metabolic ability and host metabolism of the gut microbiome species. Besides the functional role of gut microbiome species, it is also necessary to find out what types of bioactive compounds or ingredients are produced by the microbiomes in response to a number of interactions between host and microbes and study their properties so that the biopharma industries can the advantage of manufacturing commercial products useful for aquaculture farming. Gianoulis et al. (2009) while discussing microbial studies and their relationship with metagenomics mentioned that metagenomics research can provide in-depth knowledge regarding the relationship between the microbes and their hosts in understanding the symbiotic and antagonist interaction. Avila-Villa et al. (2011) reported that intracellular pathogenic bacteria are difficult to isolate from animals because some of them are intracellular obligate microorganisms that can only be cultured in semi-aqueous and/or cell culture media. Therefore, now new sequencing and bioinformatics technologies make it possible not only to investigate the diversity of gut microbiome but also to elucidate relevant genomic information from such communities.

The possibility of using microbiome species for the purpose of removing effluent effects and reducing toxicity has been reported by George et al. (2010), however, they also emphasized that the knowledge of metagenomics of such bacterial species is an added advantage for the manipulation of genomic structure involved in effluent treatment in the aquatic ecosystems. Microbial communities when they are exposed for long periods to xenobiotics and other toxic elements there is a possibility that the genetic profile of these microbiome species may get affected. In aquaculture practices, whether it is an open or closed system generates thousands of tons of xenobiotic waste each year. Therefore, the scientific communities particularly the environmentalists' take a lot of interest looking in these aspects. It has been reported that the nutrient-rich sediments generated in aquaculture farming and the effluent-receiving ecosystems possess the conditions necessary for the propagation of entire microbial communities that could be a source of new and novel genes with usable functions from a biotechnological standpoint. Lorenz and Eck (2005) mentioned that research in metagenomics of such novel genes can provide new molecules with diverse functions that may be quite useful for either therapeutic purposes in various fields or for their commercial applications to improve economic strategies. Furthermore, not only novel bioactive compounds can be found in environmental random DNA samples but also genes encoding for proteins or enzymes with higher activity for any particular biochemical reaction than those already known. For instance, Hess et al. (2011) reported the metagenomic discovery of novel genes encoding for the rapid degradation of biomass; these genes were involved in the deconstruction of cellulosic biomasses. Aquaculture farming though it deals with diverse species and diverse ecosystems, metagenomics has potential applications for all these combinations and works efficiently for enhancing aquaculture production and productivity. Most of the information regarding the microbial dynamics of profitable aquaculture systems such as salmon farming, shrimp-intensive farming systems, or bivalve culturing systems is perhaps only a small fraction of the success of the aquaculture industry. During the outbreak of diseases, most of the research studies are focused on the biology of the pathogen and the development of protocols for preventive measures. However, there is not much information on the effect of invasive pathogens on the microbial diversity of the host or any possible relationship between the microbial diversity of the environment and the host that favors the proliferation of the pathogen. Therefore, there is a need for research studies covering these aspects to understand the relations between the proliferative invasive pathogens and the microbiome diversity of the hosts.

Several workers have reported the presence of lactic acid bacteria in the digestive tract of fish and shellfish and the use of such bacterial species for the development of probiotics in aquaculture farming (Lauzon and Ringø 2012; Navarrete et al. 2010; Vazquez et al. 2005; Merrifield et al. 2010). There are other reports regarding the presence of yeast in the digestive tract of fish which helps in the decomposition of aquatic plant nutrients recycling, and biodegradation of oil and recalcitrant compounds (Kutty and Philip 2008; Song et al. 2010). Chi et al. (2009) mentioned that yeast in the digestive tract produces a number of vitamins, pigments, cell proteins, immunostimulants, and enzymes. Future research is required to find out the exact role of yeast in fish microbiomes. A number of vibrio species have been recorded in the digestive tract producing hydrolytic enzymes and helping in breaking down the dietary components in the gut system. There are also reports mentioning bacterial strains producing enzymes such as amylase, lipase, cellulose, and chitinases (Gatesoupe et al. 1997; Itoi et al. 2006; Sugita and Ito 2006). The presence of luminous Photobacterium and non-luminous Photobacterium in the digestive tract of several fish has been reported. Several workers observed that these Photobacterium species of the bacterial community in the gut system facilitate chitin digestion (Itoi et al. 2006; Ward et al. 2009; Smriga et al. 2010). While working on the microbiome of marine fish Clements et al. (2007, 2009) mentioned that the bacterial species, Clostridium found in the digestive tract of many herbivorous fish performs the function of nutrients supplying different fatty acids and vitamins. Ramirez and Dixon (2003) reported that Clostridium displays the function of producing enzymes like acid and alkaline phosphatases. Taoka et al. (2006) also mentioned that some of the species of Clostridium, viz., *C. butyricum* can be used for the manufacturing of probiotics in aquaculture for enhancing immunity in Japanese flounder.

While discussing the importance of fish and shellfish microbiomes and the role of metagenomics in promoting the aquaculture industry, a number of workers emphasized that the knowledge of metagenomics is useful in two ways, i.e., functional genomics and sequencing-based genomics. Functional genomics is essential to know the functional aspects of microbial species and sequencing-based genomics is essential to know microbial diversity (Gill et al. 2006). In the recent past, Quince et al. (2017) also reported that the application of metagenomics techniques provides in-depth analysis of gene function from microbiome communities and also helps in identifying genetic and physiological traits of the micro-organisms. Yukgehnaish et al. (2020) while working on functional metagenomics, suggested that metagenomics libraries can be constructed for future use in genomic manipulations of the bacterial species, whereas from functional metagenomics, new molecules can be discovered in the form of enzymes and other bioactive compounds which can be used for therapeutic purposes. It is also emphasized that the metagenomic research of gut microbiome is most essential for the development of appropriate probiotics, which is important to fish from the nutritional point of view. By altering the microbial composition in the gut system with the use of probiotics, one can improve the metabolic and immunological performance of fish. However, more research studies are needed on metagenomics and factors influencing the composition of the gut microbiome and also the physiological effects of this on the host fish (Yukgehnaish et al. 2020). In recent years, researchers are making efforts to understand the functional aspects of microbial species at genome levels so that by altering gene structure, desired properties can be developed in microbiome species. Such studies are very much essential and quite useful for creating innovative products for the benefit of the growth of various sectors including the pharmaceutical/bioindustry. There are several published reports also mentioning that by developing a healthy gut microbiome there will be improvement in animal's immune system and overall health and quality (Gómez and Balcázar 2008; Hanning and Diaz-Sanchez 2015; Tarnecki et al. 2017). While emphasizing the importance of metagenomics Udayangani et al. (2017) also studied the data on metagenomics in zebra fish, *Danio rerio*, after feeding them with chitosan silver Nano composites supplemented diet. They analyzed different properties of the gut microbial communities, particularly the goblet cell density, gut morphometric, and mRNA expression of immune-related and mucin-encoding genes. From the results obtained, they inferred that a chitosan silver Nano composite supplemented diet can be used as an immunostimulants for the management of disease in aquaculture farming by developing a healthy gut microbiome. Tyagi et al. (2019) while working on the gut microbial composition of freshwater carp observed that the gut contains more than three-fourths of Proteobacteria and a very low abundance of commonly found probiotic bacteria such as Bacillus, Lactobacillus, Streptococcus, and Lactococcus. Considering the herbivorous habit of fish, it has been mentioned that the gut microbiome had pathways that degrade cellulose, hemicellulose, chitin, pectin, starch, and other complex carbohydrates. A high prevalence of actinobacteria and antibiotic biosynthesis pathways have also been noticed in the gut microbiome of the carp which indicates that these microbes can produce antibiotics and this has been confirmed by identifying 51 different types of antibiotic resistance genes in microbiome species (Tyagi et al. 2019). Yukgehnaish et al. (2020) while studying metagenomics of fish gut microbiota described that the microbiome composition in the gut system has a very important role in controlling metabolism, feeding behavior, and immune response. The development of probiotics in the aquaculture industry for the management of diseases is probably the outcome of the knowledge of metagenomics of microbes that prevent infectious diseases. Furthermore, in their review paper, they also mentioned that the new methods of genome sequencing and the development of bioinformatics tools have supported the proper analysis of metagenomics data of gut microbiota and their relevant functions. In the recent past, a number of workers have attempted to carry out research studies on the metagenomics of microbial species in the fish gut system (Yukgehnaish et al. 2020). More and more studies on metagenomics covering both taxonomic diversity and functional aspects of both fish shellfish gut microbiomes are the need of the hour for augmenting aquaculture and related industries.

6.5 Gut Microbiomes Controlling the Outbreak of Diseases and the Health of the Cultivable Organisms

In aquaculture farming, farmers suffered heavy losses due to outbreaks of bacterial and viral diseases and in turn several times there was a setback for the growth of the aquaculture industry. Even today outbreaks of disease occur very frequently and there are no effective methods to control such diseases occurrence. The use of enormous amounts of antibiotics and other chemical methods further deteriorated the environment. Therefore, at present all over the world efforts are being made to replace the use of antibiotics with probiotics, bacteriophage, and biofloc technologies for controlling bacterial and viral infections in cultivable organisms. In recent times, the increasing importance of the fish and shellfish microbiome in the aquaculture industry has become a source of investor excitement. This is due to the realization of the importance of gut microbiome communities for their potential benefits and implications of the overall health of the organisms, from regulating the immune system to enhancing the growth and reproducing ability. The microbiomes found in the gut tract have the capacity to produce a number of compounds through their genes and these compounds offer several benefits. It has been already mentioned that the microbes found in the gut system produce several proteins, fatty acids, enzymes, and vitamins that help in the digestion of food, storage of nutrients, and

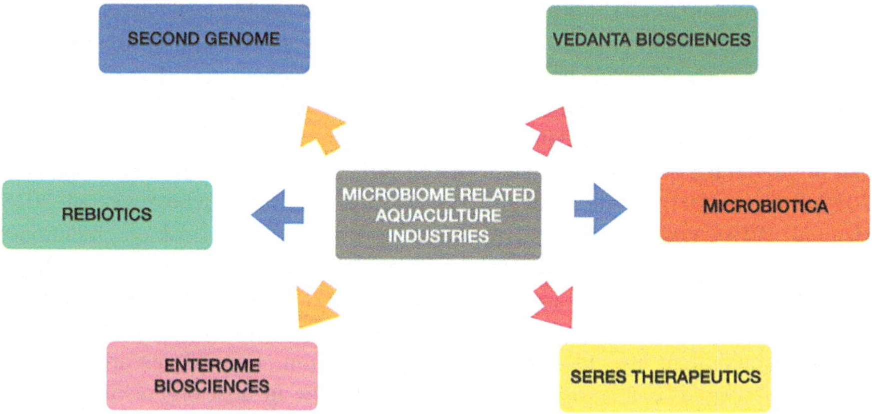

Fig. 6.2 Fish and shellfish microbiome and development of aquaculture/biopharma industries

promotion of a healthy metabolism. The microbiomes also help to maintain the integrity of cell connections in the intestinal lining, entry of preventing harmful bacteria and toxins from leaking through the intestinal wall (Diwan et al. 2021). It has been mentioned that as microbiome genes produce a number of bioactive compounds, several companies related to aquaculture industries are taking a lot of interest in manufacturing these products. All over the world, now there are several companies such as Enterome Bioscience, Second Genome, Rebiotix, Seres Therapeutic, Vedanta Biosciences, and Microbiotica that have started manufacturing drugs from microbiomes, which can be used for the treatment of various disorders not only in humans but also in other cultivable organisms (Fig. 6.2). Even some companies have started microbiome sequencing services in the last few years due to the growing support for research into genomics. This has benefited from the growing investment being driven into the life sciences sector and particularly genetic and genomic research.

There are a number of published pieces of evidence now available indicating a well-established relationship between the host's microbiome and the health status of the organisms. In recent years, more emphasis is given to research on the manipulation of the gut microbiome community to enhance the immune system and health of the host. Considering the substantial growth of the aquaculture industry, priority is now being given to manipulate the microbiome community in the digestive tract of both finfish and shellfish to improve the quality status of cultivable organisms. For manipulation of the gut microbial community, the methods adopted are changes in the feed formulations by adding more quality proteins and lipids in the diet as well as the addition of probiotics and prebiotics (Desai et al. 2012; Geurden et al. 2014; Kotzamanis et al. 2007). Several researchers have reported how the microbial composition of the gut can be changed with the manipulation of the diet. In earlier studies, different types of microbiomes in the gut tract have been reported in both cultured fish and shellfish, compared to their counterparts collected from the wild (Gilmour et al. 1976). Similarly, some workers observed that in farmed fish species there was more infection in the animals due to the poor quality of the microbiome in their gut systems (Olivier et al. 1981). By manipulating the diet with different dietary supplements such as marine protein hydrolysates, short-chain peptides, amino acids, protein hydrolysates and lipids, and fish oils, several researchers have reported that they were able to develop healthy microbiomes in the gut tract of a number of fish and improved the quality of the yield (Kiron 2012; Merrifield et al. 2011; Sila et al. 2014; Bui et al. 2014; Delcroix et al. 2015; Egerton et al. 2018; Khosravi et al. 2015; Ha et al. 2019). Yoshida et al. (2016), while working on fish microbiota, mentioned the presence of bacteria not only in the gut system which can synthesize polyunsaturated fatty acids but such bacteria can be isolated from seawater and sediments also. It is very important to investigate the use of such bacteria for the preparation of probiotics and their use for fish health and nutrition. Several researchers have reported that the presence of a beneficial microbiome in host organisms does not cause any diseases or disorders in the body system, but when there is a disturbance in the balance of the microbial community, the harmful pathogens will become more prevalent and can cause infections and diseases (Moya and Ferrer 2016; Romero et al. 2014; Turner et al. 2013). Several researchers who worked on the gut microbial composition of diseased and healthy shrimp observed that there was an imbalance in the composition of the microbiome community in the gut tract of diseased shrimp (Merrifield and Rodiles 2015; Salinas 2015; Li et al. 2017; Llewellyn et al. 2016, 2017). From these observations, it was inferred that the bacterial community in healthy shrimp was well organized and as the shrimp suffered from the severity of the disease, the bacterial community was disrupted. Several workers have made similar observations on the dysbiosis in gut microbial composition and its relationship with host diseases (Berry et al. 2012; Xiong et al. 2015). Currently, the relationship between the gut microbiota and the health of the host fish has become an important topic for research investigation. Ninawe et al. (2020) suggested the use of bacteriophage technology for controlling various bacterial and viral diseases for cultivable aquaculture species. Though phage therapy has been advocated by several researchers, the technology needs further investigation prior to its commercial applications. Earlier also several workers have attempted the use of phage therapy to cure bacterial and viral infections in fish and shellfish (Oliveira et al. 2012; Subharthi 2015). In aquaculture farming, the microbiome that is available naturally has a vital role in maintaining a healthy aquatic environment but sometimes beneficial microbiomes are added artificially also to fulfill different functional roles. Therefore, monitoring and manipulations of microbial communities in the aquaculture system hold great potential not only in the maintenance of soil and water quality but also controlling the infectious microbial pathogens in aquatic species.

6.6 Microbiome and Probiotic Development

The outbreak of bacterial and virus diseases in aquaculture farming has made huge losses to the aquaculture and seafood industries. The occurrences and the spread of diseases are still continuing and in the recent past many preventive measures are being taken to control the same. To control the spread of diseases there has been the practice of the use of enormous chemicals, medicines, and antibiotics to ensure the safety of cultivable organisms as well as enhance the productivity of the farms. The indiscriminate use of such antibiotic compounds for several years has created catastrophic effects on aquaculture farming. The continuous use of such antibiotics medicines has also affected antibiotic-resistant microorganisms in fish and shellfish and created an imbalance in the gut microbiota, which affected fish health also, and residual deposition of these toxic compounds in the fish muscle has become a potential health risk to consumers also. To replace these antibiotic medicines, several workers have suggested the use of probiotics as an alternative approach to prevent diseases in the aquaculture industry and some others have suggested its use not only for disease control but also for promoting growth and reproduction (Dawood and Koshio 2016; Fyzul et al. 2014; Hai 2015). In the recent past, many workers have emphasized the use of probiotics not only for the control of diseases but as a source of nutrients, providing enzymes for better digestion, modulating the immune system, and increasing the immune response against pathogenic bacteria (Adel et al. 2017). The most common probiotics used in aquaculture include lactic acid bacteria such as *Lactobacillus*-sp., *Bacillus*-sp., *Enterococcus*-sp. and yeast, *Saccharomyces cerevisiae* (Adel et al. 2017). The concept behind the use of probiotics is to feed an adequate amount of healthy microbes, which replace the harmful microbes in the gut flora of fish and shellfish with result that these exogenous bacteria compete with pathogens, preventing their adhesion to the intestinal wall and secreting antibacterial substances. The proliferation of healthy microbes also helps in better digestion and absorption of food in the gut resulting in increased growth performance (Ringo et al. 2010). There are several reports mentioning the better performance of the aquaculture species after feeding with probiotics as supplementary feed (Ringo et al. 2010; Williams 2017; Diwan et al. 2021). It has been reported that the beneficial effects of probiotics containing *Lactobacillus acidophilus* were significant even after the withdrawal of supplementary food like yeast and other prebiotics. In in vivo challenge studies using *Aeromonas hydrophila*, it is reported that the fish responded best in terms of resistance to infection and 100% survived when *Lactobacillus acidophilus* was the candidate probiotic compared to the use of live yeast. From these findings, it is evident that the benefits accorded by probiotics make aquaculture farming more viable compared to the use of antibiotics and chemicals. Merrifield and Carnevali (2014) and Akhter et al. (2015), while working on probiotics, mentioned the use of several microbiome species such as Gram-positive and Gram-negative bacteria, bacteriophages, microalgae, and yeasts for the preparation of probiotics. Some of the most frequently used probiotics include LAB Bacillus, Lactococcus, Shewanella, and Aeromonas genera (Merrifield and Carnevali 2014). There are several studies which indicated enhanced growth rate and improvement in the immune system in fish after manipulation of gut microbial composition using probiotics (Huang et al. 2014; Lobo et al. 2014; Cordero et al. 2015; Perry et al. 2020). Merrifield et al. (2010) reported that in aquaculture farming, the application of probiotics may not be always beneficial because of the low viability of probiotic bacteria during processing and storage and loss of bacteria from leaching in the water during feeding in culture operations. However, despite this, it has been observed that when probiotics are used, it will reduce the cost of fish farming through improvement in the fish growth and yield in a minimal culture period. It has been reported that the extensive use of probiotics is quite beneficial for the growth and development of the shellfish aquaculture industry (Ringø 2020). The use of several probiotic species such as Lactobacillus, Enterococcus, Bacillus, Aeromonas, Alteromonas, Arthrobacter, Bifidobacterium, Clostridium, Microbacterium, Paenibacillus, Phaeobacter, Pseudoalteromonas, Pseudomonas, Rhodosporidium, Roseobacter, Streptomyces, and Vibrio has been reported for sustainable aquaculture development. While working on the microbial community of the shrimp *P. vannamei* Restrepo et al. (2021) developed a probiotic using the ILI strain of *Vibrio diabolicus* bacterium which prevents the infection in the shrimp caused by the AHPND virus. Similarly, Seyed Hossein et al. (2020) also studied the effect of microbial feed additives including prebiotics, probiotics, and synbiotics on antioxidant enzymes in fish and found that adopting different dietary approaches and microbial feed additives improves the level of antioxidant enzymes and defense systems in fish. It is mentioned that probiotics are active in the small intestine and prebiotics influence the microbiomes of the large intestine. However, the use of probiotics and prebiotics in aquaculture is a fast-growing area and further research is required to investigate how these components work together in the gut system to build up adequate microbiomes.

6.7 Bioinformatics Skills in Microbiome Research for Augmenting the Aquaculture Industry

The benefits of microbiome research for the development of the aquaculture industry are getting momentum for a few years from now. In the recent past a number of probiotics have been manufactured from microbiome species and are being used for enhancing aquaculture production. However, a lot of research work is further needed in this particular niche area of fish and shellfish aquaculture to find out new resources of microbiome species for the promotion of aquaculture farming. The use of microbiomes in the form of probiotics and in the combination of prebiotics and synbiotics is being recognized for the development of aquaculture and related bio-industries also. In this context, so far not much attention has been paid to utilizing the knowledge of bioinformatics skills for aquaculture farming and also for the growth of the bioindustry. Many times in the past, development of the aquaculture industry suffered losses due to the spread of bacterial and viral diseases and several efforts have been made to prevent these pathogens entry and outbreak of diseases during fish and shellfish farming in captive conditions. In recent years, in order to understand disease dynamics and its interaction with cultivable organisms, as also the interaction of pathogens that are present in aquatic bodies and their relationship with aquatic animals, the knowledge of bioinformatics was brought to address some of the disease management issues in the aquaculture industry. For example, when the white spot syndrome virus (WSSV) invaded aquaculture farming, efforts were made to control this virus more effectively using the knowledge of bioinformatics. In such attempts, the number of workers analyzed the complete transcriptional profile of WSSV genes and also worked on the interaction mechanism between the host gene and proteins and how they react with WWSV infection (Wang et al. 2006; Zhao et al. 2007; Li et al. 2013). However, at present the information lacking on the direct interaction between WSSV and the host proteins for a proper understanding of the pathogenesis of WSSV in shrimp. In such a situation bioinformatics skills will provide a highly effective approach for identifying genes and proteins involved in WSSV/shrimp interaction based on the public protein–protein interaction (PPI) databases (Zheng et al. 2014). The exploitation of the information with regard to protein–protein interaction with bioinformatics approaches provides an effective way to analyze high-throughput experimental data in typical organisms (Pattin and Moore 2009; Remmerie et al. 2011). The predicted interaction between shrimp protein and WSSV protein which might be either independent interaction or synergistic interaction provides information on possible invasion approaches during WSSV infection to host cells. Moreover, these interactions could also lead to intracellular signaling pathways initiated by the host proteins. There are a number of research investigations on host genes and protein interaction with WSSV in different crustaceans but all these findings reported by various workers do not lead to any conclusion concerning the revealing of the mechanism of WSSV infection in the shrimp (Zhao et al. 2007; Wang et al. 2006, 2011; Li et al. 2013). Liu et al. (2009) while dealing with antiviral immunity in crustaceans mentioned that there is a cellular receptor called β-integrin which is involved in WSSV infection by interacting with WSSV envelope protein VP187. Although this particular data provide us with some useful information about the WSSV infection mechanism, it is still not clear about the molecular mechanism of WSSV infection. When whole genome sequencing information of WSS viruses is available, then the knowledge of bioinformatics will be quite useful for identifying genes and proteins involved in WSSV/host interaction. Szklarczyk et al. (2011) while discussing the bioinformatics methods mentioned the uses of different search tools for collecting protein–protein interaction (PPI) databases. The most widely used PPI databases mainly include the Search Tool for the Retrieval of Interacting Genes/Proteins (STRING), the Database of Interacting Proteins (DIP), and Reactome. STRING is a database of known and predicted protein interactions based on the sources derived from the genomic context, high-throughput experiments, co-expression, and previous knowledge. Xenarios et al. (2002) reported the use of DIP databases of interacting proteins as a search tool of bioinformatics for studying cellular networks of protein interaction. There is another database tool, mentioned by Joshi-Tope et al. (2005) called "Reactome," to find out the biological pathways related to PPI information.

6.8 Conclusions

In aquaculture farming now there is ban on the use of antibiotics and therefore all over the world efforts are being made to find out alternate methods which can replace antibiotics and augment aquaculture production safely without compromising the quality of the yield. In recent times some new and innovative methods introduced are the use of bacteriophage and biofloc technology and also the microbiome in the form of probiotics. However, there is not much information regarding the use of gut microbiome species of fish and shellfish for the development of aquaculture farming/ industry. There is a huge potential for development in this particular area, provided the knowledge pertaining to the functional aspects and the ingredient molecules that are present in the microbiome species is available. Therefore, extensive efforts are needed to develop R and D programs on the metagenomics aspects of the gut microbiome species. In order to promote the future prospects of the bio-industry, there is an urgent need for intensive research to be undertaken on fish and shellfish gut microbiomes, particularly in the areas of genomics and metagenomics. The functional aspects of the gut microbiome are an important issue for future investigation and therefore, in-depth knowledge of functional metagenomics is very much essential. Metagenomics has emerged in the last decade as a promising technology that attempts to analyze the complex genomes contained within a microbial niche.

Several attempts have been made on microbiome species related to aquaculture farming to understand the interrelations of the symbiotic and antagonistic factors with that of microbiome species and aquaculture organisms. It has been pointed out the knowledge of metagenomics can provide better insight with regard to the relationship between the microbiome species and the host inhabiting the different environments. Number of workers have suggested the use of probiotics as an alternative approach to prevent diseases in the cultivable organisms in the aquaculture industry in place of the use of antibiotic drugs some others have suggested its use not only for disease control but also for promoting growth and reproduction and other immunity related aspects. In recent years the knowledge of bioinformatics was brought in to address some of the disease management issues in the aquaculture industry and also to understand disease dynamics and its interaction with cultivable organisms. The application of bioinformatics tools will pave the way not only to understand proper interaction between the host and the pathogens at the molecular levels but also reveal the composition of the ingredients that are present in the gut microbes.

References

Adel M, El-Sayed AFM, Yeganeh S, Dadar M, Giri SS (2017) Effect of probiotic Lactococcus lactis sub sp. lactis on growth performance, intestinal microbiota, digestive enzyme activities, and disease resistance of *Litopenaeus vannamei*. Probiotics Antimicrob Proteins 9:150–156. https://doi.org/10.1007/s12602-016-9235-9

Akhter N, Bin W, Aamir MM, Muhammad M (2015) Probiotics and prebiotics associated with aquaculture: a review. Fish Shellfish Immunol 45:733–741. https://doi.org/10.1016/j.fsi.2015.05.038

Avila-Villa LA, Martınez-Porchas M, Gollas-Galvan T, Lopez-Elıas JA, Mercado L, Murguia-Lopez A et al (2011) Evaluation of different microalgae species and Artemia (*Artemia franciscana*) as possible vectors of necrotizing hepatopancreatic bacteria. Aquaculture 318: 273–276

Balcázar JL, de Blas I, Ruiz-Zarzuela I, Vendrell D, Giron ESO, Muzquiz JL (2007) Enhancement of the immune response and protection induced by probiotic lactic acid bacteria against furunculosis in rainbow trout (*Oncorhynchus mykiss*). FEMS Immunol Med Microbiol 51: 185–193

Baldo L, Riera JL, Tooming KA, Alba MM, Salzburger W (2015) Gut microbiota dynamics during dietary shift in eastern African cichlid fishes. PLoS One 10(5):e0127462. https://doi.org/10.1371/Journal.pone.0127462

Berry D, Schwab C, Milinovich G, Reichert J, Mahfoudh KB, Decker T, Engel M, Hai B, Hainzl E, Heider S, Kenner L, Muller M, Rauch I, Strobl B, Wagner M, Schleper C, Urich T, Loy A (2012) Phylotype-level 16S rRNA analysis reveals new bacterial indicators of health state in acute murine colitis. ISME J 6:2091–2106. https://doi.org/10.1038/ismej.2012.39

Bordenstein SR, Theis KR (2015) Host biology in light of the microbiome: ten principles of holobionts and hologenomes. PLoS Biol 13(8):e1002226. https://doi.org/10.1371/journal.pbio.1002226

Bui HTD, Khosravi S, Fournier V, Herault M, Lee KJ (2014) Growth performance, feed utilization, innate immunity, digestibility and disease resistance of juvenile red sea bream (Pagrus major) fed diets supplemented with protein hydrolysates. Aquaculture 418–419:11–16. https://doi.org/10.1016/j.aquaculture.2013.09.046

Caruso G (2013) Microbes and their use as indicators of pollution. J Pollut Eff Control 1:1–3

Chi Z, Chi Z, Zhang T, Liu G, Li J, Wang X (2009) Production, characterization and gene cloning of the extracellular enzymes from the marine-derived yeasts and their potential applications. Biotechnol Adv 27(3):236–255. https://doi.org/10.1016/j.biotechadv.2009.01.002

Clements KD, Choat JH (1995) Fermentation in tropical marine herbivorous fishes. Physiol Zool 68:355–378

Clements KD, Pasch IB, Moran D, Turner SJ (2007) Clostridia dominate 16S rRNA gene libraries prepared from the hindgut of temperate marine herbivorous fishes. Mar Biol 150:1431–1440. https://doi.org/10.1007/s00227-006-0443-9

Clements KD, Raubenheimer D, Choat JH (2009) Nutritional ecology of marine herbivorous fishes: ten years on. Funct Ecol 23:79–92. https://doi.org/10.1111/j.1365-2435.2008.01524.x

Cordero H, Guardiola FA, Tapia-Paniagua ST, Cuesta A, Meseguer J, Balebona MC, Moriñigo MÁ, Esteban MÁ (2015) Modulation of immunity and gut microbiota after dietary administration of alginate encapsulated Shewanella putrefaciens Pdp11 to gilthead sea bream (*Sparus aurata* L.). Fish Shellfish Immunol 45:608–618. https://doi.org/10.1016/j.fsi.2015.05.010

Dawood MAO, Koshio S (2016) Recent advances in the role of probiotics and prebiotics in carp aquaculture: a review. Aquaculture 454:243–251. https://doi.org/10.1016/J.AQUACULTURE.2015.12.033

Delcroix J, Gatesoupe FJ, Desbruyères E, Huelvan C, Le Delliou H, Le Gall MM, Quazuguel P (2015) The effects of dietary marine protein hydrolysates on the development of sea bass larvae, Dicentrarchus labrax, and associated microbiota. Aquac Nutr 21:98–104. https://doi.org/10.1111/anu.12139

Desai AR, Links MG, Collins SA, Mansfield GS, Drew MD, Van Kessel AG, Hill JE (2012) Effects of plant-based diets on the distal gut microbiome of rainbow trout (*Oncorhynchus mykiss*). Aquaculture 350:134–142. https://doi.org/10.1016/j.aquaculture.2012.04.005

Diwan AD, Harke SN, Gopalkrishna, Archana NP (2021) Aquaculture industry prospective from the gut microbiome of fish and shellfish: an overview. J Anim Physiol Anim Nutr 106:441–469. https://doi.org/10.1111/jpn.13619

Egerton S, Culloty S, Whooley J, Stanton C, Ross RP (2018) Characterization of protein hydrolysates from blue whiting (*Micromesistius poutassou*) and their application in beverage fortification. Food Chem 245:698–706. https://doi.org/10.1016/j.foodchem.2017.10.107

Eichmiller JJ, Hamilton MJ, Staley C, Sadowsky MJ, Sorensen PW (2016) Environment shapes the faecal microbiome of invasive carp species. Microbiome 4:44. https://doi.org/10.1186/s40168-016-0190-1

Ezemonye LIN, Ogeleka DF, Okieimen FE (2009) Lethal toxicity of industrial detergent on bottom dwelling sentinels. Int J Sediment Res 24:479–474

FAO (2020) The state of World's fisheries and aquaculture (SOFIA), Sustainable fisheries; sustainable aquaculture; Code of conduct for responsible fisheries; new technology; biosecurity; Sustainable development goals. FAO, Rome. ISBN: 978–92–5-132692-3

Foster KR, Schluter J, Coyte KZ, Rakoff-Nahoum S (2017) The evolution of the host microbiome as an ecosystem on a leash. Nature 548:43

Fyzul AN, Alharbi AH, Austin B (2014) Review: development in the use of probiotics for disease control in aquaculture. Aquaculture 431:1–11

Gatesoupe FJ, Infante JLZ, Cahu C, Quazuguel P (1997) Early weaning of sea bass larvae, Dicentrarchus labrax: the effect on microbiota, with particular attention to iron supply and exoenzymes. Aquaculture 158:117–127. https://doi.org/10.1016/S0044-8486(97)00179-8

George I, Stenuit B, Agathos S, Marco D (2010) Application of metagenomics to bioremediation. In: Metagenomics: theory, methods and applications, vol 1. Caister Academic Press, Norfolk, pp 119–140

Geurden I, Mennigen J, Plagnes-Juan E, Veron V, Cerezo T, Mazurais D, Zambonino-Infante J, Gatesoupe J, Skiba-Cassy S, Panserat S (2014) High or low dietary carbohydrate: protein ratios during first-feeding affect glucose metabolism and intestinal microbiota in juvenile rainbow trout. J Exp Biol 217:3396–3406. https://doi.org/10.1242/jeb.106062

Gianoulis TA, Raes J, Patel PV, Bjornson R, Korbel JO, Letunic I et al (2009) Quantifying environmental adaptation of metabolic pathways in metagenomics. Proc Natl Acad Sci 106: 1374–1379

Gilbert JA, Dupont CL (2011) Microbial metagenomics: beyond the genome. Ann Rev Mar Sci 3: 347–371

Gilbert JA, Steele JA, Caporaso JG, Steinbruck L, Reeder J, Temperton B et al (2012) Defining seasonal marine microbial community dynamics. ISME J 6:298–308. https://doi.org/10.1038/ismej.2011.107

Gill SR, Pop M, DeBoy RT, Eckburg PB, Turnbaugh PJ, Samuel BS (2006) Metagenomics analysis of the human distal gut microbiome. Science 312:1355

Gilmour A, McCallum MF, Allan M (1976) A study of the bacterial types occurring on the skin and in the intestines of farmed plaice (*Pleuronectes platessa* L.). Aquaculture 7:161–172. https://doi.org/10.1016/0044-8486(76)90005-3

Gómez GD, Balcázar JL (2008) A review on the interactions between gut microbiota and innate immunity of fish. FEMS Immunol Med Microbiol 52:145–154. https://doi.org/10.1111/j.1574-695X.2007.00343.x

Guivier E, Nicolas P, Rémi C, André G (2019) Microbiota associated with the skin, gills, and gut of the fish *Parachondrostoma toxostoma* from the Rhône basin. Freshw Biol 65(3):446–459. https://doi.org/10.1111/fwb.13437

Ha NG, Fernandes A, Jesus A, Fernando N, Gonçalves N, Soaresde O, Juliet K, Sugai M, Luiz P, José L, Pedreira M, Thiago EL, Hadi PF (2019) Sardine (*Sardinella* spp.) protein hydrolysates as growth promoter in south American catfish (*Rhamdia quelen*) feeding: productive performance, digestive enzymes activity, morphometry and intestinal microbiology. Aquaculture 500(1):9

Hai NV (2015) The use of probiotics in aquaculture. J Appl Microbiol 119:917–935. https://doi.org/10.1111/jam.12886

Hanning I, Diaz-Sanchez S (2015) The functionality of the gastrointestinal microbiome in non-human animals. Microbiome 3:51. https://doi.org/10.1186/s40168-015-0113-6

Hess M, Sczyrba A, Egan R, Kim T-W, Chokhawala H, Schroth G et al (2011) Metagenomic discovery of biomass-degrading genes and genomes from cow rumen. Science 331:463–467

Holt CC, David B, Stentiford GD, Giezen M (2020) Understanding the role of the shrimp gut microbiome in health and disease. J Invertebr Pathol 4(1):1–16. https://doi.org/10.1016/j.jip.2020.107387

Huang Z, Li X, Wang L, Shao Z (2014) Changes in the intestinal bacterial community during the growth of white shrimp, Litopenaeus vannamei. Aquac Res 47:1737–1746

Itoi S, Okamura T, Koyama Y, Sugita H (2006) Chitinolytic bacteria in the intestinal tract of Japanese coastal fishes. Can J Microbiol 52:1158–1163. https://doi.org/10.1139/w06-082

Joshi-Tope G, Gillespie M, Vastrik I, D'Eustachio P, Schmidt E, de Bono B, Jassal B, Gopinath GR, Wu GR, Matthews L, Lewis S, Birney E, Stein L (2005) Reactome: a knowledgebase of biological pathways. Nucleic Acids Res 33:D428–D432. https://doi.org/10.1093/nar/gki072

Khosravi S, Rahimnejad S, Herault M, Fournier V, Lee CR, Dio Bui HT (2015) Effects of protein hydrolysates supplementation in low fish meal diets on growth performance, innate immunity and disease resistance of red sea bream Pagrus major. Fish Shellfish Immunol 45:858–868. https://doi.org/10.1016/j.fsi.2015.05.039

Kiron V (2012) Fish immune system and its nutritional modulation for preventive health care. Anim Feed Sci Technol 173:111–133. https://doi.org/10.1016/j.anifeedsci.2011.12.015

Kotzamanis YP, Gisbert E, Gatesoupe FJ, Zambonino Infante J, Cahu C (2007) Effects of different dietary levels of fish protein hydrolysates on growth, digestive enzymes, gut microbiota, and resistance to vibrio anguillarum in European sea bass (*Dicentrarchus labrax*) larvae. Comp Biochem Physiol A Mol Integr Physiol 147:205–214. https://doi.org/10.1016/j.cbpa.2006.12.037

Kutty SN, Philip R (2008) Marine yeast-a review. Yeast 25(7):465–483

Lauzon HL, Ringø E (2012) Prevalence and application of lactic acid bacteria in aquatic environments. In: Lahtinen S, Ouwehand A, Salminen S, von Wright A (eds) Lactic acid bacteria: microbiological and functional aspects, 4th edn. CRC, Boca Raton, FL, pp 593–631

Li SH, Zhang XJZ, Sun F, Li H, Xiang JH (2013) Transcriptome analysis on Chinese shrimp *Fenneropenaeus chinensis* during WSSV acute infection. PLoS One 8:e58627. https://doi.org/10.1371/journal.pone.0058627

Li X, Zhou L, Yu Y, Ni J, Xu W, Yan Q (2017) Composition of gut microbiota in the gibelio carp (*Carassius auratus gibelio*) varies with host development. Microb Ecol 74:239–249. https://doi.org/10.1007/s00248-016-0924-4

Li W, Zhou Z, Li H, Wang S, Ren L, Hu J et al (2022) Successional changes of microbial communities and host-microbiota interactions contribute to dietary adaptation in allodiploid hybrid fish. Microb Ecol 85:1190. https://doi.org/10.1007/s00248-022-01993-y

Liu H, Soderhall K, Jiravanichpaisal P (2009) Antiviral immunity in crustaceans. Fish Shellfish Immunol 27(2):79–88. https://doi.org/10.1016/j.fsi.2009.02.009

Liu H, Guo X, Gooneratne R, Lai R, Zeng C, Zhan F, Wang W (2016) The gut microbiome and degradation enzyme activity of wild freshwater fishes influenced by their trophic levels. Sci Rep 6:24340. https://doi.org/10.1038/srep24340

Llewellyn MS, McGinnity P, Dionne M, Letourneau J, Thonier F, Carvalho GR, Creer S, Derome N (2016) The biogeography of the Atlantic salmon (*Salmo salar*) gut microbiome. ISME J 10: 1280–1284. https://doi.org/10.1038/ismej.2015.189

Llewellyn MS, Leadbeater S, Garcia C, Sylvain F-E, Custodio M, Ang KP, Powell F, Carvalho GR, Creer S, Elliot J, Derome N (2017) Parasitism perturbs the mucosal microbiome of Atlantic salmon. Sci Rep 7:43465. https://doi.org/10.1038/srep43465

Lobo C, Moreno-Ventas X, Tapia-Paniagua S, Rodrıguez C, Morinigo MA, de La Banda IG (2014) Dietary probiotic supplementation (Shewanella putrefaciens Pdp11) modulates gut microbiota and promotes growth and condition in Senegalese sole larviculture. Fish Physiol Biochem 40: 295–309. https://doi.org/10.1007/s10695-013-9844-0

Lorenz P, Eck J (2005) Metagenomics and industrial applications. Nat Rev Microbiol 3:510–516

Martınez-Cordova LR, Emerenciano M, Miranda-Baeza A, Martınez-Porchas M (2015) Microbial-based systems for aquaculture of fish and shrimp: an updated review. Rev Aquac 7:131–148

McFall-Ngai M et al (2013) Animals in a bacterial world, a new imperative for the life sciences. Proc Natl Acad Sci U S A 110:3229–3236. https://doi.org/10.1073/pnas.1218525110

Merrifield DL, Carnevali O (2014) Probiotic modulation of the gut microbiota of fish. In: Merrifield D, Ringø E (eds) *Aquaculture nutrition*: gut health, probiotics and prebiotics. Wiley, Hoboken, NJ, pp 185–222

Merrifield DL, Rodiles A (2015) The fish microbiome and its interactions with mucosal tissues. In: Benjamin H, Peatman E (eds) Mucosal health in aquaculture. Academic, Cambridge, MA, pp 273–295. https://doi.org/10.1016/B978-0-12-417186-2.00010-8

Merrifield DL, Dimitroglou A, Foey A, Simon J, Remi D, Baker TM, Jarl B, Castex M, Ringø E (2010) The current status and future focus of probiotic and prebiotic applications for salmonids. Aquaculture 302(1–2):1–18. https://doi.org/10.1016/j.aquaculture.2010.02.007

Merrifield DL, Harper GM, Mustafa S, Carnevali O, Picchietti S, Davies SJ (2011) Effect of dietary alginic acid on juvenile tilapia (*Oreochromis niloticus*) intestinal microbial balance, intestinal histology and growth performance. Cell Tissue Res 344:135–146. https://doi.org/10.1007/s00441-010-1125-y

Monchy S, Sanciu G, Jobard M, Rasconi S, Gerphagnon M (2011) Exploring and quantifying fungal diversity in freshwater lake ecosystems using rDNA cloning/sequencing and SSU tag pyro sequencing. Environ Microbiol 13:1433–1453

Mountfort DO, Campbell J, Clements KD (2002) Hindgut fermentation in three species of marine herbivorous fish. Appl Environ Microbiol 68:1374–1380. https://doi.org/10.1128/AEM.68.3.1374-1380.2002

Moya A, Ferrer M (2016) Functional redundancy-induced stability of gut microbiota subjected to disturbance. Trends Microbiol 24(5):402–513. https://doi.org/10.1016/j.tim.2016.02.002

Navarrete P, Mardones P, Opazo R, Espejo R, Romero J (2010) Oxytetracycline treatment reduces bacterial diversity of intestinal microbiota of Atlantic salmon. J Aquat Anim Health 20:177–183. https://doi.org/10.1577/H07-043.1

Nielsen S, Walburn JW, Verge SA, Thomas T, Egan S (2017) Microbiome patterns across the gastrointestinal tract of the rabbit fish Siganus fuscescens. Peer J 5:e3317. https://doi.org/10.7717/peerj.3317111

Ninawe AS, Sivasankari S, Ramasamy P, Seghal KG, Selvin J (2020) Bacteriophages for aquaculture disease control. Aquac Int 28(5):1925–1938. https://doi.org/10.1007/s10499-020-00567-4

Oliveira J, Castilho F, Cunha A, Pereira MJ (2012) Bacteriophage therapy as a bacterial control strategy in aquaculture. Aquac Int 20:879–910. https://doi.org/10.1007/s10499-012-9515-7

Olivier G, Lallier R, Lariviere S (1981) A toxigenic profile of *Aeromonas hydrophila* and *Aeromonas sobria* isolated from fish. Can J Microbiol 27:330–333. https://doi.org/10.1139/m81-050

Pattin KA, Moore JH (2009) Role for protein-protein interaction databases in human genetics. Expert Rev Proteomics 6(6):647–659. https://doi.org/10.1586/epr.09.86

Pérez T, Balcázar JL, Ruiz-Zarzuela I, Halaihel N, Vendrell D, de Blas I, ú Zquiz JLM (2010) Host–microbiota interactions within the fish intestinal ecosystem. Mucosal Immunol 3(4):355–360. https://doi.org/10.1038/mi.2010.12

Pérez-Sánchez T, Balcázar JL, Merrifield DL, Carnevali O, Gioacchini G, de Blas I, Ruiz-Zarzuela I (2011) Expression of immune-related genes in rainbow trout (*Oncorhynchus mykiss*) induced by probiotic bacteria during *Lactococcus garvieae* infection. Fish Shellfish Immunol 31:196–201

Perry WB, Lindsay E, Payne CJ, Brodie C, Kazlauskaite R (2020) The role of the gut microbiome in sustainable teleost aquaculture. Proc R Soc B Biol Sci 287:20200184. https://doi.org/10.1098/rspb.2020.0184

Piazzon MC, Naya-Català F, Perera E, Palenzuela O, Sitjà-Bobadilla A, Pérez-Sánchez J (2020) Genetic selection for growth drives differences in intestinal microbiota composition and 559 parasite disease resistance in gilthead sea bream. Microbiome 8:168

Porreca GJ, Shendure J, Church GM, Pace B (2006) Polony DNA sequencing. In: Ausubel FM (ed) Current protocols in molecular biology, vol 76. Wiley, Hoboken, NJ

Quince C, Lundin EE (2013) The impact of Crohn's disease genes on healthy human gut microbiota: a pilot study. Gut 62:952

Quince C, Walker AW, Simpson JT, Loman N, Segata N (2017) Shotgun metagenomics, from sampling to analysis. Nat Biotechnol 35:833–844. https://doi.org/10.1038/nbt.3935

Rajeev R, Adithya KK, Kiran GS, Joseph S (2021) Healthy microbiome: a key to successful and sustainable shrimp aquaculture. Rev Aquac 13:238–258. https://doi.org/10.1111/raq.12471

Ramirez RF, Dixon BA (2003) Enzyme production by obligate intestinal anaerobic bacteria isolated from Oscars (Astronotus ocellatus), angelfish (Pterophyllum scalare) and southern flounder (Paralichthys lethostigma). Aquaculture 227:417–426. https://doi.org/10.1016/S0044-8486(03)00520-9

Rawls JF, Mahowald MA, Ley RE, Gordon JI (2006) Reciprocal gut microbiota transplants from zebrafish and mice to germ-free recipients reveal host habitat selection. Cell 127:423–433. https://doi.org/10.1016/j.cell.2006.08.043

Ray AK, Ghosh K, Ringø E (2012) Enzyme-producing bacteria isolated from the fish gut: a review. Aquac Nutr 18:465–492. https://doi.org/10.1111/j.1365-2095.2012.00943.X

Redfern LK, Jayasundara N, Singleton DR, Di Giulio RT, Carlson J, Sumner SJ et al (2021) The role of gut microbial community and metabolomic shifts in adaptive resistance of Atlantic killifish (*Fundulus heteroclitus*) to polycyclic aromatic hydrocarbons. Sci Total Environ 776: 145955

Remmerie N, de Vijlder T, Laukens K, Dang TH, Lemière F, Mertens I, Valkenborg D, Blust R, Witters E (2011) Next generation functional proteomics in non-model plants: a survey on techniques and applications for the analysis of protein complexes and post-translational modifications. Phytochemistry 72(10):1192–1218. https://doi.org/10.1016/j.phytochem.2011.01.003

Restrepo L, Cristóbal DB, Leandro B, Irma B, Jenny R, Bonny B, Alejandro R (2021) Microbial community characterization of shrimp survivors to AHPND challenge test treated with an effective shrimp probiotic (Vibrio diabolicus). Microbiome 9:88. https://doi.org/10.1186/s40168-021-01043

Ringø E (2020) Probiotics in shellfish aquaculture. Aquac Fish 5(1):1–27. https://doi.org/10.1016/j.aaf.2019.12.001

Ringo E, Olsen RE, Gifstad TO, Dalmo RA, Amlund H, Hemre GI, Bakke AM (2010) Prebiotics in aquaculture: a review. Aquac Nutr 16:117–136. https://doi.org/10.1111/j.1365-2095.2009.00731.x

Roeselers G, Mittge EK, Stephens WZ, Parichy DM, Cavanaugh CM (2011) Evidence for a core gut microbiota in the zebra fish. ISME J 10:1595–1608. https://doi.org/10.1038/ismeJ2011.38

Romero J, Ringø E, Merrifield DL (2014) The gut microbiota of fish. In: Merrifield D, Ringø E (eds) *Aquaculture nutrition*: gut health, probiotics and prebiotics. Wiley Blackwell Publishing, Oxford, OX, pp 75–100

Salinas I (2015) The mucosal immune system of teleost fish. Biology 4:525. https://doi.org/10.3390/biology4030525

Sanchez LM, Wong WR, Romina MR, Christopher JS, Roger GL (2012) Examining the fish microbiome: vertebrate-derived bacteria as an environmental niche for the discovery of unique marine natural products. PLoS One 7(5):e35398. https://doi.org/10.1371/journal.pone.0035398

Seyed Hossein H, Samira Y, Hien VD, Ghasem A, Giorgia G, Francesca M, Oliana C (2020) Oxidative stress and antioxidant defense in fish: the implications of probiotic, prebiotic, and synbiotics. Rev Fish Sci Aquac 29(2):198–217. https://doi.org/10.1080/23308249.2020.1795616

Sharma PV, Thaiss CA (2020) Host-microbiome interactions in the era of single-cell biology. Front Cell Infect Microbiol 10:569070. https://doi.org/10.3389/fcimb.2020.569070

Sila A, Nedjar-Arroume N, Hedhili K, Chataigné G, Balti R, Nasri M, Dhulster P, Bougatef A (2014) Antibacterial peptides from barbel muscle protein hydrolysates: activity against some pathogenic bacteria. LWT Food Sci Technol 55:183–188. https://doi.org/10.1016/j.lwt.2013.07.021

Simon JC, Julian RM, Christophe M, Marc AS (2019) Host-microbiota interactions: from holobiont theory to analysis. Microbiome 7:5. https://doi.org/10.1186/s40168-019-0619-4

Smriga S, Sanding SA, Azam F (2010) Abundance, diversity, and activity of microbial assemblages associated with coral reef fish guts and faces. FEMS Microbiol Ecol 73:31–42. https://doi.org/10.1111/j.1574-6941.2010.00879.x

Song C, Zhenming C, Li J, Wang X (2010) Galactosidase production by the psychrotolerant yeast Guehomyces pullulans 17–1 isolated from sea sediment in Antarctica and lactose hydrolysis. Bioprocess Biosyst Eng 33(9):1025–1031

Streit WR, Schmitz RA (2004) Metagenomics—the key to the unculturable microbes. Curr Opin Microbiol 7:492–498

Subharthi P (2015) Phage therapy an alternate disease control in aquaculture: a review on recent advancements. IOSR J Agric Vet Sci 8(9):68–81

Sugita H, Ito Y (2006) Identification of intestinal bacteria from Japanese flounder (*Paralichthys olivaceus*) and their ability to digest chitin. Lett Appl Microbiol 43:336–342. https://doi.org/10.1111/j.1472-765X.2006.01943.x

Suttle CA (2007) Marine viruses—major players in the global ecosystem. Nat Rev Microbiol 5:801–812

Szklarczyk D, Franceschini A, Kuhn M, Simonovic M, Roth A, Minguez P, Doerks T, Stark M, Muller J, Bork P, Jensen LJ, Mering CV (2011) The STRING database in 2011: functional interaction networks of proteins, globally integrated and scored. Nucleic Acids Res 39(1):D561–D568. https://doi.org/10.1093/nar/gkq973

Taoka Y, Maeda H, Jo J-Y, Jeon M-J, Bai SC, Lee W-J, Yuge K, Koshio S (2006) Growth, stress tolerance and non-specific immune response of Japanese flounder *Paralichthys olivaceus* to probiotics in a closed recirculating system. Fish Sci 72:310–321. https://doi.org/10.1111/j.1444-2906.2006.01152.x

Tarnecki AM, Burgos FA, Ray CL, Arias CR (2017) Fish intestinal microbiome: diversity and symbiosis unravelled by metagenomics. J Appl Microbiol 123(1):2–17. https://doi.org/10.1111/jam.13415

Trindade M, Zyl LJ, Navarro FJ, Abd Elrazak A (2015) Targeted metagenomics as a tool to tap into marine natural product diversity for the discovery and production of drug candidates. Front Microbiol 6:890. https://doi.org/10.3389/fmicb.2015.00890

Tsuchiya C, Sakata T, Sugita H (2008) Novel ecological niche of *Cetobacterium somerae*, an anaerobic bacterium in the intestinal tracts of freshwater fish. Lett Appl Microbiol 46:43–48

Turner TR, James EK, Poole PS (2013) The plant microbiome. Genome Biol 14:209. https://doi.org/10.1186/gb-2013-14-6-209

Tyagi A, Singh B, Billekallu TNK, Singh NK (2019) Shotgun metagenomics offers novel insights into taxonomic compositions, metabolic pathways and antibiotic resistance genes in fish gut microbiome. Arch Microbiol 201(3):295–303. https://doi.org/10.1007/s00203-018-1615-y

Udayangani RMC, Dananjaya SHS, Nikapitiya C, Heo GJ, Lee J, De Zoysa M (2017) Metagenomics analysis of gut microbiota and immune modulation in zebra fish (*Danio rerio*) fed chitosan silver nanocomposites. Fish Shellfish Immunol 66:173–184. https://doi.org/10.1016/j.fsi.2017.05.018

Van Kessel MAHJ, Dutilh BE, Neveling K, Kwint MP, Veltman JA, Flik G, Jetten MSM, Klaren PHM, Op den Camp HJM (2011) Pyro sequencing of 16S rRNA gene amplicons to study the microbiome in the gastrointestinal tract of carp (*Cyprinus carpio* L.). AMB Express 1:41

Vazquez J, Gonzalez M, Murado M (2005) Effects of lactic acid bacteria cultures on pathogenic microbiota from fish. Aquaculture 245(1–4):149–161. https://doi.org/10.1016/j.aquaculture.2004.12.008

Wang B, Li F, Dong B, Zhang X, Zhang C, Xiang J (2006) Discovery of the genes in response to white spot syndrome virus (WSSV) infection in Fenneropenaeus chinensis through cDNA microarray. Mar Biotechnol 8(5):491–500. https://doi.org/10.1007/s10126-005-6136-4

Wang H, Jin L, Zhang H (2011) Comparison of the diversity of the bacterial communities in the intestinal tract of adult Bactrocera dorsalis from three different populations. J Appl Microbiol 110:1390–1401

Ward N, Steven B, Penn K, Methé B, Detrich W (2009) Characterization of the intestinal microbiota of two Antarctic notothenioid fish species. Extremophiles 13(4):679–685. https://doi.org/10.1007/s00792-009-0252-4

Williams J (2017) The role of prebiotics and probiotics in aquaculture. https://www.global.engage.com

Wooley JC, Godzik A, Friedberg I (2010) A primer on metagenomics. PLoS One 6:e100066

Xenarios I, Salwınski Ł, Duan XJ, Higney P, Kim DE (2002) DIP, the data base of interacting proteins: a research tool for S. studying cellular networks of protein interactions. Nucleic Acids Res 30(1):303–305

Xing MX, Hou ZH, Yuan JB, Liu Y, Qu Y, Liu B (2013) Taxonomic and functional metagenomic profiling of gastrointestinal tract microbiome of the farmed adult turbot (*Scophthalmus maximus*). FEMS Microbiol Ecol 86:432–443. https://doi.org/10.1111/1574-6941.12174

Xiong J, Wang K, Wu J, Qiu L, Yang K, Qian Y, Zhang D (2015) Changes in intestinal bacterial communities are closely associated with shrimp disease severity. Appl Microbiol Biotechnol 99: 6911–6919. https://doi.org/10.1007/s00253-015-6632-z

Ye L, Amberg J, Chapman D, Gaikowski M, Liu WT (2014) Fish gut microbiome analysis differentiates physiology and behavior of invasive Asian carp and indigenous American fish. ISME J 8:541–551

Yoshida K, Hashimoto M, Hori R, Adachi T, Okuyama H, Orikasa Y, Nagamine T, Shimizu S, Ueno A, Morita N (2016) Bacterial long-chain polyunsaturated fatty acids: their biosynthetic genes, functions, and practical use. Mar Drugs 14:E94. https://doi.org/10.3390/md14050094

Yukgehnaish K, Praveen K, Parimannan S, Marimuthu K, Aziz A, Bilal AP, Jesu A (2020) Gut microbiota metagenomics in aquaculture: factors influencing gut microbiome and its physiological role in fish. Rev Aquac 1:1–25. https://doi.org/10.1111/raq.12416

Zhang S, Sun X (2022) Core gut microbiota of shrimp function as a regulator to maintain immune homeostasis in response to WSSV infection. Microbiol Spectr 10(2):e0246521. https://doi.org/10.1128/spectrum.02465-21

Zhang M, Shan C, Tan F, Limbu SM, Chen L, Du ZY (2021) Gnotobiotic models: powerful tools for deeply understanding intestinal microbiota-host interactions in aquaculture. Aquaculture 517:734800. https://doi.org/10.1016/j.aquaculture.2019.734800

Zhao Z, Yin Z, Weng S (2007) Profiling of differentially expressed genes in hepatopancreas of white spot syndrome virus-resistant shrimp (Litopenaeus vannamei) by suppression subtractive hybridization. Fish Shellfish Immunol 22(5):520–534

Zheng S, Li S, Li F, Xiang J (2014) Bioinformatics prediction of WSSV-host protein-protein interaction. Biomed Res Int 2014:416543. https://doi.org/10.1155/2014/416543

Part II

Microbiome of Shellfish

Shellfish Microbiome 7

Aifa Fathima, Yaser Arafath, Saqib Hassan, Aaron Lipton,
Arun Ninawe, George Seghal Kiran, and Joseph Selvin

Abstract

The presence of a microbiome has an immense effect on the host system. It varies from terrestrial to marine environments. Also based on the feeding habitat and place of colonization one will be richer than the other one. In this chapter, we will discuss about the shellfish types and the variation in the microbiome community of different species. The mitten crab microbiome comprises of *Proteobacteria, Bacteroidetes, Chloroflexi, Tenericutes,* and other phyla. The comparative study reveals the difference between the Horseshoe crab microbiome from shark reef and Florida sediment, whereas the shark reef shows a richer microbiome

A. Fathima · Y. Arafath · J. Selvin (✉)
Department of Microbiology, School of Life Sciences Pondicherry University, Puducherry, India

S. Hassan
Department of Microbiology, School of Life Sciences Pondicherry University, Puducherry, India

Future Leaders Mentoring Fellow, American Society for Microbiology, Washington, USA

Department of Biotechnology, School of Bio and Chemical Engineering, Sathyabama Institute of Science and Technology, Chennai, India

A. Lipton
Centre for Marine Science and Technology, Kanyakumari, India

A. Ninawe
Department of Biotechnology, Ministry of Science and Technology, Government of India, New Delhi, India

G. S. Kiran
Centre for Marine Science and Technology, Kanyakumari, India

Department of Food Science and Technology, School of Life Sciences Pondicherry University, Puducherry, India

A. Diwan et al. (eds.), *Microbiome of Finfish and Shellfish*,
https://doi.org/10.1007/978-981-99-0852-3_7

compared to Florida. In this chapter will see about the types of shellfish and their microbiome composition.

Keywords

Shellfish · Microbiome · Crustaceans · Mollusks · Echinoderms

7.1 Introduction

The existing knowledge of fish and shellfish microbiome composition is gained from multiple investigations carried out by various researchers as well as a large number of scientific publications. According to these reports, the microbiome focus appears to be on farmed finfish and shellfish. In the past, the skin, gills, and digestive tract were the primary areas of focus for studies on microbes. Several researchers have made an effort to look into the fish microbiome (Harrison 1929). This is because fish and shellfish's primary entry points for microbes are these body parts (Diwan et al. 2022). The intestinal tract microbiota, among other parts of the body, has an impact on fish health, according to research by Li et al. (2019). It is emphasized that mucosal adhesion, particularly in the intestinal tract, is crucial for the development of early bacterial infection in fish (Diwan et al. 2022).

But current research trends are concentrated on the digestive tract microbes, which are connected to roles in a variety of physiological processes such as disease defense, immune response modulation, nutrient absorption, metabolic regulation, and vitamin synthesis (Diwan et al. 2022). A microbiome is crucial for maintaining an organism's optimal health and well-being for a number of additional reasons. Without the gut microbiome, it is impossible to digest and break down the complex molecules in food, which are necessary for energy production throughout the day and the absorption of essential nutrients. Dietary requirements, food cravings, and feelings of fullness are all influenced by the metabolic activity of the gut microbiome, which also determines the nutritional pattern (Diwan et al. 2022). Previously, microbiome species were identified using phenotypic and biochemical traits, and bacterial levels were counted using different types of cultural media. Due to the accessibility of cutting-edge genetic engineering tools like next-generation sequencing (NGS) technologies, researchers have recently been able to analyze the microbiome community in greater detail (Diwan et al. 2022). As a result, many studies on the microbial species' genome structural composition in finfish and shellfish have been conducted. It will be simpler to manipulate microbes with these genetic details in order to increase their necessary functional characteristics, such as feed effectiveness and disease resistance. To cultivate a healthy microbiome environment in systems for aquaculture farming as well as cultivable organisms, it is essential to understand host–microbe relationships and to recognize the functions of microbial communities (Tarnecki et al. 2017). While reviewing the fish gut microbiome, Talwar et al. (2018) conducted an extensive analysis of the diversity and functional characteristics of the various microorganisms. Furthermore, they emphasized that studying the functional characteristics of the microbiome has the potential to improve aquaculture production on a long-term basis. Recent studies on these topics have also been discussed in various papers (Diwan et al. 2022). While conducting an extensive analysis of the interaction between the host and microbial system in fish, Legrand et al. (Legrand et al. 2020) reported the functional involvement of microbiota with the host and their interplay mechanisms. Fish microbiome diversity is influenced by a variety of environmental, dietary, and genetic factors, among others. The significance of more cutting-edge techniques such as metagenomics, meta-transcriptomics, meta-proteomics, and metabolomics was emphasized in order to comprehend the functional interactions and associations between bacteria and other significant microbes and hosts. These methods could offer important details about the functional interactions between microorganisms and their hosts.

Microbiome studies on shellfish were previously conducted using conventional methods. Due to the culture-based techniques used in the shrimp farming system, microbiome diversity was present in the majority of studies. There have been several previous attempts to examine the microbiome composition of penaeid shrimp species using these conventional methods. The primary target organs for microbiota analysis were hepatopancreatic tissue and the digestive tract, and a comparison between shrimp collected in the wild and those raised in grow-out ponds were made (Diwan et al. 2022). Due to the continuous nature of the digestive tract and its strong connections to hepatopancreatic tissues, little is known about the microbiota of shrimp organs other than the intestine (Cheung et al. 2015). To date, only one study has looked at the microbiome of hepatopancreatic tissue in the shrimp *Neocaridina denticulate* (Cheung et al. 2015). Shrimp aquaculture has experienced significant losses due to the occurrence of diseases brought on by acute hepatopancreatic necrosis in the shrimp (Joshi et al. 2014). Work is being done to comprehend the pathogens responsible for the disease and its molecular basis in order to lessen these losses (Gomez-Jimenez et al. 2014). Several studies have been conducted to better understand how gut microbiota affects shellfish health. In this chapter, we summarized the various types of shellfish and their associated gut microbiota.

7.2 Shellfish

The term “shellfish” refers to aquatic invertebrates that have exoskeletons and are consumed, including mollusks, crustaceans, and echinoderms. Some shellfish can be found in freshwater, despite the fact that most are harvested in saltwater. 75–80% of a shellfish’s weight is water, followed by 15–25% protein, 2.5–5% fat, and 1.2–2% minerals. With white flesh and connective tissue that contains more collagen than fish, crustacean meat like lobster, shrimp, and crab is less tender. Clams and oysters have an adductor muscle that enables them to open and close their shells, whereas mollusks like abalone have a muscular foot that allows them to move around. Mollusks that have been turned inside out and have tough, collagen-rich muscles include squid and octopus. Crustaceans, such as shrimp, crab, and lobster have exoskeletons that protect their bodies with hard outer shells. As they get bigger, they shed their shells through molting. Their segmented bodies serve as structural support and protection for their muscles and organs. Mollusks have hard shells made of a chalky material called calcium carbonate that protect them from predators.

7.3 Types of Shellfish

Crustaceans and mollusks are the two groups into which shellfish are divided. The hard, segmented bodies of crustaceans, which are arthropods and closely related to insects, set them apart from other animals. Mollusks are divided into three groups: cephalopods, which lack shells like octopus, bivalves, which have hinged shells like clams, and gastropods, which have a single shell resembling a snail. The spiny sea urchin belongs to the small group of shellfish known as echinoderms. It has a hard shell and long, sharp spines that serve as a defense against predators (Fig. 7.1).

7.3.1 Crustaceans

A few examples of crustaceans are lobster, shrimp, and crab as well as freshwater species like crawfish. Shrimp and lobster meat is most plentiful in the tail, with the exception of the American lobster, which has large claws and a fair amount of meat. Because a crab's body is shorter and it does not move around much, most of the meat in its tentacles and claws is edible. Soft shell crabs are molting crabs that are cooked and eaten whole. Crustaceans will deteriorate quickly if not kept alive because the mid-gut glands, liver, and intestines will break down and turn the muscles into mush. Because of this rapid deterioration, most crabs are pre-cooked, whereas lobster and

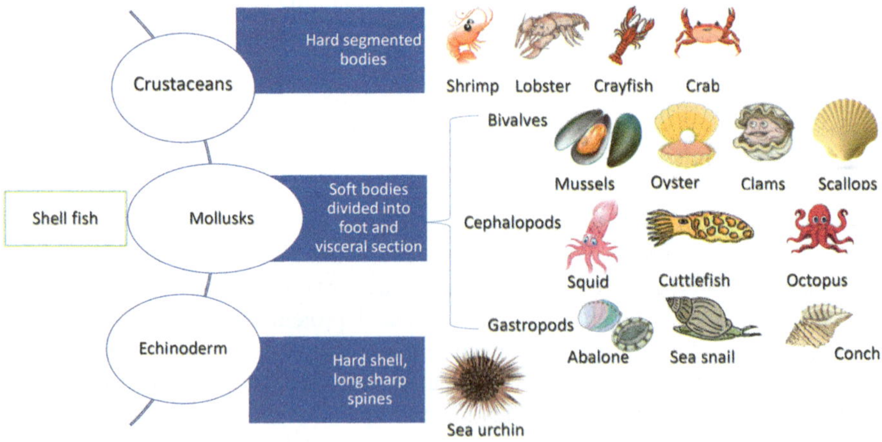

Fig. 7.1 A schematic representation of classification of shellfish

shrimp tails are commonly separated and frozen in their raw state. Crustaceans are a dark green-to-blue color that blends in with their environment, but when cooked, they turn a bright orange or red color due to carotenoid pigments caused by their diet's plankton.

7.3.1.1 Shrimp

Shrimp are aquatic crustaceans that can reach lengths of up to 10 in/25 cm. There are about 300 varieties of edible freshwater and saltwater shrimp. In the UK, the word "prawn" is frequently used to describe shrimp, but in the USA, it can also be used to describe a large shrimp.

7.3.1.2 Lobsters

Crustaceans that scurry across the ocean floor include lobsters. The majority of marine species used for commercial lobster fishing are taken from tropical to cold temperate waters. Due to their typically lengthy and complex larval stages—some spiny lobster species take up to a year to reach the puerulus or stage that settles onto the sea floor—marine lobster species are less frequently raised in aquaculture than many other crustacean groups (Penn et al. 2018). They are divided into several species in order to serve commercial needs. The American lobster, which can be found along the eastern seaboard of the USA and Canada, is known for its distinctive claws. For a lobster to reach an edible size of 1–2 Lb/450–900 g, it takes 6–7 years. Spiny or rock lobsters live in warmer waters like those in the Caribbean, Mediterranean, Australia, Asia, and South Africa. They lack the American Lobster's large claws but have long, spiny antennae and large tails instead. In Australia, New Zealand, and South Africa, they are referred to as crayfish. As far north as Iceland and northern Norway and as far south as Portugal and portions of the Mediterranean Sea, langoustine, also known as Dublin Bay Prawns, Norway Lobster, or Scampi, can be found in the northern Atlantic Ocean and the North Sea. They can grow up to 7–8 in tall/18–20 cm.

Cold water lobsters from the north Atlantic and off the coasts of New Zealand, Australia, and South Africa are sold alongside warm water lobsters from California, Florida, the Mediterranean, the South Pacific, the Caribbean, New Guinea, and New Zealand. Cold water tails have firmer, more tender meat and a sweeter flavor than warm water tails, which can be softer in texture. Yellow spots and a yellow band across the tail distinguish Caribbean warm water tails. Because they grow more slowly in colder winters, Maine Lobster Tails of the American Lobster variety have whiter meat and are considered more tender.

7.3.1.3 Crab

Within the order Decapoda, the majority of crabs with significant commercial value are either "Brachyura (true crabs) or Anomura (hermit crabs and king crabs)" (Penn et al. 2018). They typically have four pairs of walking legs, two pairs of claws, and a broad, flattened body. Of all the crustaceans, the most sophisticated, successful, and diverse crustaceans are probably crabs. They live in a variety of habitats, including freshwater, estuaries, shallow tropical seas, and deep ocean trenches. Several tropical species live their entire lives on land, only returning to the water to reproduce (Penn et al. 2018).

Crabs have a variety of reproductive habits, including complex courtship behaviors in which the male defends the female before she molts and mating takes place. Female crabs store spermatozoa after copulation until spawning when fertilization takes place as the eggs are extruded. For extended periods of time—more than a year in some species—the female can hold onto spermatozoa in a viable state. The important swimming crab species are typically resistant to heavy fishing pressure due to their frequently complex but effective reproductive behavior and high levels of fecundity; however, more recently, some stocks of these species have been overfished due to recruitment (Caputi et al. 2014). During the early stages of development, some species extrude, fertilize, and attach multiple broods of eggs to the underside of the female. Each female typically produces between 50,000 and 500,000 eggs per year, and some species can produce over a million eggs. The four or five zoeal stages that most marine species go through before molting into a megalopa, which typically settles out of its planktonic existence onto the sea floor, are referred to as crab larval stages or zoea (Penn et al. 2018).

The three most significant species of commercial crab are all quickly expanding "swimming crabs," which are found in shallow tropical or temperate waters and bays. These crabs belong to a family that uses its flattened, paddle-like hind leg to propel itself through the water on rare occasions when it needs to "swim" a short distance through sand or mud. Between 1 and 3 years of age is when they mature and are harvested. The most common crab species caught worldwide is the gazami crab, which is widely distributed throughout the western Pacific and dwells in shallow inshore waters in safe coastal bays. The reported landings may, however, include a sizeable but unidentified portion from aquaculture operations. It has become common practice, especially off the coast of Japan, to restock waters with young gazami crabs, and this practice is regarded as an effective economic tactic (Penn et al. 2018).

The western and central western Atlantic is home to blue crabs. States in the Gulf of Mexico and the mid-Atlantic make up the vast majority of the states from which landings are made off the coast of the USA. Both hard and peeler/soft crabs are targeted by commercial fishery, and soft-shelled crabs are prized in the USA. They have just frequently molted, so their shells are still soft (Penn et al. 2018).

7.3.2 Mollusks

Mollusks are divided into three groups: cephalopods, bivalves, and gastropods, also referred to as univalves. Gastropods, which include abalone, conch, and snails (escargot), all have just one shell. These mollusks can become very tough when cooked because of their connective tissue. They are frequently thinly sliced, mechanically tenderized, and either quickly cooked or stewed for several hours. In order to help them move through the water, oysters, mussels, clams, and scallops have hinged shells that open and close. While mussels are frequently served lightly steamed for a few minutes, clams and oysters are frequently served raw. The white adductor muscles of scallops, which are tender even when cooked briefly, are prized.

7.3.2.1 Bivalves

Clams

Clams are bivalve mollusks that can be found all over the world in a range of sizes. There are more than 2000 species of clams, and there are about 150 edible clam varieties. Clams burrow in sand or sediment as opposed to mussels and oysters, which cling to rocks and reefs.

Clams can either have a hard shell or a soft shell. While hard shell clams are larger and live in deeper waters, soft shell clams are found in shallower waters near the coast. Soft shell steamer clams have a size range of 1.5 in. (4 cm) and hard shell Geoduck clams have a size range of 8 in. (20 cm). Among the species raised in North America are soft shell steamers, littleneck, cherrystone, manila razor, quahog, and geoduck.

Mussels

Bivalve mollusks known as mussels are widespread throughout the world. In Canada, the New Zealand Green-Lipped Mussel and the Prince Edward Island Blue Mussel are the two most frequently cultivated species. Mussels have a large organ called a foot that is connected to the byssus thread, also called the "beard," which is removed prior to cooking, like the majority of bivalves. The beard is now removed in modern farm-raised mussel processing.

Oyster

Bivalve mollusks called oysters are found all over the world, typically in the beds and reefs of estuaries and shallow waters. Salinity, mineral content, and nutrient content all affect the flavor profile of water. They can procreate on their own and will transform from male to female. The two most prevalent oysters are the Eastern American Oyster, which can be found off the coasts of Japan, Washington State, and as far south as Australia, and the Pacific Oyster, which can be found along the eastern Atlantic coast from Canada to Argentina. The Pacific oyster, which includes the Kumamoto and Olympia, is the most widely cultured in the world. On the Atlantic coast, North American varieties include Malpeque, Cotuit, and Blue Point. The Belon oyster is a European Flat variety from the Brittany region of France.

Scallops

Scallops are marine bivalves that can be found all over the world, but they are primarily harvested off the east coasts of countries such as the USA, Canada, Japan, and China. Although the roe is occasionally sold whole, scallops are typically processed for their abductor meat. The two most popular market varieties are the larger variety of sea scallops and the smaller variety of bay scallops. Sea scallops are sorted and sold by weight count-per-pounder pound system. Sea scallops can cost anywhere from $10 to $40 per pound/450 g, with the most common sizes found in markets being 10–20, 20–30, or 30–40.

7.3.2.2 Cephalopods

Squid (calamari), octopuses, and cuttlefish are examples of cephalopods, which also contain an internal shell known as a pen or cuttlebone. Cooking these creatures is challenging due to their extremely tough connective tissue, which is made up of a network of collagen. They work best when mechanically tenderized and cooked quickly or slowly for a long time.

Octopus

There are about 300 different species of octopuses, a type of cephalopod (mollusk without a shell) that can be found in the majority of the world's oceans. In the Mediterranean region, especially in Greece, Italy, and Spain, as well as in Asia, particularly in Japan and Korea, and in Australia, octopuses are a favorite food. They share a gland with squids and cuttlefish that releases a black ink when threatened.

Squid

The cephalopod mollusk squid, also known as calamari in Italian, has a soft body supported by an internal shell. While some of the 350 species of squid thrive in the ocean's depths, many of them live in shallow coastal waters or close to the surface of the open sea. They are well-known in the USA, Australia, Japan, and Europe.

Cuttlefish

A cephalopod mollusk with a flatter body than the squid, the cuttlefish is related to the squid and popular in the Mediterranean and Asia. As camouflage, the cuttlefish mimics the color of its surroundings. The cuttlefish's white flesh is firm and similar in texture to squid.

7.3.2.3 Gastropods

Gastropods are mollusks that live in freshwater, saltwater, and terrestrial environments. They are also known as univalves because of their single shell, and they include snails, conch, and abalone.

Abalone

Abalone has soft white bodies with flattened, slightly spiraled shells and muscular foot that clings to rocks in chilly water. Off the coasts of Western North America, Japan, South Africa, Australia, New Zealand, and South Africa. There are more than 100 different varieties, but many are endangered due to overfishing. Although farming of abalone is practiced in California, Hawaii, and Japan, commercial abalone fishing is prohibited in the USA. The tough white meat is frequently pounded with a mallet to make it more tender before being sliced thinly for eating raw, sautéing, or grilling. The flavor and texture are similar to calamari.

Conch

Large conch snails can be found from Florida to Brazil in the warm Atlantic and Caribbean waters. In the Florida Keys, Puerto Rico, and other Caribbean islands, it is a favorite. Conch must be imported from other Caribbean regions because it is an endangered species in the USA. Like abalone, conch meat has a mild, sweet clam-like flavor but is also extremely tough and needs to be tenderized before cooking.

Snails

Various land and water snails, better known as escargot in French, are consumed in various cultures. The most common is the French "petit gris" (for little gray) and the larger "très gro" or Roman Snails. Snails are rarely sold fresh, but rather in canned form that requires little preparation. Along with Greece, Turkey, and Nigeria, Indonesia is a major producer of snails.

7.3.3 Echinoderms

7.3.3.1 Sea Urchin

The most famous edible item in this category is the sea urchin, which gets its name from the Greek for "Spiny Skin." Sea urchins are typically eaten raw or in ceviche. A small marine invertebrate with a tough shell that lives in coastal waters is the sea urchin. It has a spherical shell with spines all over it. The edible parts of this unisexual animal are the ovaries, also referred to as the "coral," and the fluid that surrounds them. Coral has an orange hue, resembling a scallop. Most seas contain sea urchins, but many of the nearly 500 species are poisonous and cannot be eaten. Sea urchins are extremely perishable and are sold live, whole, or as corals only. When handling or opening sea urchins, thick, sturdy gloves are required.

7.4 Shellfish Microbiome

7.4.1 Microbiome

The primary colonization of the microbiome in fish begins at the early larval stage, followed by the surrounding water system and from the first feed (Bell et al. 1971). There are notable inter-species variations seen in some fishes like *Gadus morhua L.*, and halibut, *Hippoglossus hippoglossus*, eggs (Hansen and Olafsen 1989). Some of the bacteria were accepted as species-specific based on the binding capacity over the glycoprotein present on the egg surface (Larsen 2014; Hansen and Olafsen 1999). Interestingly like humans, the GI tract of fish contains only a few bacteria, which tend to get diversified in various feeding habits (Yatsunenko et al. 2012). Ringø and Birkbeck's in 1999 analyzed intestinal microflora of fish larvae and fry to contain Vibrio sps, Pseudomonas sps, Cytophaga sps, and Flavobacterium sps. The microbiome becomes stable within the first 50 days in many fish species (McIntosh et al. 2008).

7.4.2 Crab

The terrestrial biodiversity is occupied by the phylum Arthropoda which dated back to Cambrian and Silurian periods (Lozano-Fernandez et al. 2016). There is a difference between aquatic habitat and terrestrial habitat. At the time of transition, the crabs body has to adapt to physiological as well as morphological changes including movement, excretion, gaseous changes, reproduction, and salinity level (Anger 1995). This is due to the huge difference in the physical nature of air and water (Little 1990). During the transition, major changes can be seen in the gills which serve as the organ of respiration, pH regulation, and nitrogenous waste excretion. At the transitional stage, some of the intertidal and terrestrial brachyurans lost their respiratory function with the evolvement of branchiostegal lungs (Farrelly and Greenaway 1994) or tympana on the legs (Maitland 1986). Crabs that live in terrestrial and semi-terrestrial land feed on vascular plant tissue for food (Linton and Greenaway 2007). Due to the low nitrogen/carbon ratio in land feed to aquatic feed physiological and morphological changes are evident (Linton and Greenaway 2007).

Darwinian evolution has an impact on the genes of both host and commensal microbes. In arthropoda, the host–microbiome helps in the adaptation and evolution of the particular organism. A study conducted by Cuellar-Gempeler and Leibold in 2018 and 2019 analyzed the microbial community associated with fiddler crabs and found that the carapace was colonized by the microbiome from sediment and the gut consists of the microbiome from the burrow sediment as well as the surface sediment. The metatranscriptomic analysis gives insight of the functions of microbiomes (Shakya et al. 2019).

7.4.3 Chinese Mitten Crab (*Eriocheir sinensis*)

The Chinese mitten crab forms the major income according to Chinese fisheries Bureau (~750,945 tons in 2017) (Liu et al. 2018). Pathogens that affecting crab enter through the mouth and ends up in gut eventually misbalancing the gut microbiota. The composition of the bacterial community present in the aquaculture ponds affects the bacterial community that colonize the gut of the farming animals. The microbiome that is present plays an important role in immune functioning, nutrition, resistance against disease. When the crabs are fed with iced fresh fish the following bacterial communities were seen in the microbiome analysis were mentioned below (Tables 7.1 and 7.2).

When fed with iced fresh trash fish feed the sequencing results showed 86 shared Operational Taxonomic units (OTUs) between sediment (6.95%), water (2.77%), and crabs gut (25.60%) (Sun et al. 2020). Whereas the formulated feed showed shared Operational Taxonomic Units of 8.85% (water), 2.08% (sediment), and 18.99% (Intestine). There were 174, 305, and 2632 unique OTUs from crab intestine, water, and sediment respectively. Li et al. showed that Proteobacteria and Bacteroidetes formed the dominant phyla compared to other genera found in *Eriocheir sinensis*.

Table 7.1 The composition of different microbiomes when Horseshoe crabs are fed with iced trash fresh fish

Phylum	Total count of genera	Genera	Reference
Proteobacteria	318	*Rheinheimera spp., Limnohabitans* spp., *Polynucleobacter* spp., *Pragia* spp., *Thiobacillus* spp.	Friel et al. (2020)
Bacteroidetes	107	*Bacteriodes* spp., *Roseimarinus spp., Dysgonomonas spp., Pseudarcicella spp., Chlorobaculum spp., Fluviicola spp.*	Friel et al. (2020)
Chloroflexi	65	*Chloroflexi bacterium RBG-16-58-14, Oscillochloris* spp., *Anaerolinea* spp., *Leptolinea* spp., *Anaerolineae bacterium* UTCFX1.	Friel et al. (2020)
Tenericutes	5	*Candidatus Bacilloplasma, Candidatus* spp., *Hepatoplasma* spp., *Mycoplasmataceae_uncultured, Mollicutes* spp., *Mycoplasmataceae_Unclassified*	Friel et al. (2020)
Other phylum	354	*Deinococcus* spp., *Truepera* spp., *Meiothermus* spp., *Candidatus Omnitrophus, Gemmatimonas spp., Gemmatirosa spp.*	Friel et al. (2020)

Table 7.2 The composition of different microbiomes when Horseshoe crabs are fed with formulated diet

Phylum	Total count of genera	Genera	Reference
Proteobacteria	316	*Polynucleobacter* spp., *Thiobacillus* spp., *Limnohabitans* spp., *Acinetobacter spp., Shewanella spp.*	Friel et al. (2020)
Bacteroidetes	103	*Dysgonomonas* spp., *Pseudarcicella* spp., *Flavobacterium* spp., *Sediminibacterium* spp., *Fluviicola* spp.	Friel et al. (2020)
Actinobacteria	46	*Candidatus Limnoluna, Candidatus Aquiluna*	Friel et al. (2020)
Tenericutes	5	*Candidatus Bacilloplasma, Candidatus Hepatoplasma, Mycoplasmataceae_uncultured, Mollicutes RF39_norank, Mycoplasmataceae_Unclassified*	Friel et al. (2020)
Other phylum	379	*Armatimonas,Opitutus,IMCC26134, ADurb. Bin063–1, Lacunisphaera, Cephaloticoccus, etc.*	Friel et al. (2020)

7.4.4 Horseshoe Crab *(Limulus polyphemus)*

The Horseshoe crab belongs to the order *Xiphosurida*. They have dated back 400 million years ago and are the only species seen in North America (Rudkin and Young 2009). They are widely used in various research fields like pharmacology to

Fig. 7.2 Diagrammatic representation of the difference in microbiome composition from shark reef and Florida sediment horseshoe crab

produce *Limulus* amebocyte lysate. It can be used as a model organism in embryology and vision research (Williams 2019). Both the wild type and tank grown horseshoe have been tested for their microbiome. The bacterial microbiome formed 99%, *Archaea* microbiome was more in shark reef (Florida) (Fig. 7.2).

The microbiome from four parts (carapace, oral, gills, and cloaca) of the Horseshoe crab was analyzed. The dominant species were found to be *Proteobacteria* (*Gammaproteobacteria* and *Alphaproteobacteria*) and *Bacteroidetes* (Bacteroidia) along with other genera such as *Planctomycetes, Patescibacteria, Verrucomicrobia*, and *Actinobacteria* (Friel et al. 2020).

Once they are taken into captivity (14 weeks) the dominant species of microbiomes are *Firmicutes, Tenericutes, Epsilonbacteraeota, and Fusobacteria,* whereas the unclassified phylum contributes to 8–70% of the microbiome population (Friel et al. 2020). The wild horseshoe crab contains more than 50% of unclassified members of *Gammaproteobacteria* spp., *Chitinophagales* spp., *Rhodobacteraceae* spp., *Saprospiraceae* spp., *Flavobacteriaceae* spp., also *Vibrio spp.*, *Tenacibaculum spp., Thiothrix spp.*, and *Rubritalea spp*. Apart from these the dominant species found in wild type after 14 months in captivity were unclassified *Oceanospirillales spp.*, unclassified *Rhodobacteraceae spp*. The cloaca and the outer shell (carapace) showed abundance in *Rubritalea spp., Spongiimonas* spp. in oral cavity, gills, cloaca except carapace (Friel et al. 2020). In Gills unclassified *Gammaproteobacteria,*

Flavobacteriaceae, and *BD1–7 clade* were seen. The tank grown samples showed 75% *Aeromonas, Pseudomonas, Cocleimonas* in the carapace, oral cavity, and gills. The Cloaca microbiome was much richer and more diverse than other body parts showing *Aeromonas* spp., *Shewanella* spp., *Vagococcus* spp., *Enterococcus* spp., *Lactococcus* spp., and *Proteus spp.*

7.4.5 White Shrimp (*Litopenaeus vannamei*)

The white shrimp is a highly farmed animal among shrimps all over the world (Zhang et al. 2014). The gut microbiome of the shrimp plays important role in nutrient absorption and digestion. Throughout the developmental stage, the microbiome varies along with the growth of the shrimp and based on the mode of feed. Among the 144 samples analyzed by Fan et al. 2019, it was observed to that the samples comprised of 22.6% Unclassified bacteria, *Proteobacteria* 40.8%, *Bacteriodetes* 19.96%, *Verrucomicrobia* 8.26%, *Firmicutes* 6.17% and *Actinobacteria* 1.59%. The Operational Taxonomic Unit forms 85.5% of the total sequencing where *Vibrionaceae* forms the second largest bacterial proportion 15.7% (Zhang et al. 2014). The OTU *Haloferula* occupied 7.04%, *Bifidobacterium* 1.07%, and *Bacteroidetes* 4.12%. The presence of OTU in the gut of the shrimp varies with age. The *Proteobacteria* has 32 OTUs forming 34.4% of the sequences among them four OTUs were from *Vibrionaceae*.

7.5 Conclusion

The benefits of the microbiome in the health and wellness of marine organisms were immense and have lots of potential benefits to the host. The composition varies with changes in sediment, feed, and surrounding water source. They can be used as model organisms for humans in the pharmacological field.

References

Anger K (1995) The conquest of freshwater and land by marine crabs: adaptations in life-history patterns and larval bioenergetics. J Exp Mar Biol Ecol 193(1–2):119–145

Bell GR, Hoskins GE, Hodgkiss W (1971) Aspects of the characterization, identification, and ecology of the bacterial flora associated with the surface of stream-incubating Pacific salmon (Oncorhynchus) eggs. J Fish Board Can 28(10):1511–1525

Caputi N, de Lestang S, Hart A, Kangas M, Johnston D, Penn J (2014) Catch predictions in stock assessment and management of invertebrate fisheries using pre-recruit abundance—case studies from Western Australia. Rev Fish Sci Aquac 22(1):36–54

Cheung MK, Yip HY, Nong W, Law PTW, Chu KH, Kwan HS, Hui JHL (2015) Rapid change of microbiota diversity in the gut but not the hepatopancreas during gonadal development of the new shrimp model Neocaridina denticulata. Mar Biotechnol 17(6):811–819

Cuellar-Gempeler C, Leibold MA (2018) Multiple colonist pools shape fiddler crab-associated bacterial communities. ISME J 12(3):825–837

Cuellar-Gempeler C, Leibold MA (2019) Key colonist pools and habitat filters mediate the composition of fiddler crab–associated bacterial communities. Ecology 100(4):e02628

Diwan AD, Harke SN, Panche AN (2022) Aquaculture industry prospective from gut microbiome of fish and shellfish: an overview. J Anim Physiol Anim Nutr 106(2):441–469

Fan J, Chen L, Mai G, Zhang H, Yang J, Deng D, Ma Y (2019) Dynamics of the gut microbiota in developmental stages of Litopenaeus vannamei reveal its association with body weight. Sci Rep 9(1):1–10

Farrelly CA, Greenaway P (1994) Gas exchange through the lungs and gills in air-breathing crabs. J Exp Biol 187:113–130

Friel AD, Neiswenter SA, Seymour CO, Bali LR, McNamara G, Leija F et al (2020) Microbiome shifts associated with the introduction of wild Atlantic horseshoe crabs (Limulus polyphemus) into a touch-tank exhibit. Front Microbiol 11:1398

Gomez-Jimenez S, Noriega-Orozco L, Sotelo-Mundo RR, Cantu-Robles VA, Cobian-Guemes AG, Cota-Verdugo RG et al (2014) High-quality draft genomes of two Vibrio parahaemolyticus strains aid in understanding acute hepatopancreatic necrosis disease of cultured shrimps in Mexico. Genome Announc 2(4):e00800–e00814

Hansen GH, Olafsen JA (1989) Bacterial colonization of cod (Gadus morhua L.) and halibut (Hippoglossus hippoglossus) eggs in marine aquaculture. Appl Environ Microbiol 55(6): 1435–1446

Hansen G, Olafsen J (1999) Bacterial interaction in early life stages of marine cold water fish. Microb Ecol 38(1):1–26

Harrison FC (1929) The discoloration of halibut. Can J Res 1(3):214–239

Joshi J, Srisala J, Truong VH, Chen IT, Nuangsaeng B, Suthienkul O et al (2014) Variation in vibrio parahaemolyticus isolates from a single Thai shrimp farm experiencing an outbreak of acute hepatopancreatic necrosis disease (AHPND). Aquaculture 428:297–302

LarsenA (2014) Studies on the microbiota of fishes and the factors influencing their composition. (Doctoral dissertation)

Legrand TP, Wynne JW, Weyrich LS, Oxley AP (2020) A microbial sea of possibilities: current knowledge and prospects for an improved understanding of the fish microbiome. Rev Aquac 12(2):1101–1134

Li X, Ringø E, Hoseinifar SH, Lauzon HL, Birkbeck H, Yang D (2019) The adherence and colonization of microorganisms in fish gastrointestinal tract. Rev Aquac 11(3):603–618

Linton SM, Greenaway P (2007) A review of feeding and nutrition of herbivorous land crabs: adaptations to low quality plant diets. J Comp Physiol B 177(3):269–286

Little C (1990) The terrestrial invasion: an ecophysiological approach to the origins of land animals. CUP Archive, Cambridge

Liu L, Long X, Deng D, Cheng Y, Wu X (2018) Molecular characterization and tissue distribution of carnitine palmitoyltransferases in Chinese mitten crab Eriocheir sinensis and the effect of dietary fish oil replacement on their expression in the hepatopancreas. PLoS One 13(8): e0201324

Lozano-Fernandez J, Carton R, Tanner AR, Puttick MN, Blaxter M, Vinther J et al (2016) A molecular palaeobiological exploration of arthropod terrestrialization. Philos Trans R Soc B Biol Sci 371(1699):20150133

Maitland DP (1986) Crabs that breathe air with their legs-Scopimera and Dotilla. Nature 319(6053): 493–495

McIntosh D, Ji B, Forward BS, Puvanendran V, Boyce D, Ritchie R (2008) Culture-independent characterization of the bacterial populations associated with cod (Gadus morhua L.) and live feed at an experimental hatchery facility using denaturing gradient gel electrophoresis. Aquaculture 275(1–4):42–50

Penn JW, Caputi N, de Lestang S, Johnston D, Kangas M, Bopp J (2018) Crustacean fisheries. In: Reference module in earth systems and environmental sciences. Elsevier, Amsterdam, pp 1–12

Rudkin DM, Young GA (2009) Horseshoe crabs–an ancient ancestry revealed. In: Biology and conservation of horseshoe crabs. Springer, Boston, MA, pp 25–44

Shakya M, Lo CC, Chain PS (2019) Advances and challenges in metatranscriptomic analysis. Front Genet 10:904

Sun Y, Han W, Liu J, Huang X, Zhou W, Zhang J, Cheng Y (2020) Bacterial community compositions of crab intestine, surrounding water, and sediment in two different feeding modes of Eriocheir sinensis. Aquac Rep 16:100236

Talwar C, Nagar S, Lal R, Negi RK (2018) Fish gut microbiome: current approaches and future perspectives. Indian J Microbiol 58(4):397–414

Tarnecki AM, Burgos FA, Ray CL, Arias CR (2017) Fish intestinal microbiome: diversity and symbiosis unravelled by metagenomics. J Appl Microbiol 123(1):2–17

Williams KL (ed) (2019) Endotoxin detection and control in pharma, limulus, and mammalian systems. Springer, Cham

Yatsunenko T, Rey FE, Manary MJ, Trehan I, Dominguez-Bello MG, Contreras M et al (2012) Human gut microbiome viewed across age and geography. Nature 486(7402):222–227

Zhang M, Sun Y, Chen K, Yu N, Zhou Z, Chen L et al (2014) Characterization of the intestinal microbiota in Pacific white shrimp, Litopenaeus vannamei, fed diets with different lipid sources. Aquaculture 434:449–455

Shellfish Microbiome and Its Significant Role in Immune Response and Health Management

8

A. Panigrahi, T. N. Vinay, P. K. Patil, M. S. Nithin, and A. D. Diwan

Abstract

Gut microbiota is known to influence the health, nutrient absorption, reproduction, and other metabolic activities in both terrestrial and aquatic animals and can be used as biomarker to predict the performance in selective breeding programs. Several studies have generated information on microbial community structures associated with, developmental stages, dietary modifications, osmotic stress, and wild and captive production conditions. The analysis of wild and captive reared *P. indicus* gut microbiota revealed that *Proteobacteria, Fusobacteria, Tenericutes*, and *Bacteroidetes,* were the dominant phyla, indicating host is the major determinant factor in theformation of gut microbiota. Microbes associated with the developmental stages (egg, nauplii, zoea, mysis, PL1, PL6 and PL12) showed that *Proteobacteria* and *Bacteroidetes*, were the two dominant phyla. To study the effects of the probiotics on the gut microbiota, a combination of *Bacillus* spp. were added into the rearing environment of shrimp larvae. The analysis showed that the abundance of *Proteobacteria* was higher in probiotic treated group compared to control, indicating the possibilities for microbiota modulation at early stages. Further, understanding how environmental factors influence the shaping of gut microbiota will aid in using gut microbiota as a tool to prevent and control diseases in aquaculture. Intestine of shrimps adapted to two different salinities showed that the dominant phyla recorded was *Proteobacteria* followed

A. Panigrahi (✉) · T. N. Vinay · P. K. Patil
ICAR-Central Institute of Brackishwater Aquaculture, Chennai, Tamil Nadu, India

M. S. Nithin
College of Fisheries, Karnataka Veterinary Animal and Fisheries Sciences University, Mangalore, Karnataka, India

A. D. Diwan
Institute of Biosciences and Technology, MGM University, Aurangabad, Maharashtra, India

A. Diwan et al. (eds.), *Microbiome of Finfish and Shellfish*,
https://doi.org/10.1007/978-981-99-0852-3_8

by *Bacteroidetes* and *Fusobacteria,* though the abundance of these microbes varied significantly due change in salinity. These observations provide valuable information to device intervention strategies for health and production.

Keywords

Aquaculture · Crustaceans · Microbial diversity · Gut microbiota · Metagenomics

8.1 Introduction

Aquaculture is emerging as a sustainable alternative for the production of food for the growing global population. India ranks second in global total fish production which stands at 16.187 million tons (2021–2022) and is about 7.7% of total global fish production (PIB, GOI). Indian coastal aquaculture is majorly dominated by shrimp farming. Pacific white leg, *Penaeus vannamei,* is the major species cultured and it accounts for more than 90% of Indian and more than 76% globally (FAO 2020) shrimp production. In 2021, India produced 8.43 metric tons of shrimp and among the species cultured, *P. vannamei* farming increased by around 15%, from 6,18, 678 tons to 8,15,745 tons. Furthermore, *P. vannamei* farming in India has started facing production losses due to serious disease problems including running mortality syndrome, white fecal syndrome, and hepatopancreatic microsporidiosis (Otta et al. 2014; Rajendran et al. 2016; Alavandi et al. 2019). Disease risks can be minimized or even completely avoided by using best management practices, judicious use of chemical additives, appropriate effluent discharge, etc. Improving productivity through sustainable approach is the need of the hour and one of the ways is through the microbial association technologies for which it is inevitable to understand the role of microbes associated with the intestine of the species. Increasing growth and productivity through beneficial bacterial interventions including alterations in gut microbiota are becoming a reality through various technologies such as probiotic and biofloc based ones. These innovative practices among other things influences the colonization of heterotrophic bacterial and related microbial aggregation which in turn uses twice the proteins (feed and converted waste), improving digestion (with enzymes and growth promoters), and other advantages.

Now it is increasingly evident in various terrestrial and aquatic animals that the intestinal microbiota plays a huge role in the overall functioning of the organism including the nutrient absorption, host development, physiological activities, health, and several phenotypic traits. A continuous effort on enhancing the knowledge on aquatic animal microbiome is underway. Globally, penaeid shrimps *(P. vannamei. P. monodon, P. japonicus, P. indicus, P. chinensis, P. stylirostris, and P. merguiensis)* are farmed in commercial scale for food, nutrition, and culinary preferences. The intestinal microbiome of these commercially important shrimps, *P. merguiensis, P. chinensis,* and *P. penicillatus*, has been reported from various environments using conventional culture-dependent methods. Recently, next-generation sequencing based microbial culture-independent approaches were applied to study the intestinal microbiome of farmed, wild, diseased fish/shrimp and when they are grown under different environmental conditions. A highly complex microbiota with symbiosis inhabits the intestine of all the organisms; these interactions play a major role in various physiological functions of all the organisms (Kostic et al. 2013). It is evident from several studies that the microbiota influences the growth, nutrient absorption, reproductive status, immune response, and disease resistance (Khanjani et al. 2022), which is reported to positively contribute to the physiological and metabolic status of any organism. Imbalance in the host–microbiota homeostasis can lead to several pathophysiological conditions.

8.2 Microbial Diversity and Microbiome Study

The conventional microbial culture methods are not reliable to study the microbial diversity, as the majority (>99%) of the bacteria are uncultivable. Hence, the culture-independent molecular approaches are developed as reliable methods to characterize the diversity and abundance from various sources. The 16S rRNA gene amplicon high-throughput sequencing, a next-generation sequencing (NGS) technology has become popular approach to study the microbiota from various environmental samples (Frémont et al. 2013; Patil et al. 2021). Furthermore, several methods like extracellular enzymes production using specific agar plates, antibiofilm activity of culture supernatants (Panigrahi et al. 2019a, b, c) using microtiter-plate assay, cell aggregation, biofilm, and virulence-related genes analysis using real-time PCR and microscopic analysis to characterize gut microbiome.

8.3 Gut Microbiome Composition in Shellfish

From our previous studies, it is evident that the core gut microbiota of *P. indicus* from wild was mainly constituted of *Proteobacteria, Fusobacteria, Bacteroidetes,* Planctomycetes, and *Epsilonbacteraeota* at the phylum level. While, in farmed shrimps the core microbiota observed was of *Proteobacteria, Fusobacteria, Tenericutes, Firmicutes,* and *Cyanobacteria. Proteobacteria* is reported to be the major phyla in all other important penaeid shrimps *(P. vannamei, P. monodon,* and *P. stylirostris)* and other aquatic organisms. Furthermore, it is also demonstrated that *Proteobacteria* is the major phylum even after dietary modifications and any other physiological alterations (Diwan et al. 2022). In *P. indicus* the abundance of microbiome varied significantly between the wild and farmed shrimps, while the prevalence remained similar in both the environments. The microbiota is observed to change according to the host developmental stages, food habits, disease condition, and species (Tepaamorndech et al. 2020). Furthermore, the role of gut microbiome in improving the health and production has been reported for several aquatic organisms. The gut microbiota of crab *(S. paramamosain, Eriocheir sinensis,* and

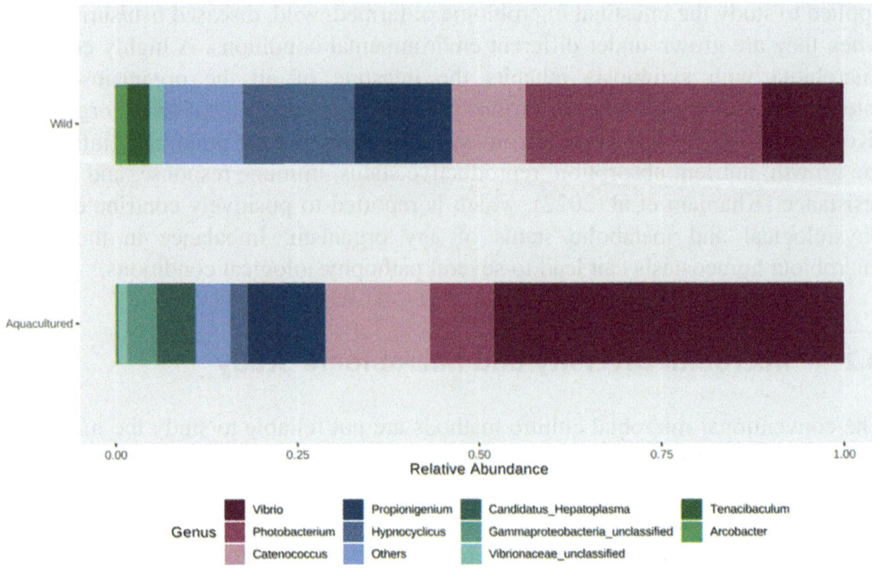

Fig. 8.1 Taxonomy of intestinal microbiota at phylum and genus levels in the wild and farmed *P. indicus* (Patil et al. 2021)

Callinectes sapidus) sourced from different conditions is reported employing high-throughput sequencing approach (Fig. 8.1).

8.4 Factors Effecting Microbial Diversity

The gut microbial composition is reported to be species specific, and also it changes according to developmental stages, early feeding, disease conditions, physiological stress, salinity, and geographical location, and it is evident from several reports that the host species plays a major role in shaping the gut microbiota. Hence, the gut microbial composition of each species should be characterized to develop species specific intervention strategies. Intestinal microbiota is an important and versatile organ of the host, involved in numerous physiological functions including growth, development, and health of several aquatic organisms (Saeed et al. 2021). Little emphasis is given on the effect of salinity on intestinal microbiota of aquatic organisms. Recently, the effect of salinity on intestinal microbiota of *P. vannamei*, *P. monodon* and several other finfish species were reported (Zhou et al. 2020) (Fig. 8.2).

In one of the studies, the role of salinity adaptation on intestinal microbiome of juvenile Indian white shrimp, *P. indicus* was studied. Intestine of shrimps adapted to two different salinities were analyzed using 16S rRNA V3-V4 hypervariable region amplicon sequencing (Chaiyapechara et al. 2022). At all salinity levels, the dominant phylum recorded was *Proteobacteria* followed by *Bacteroidota* and *Fusobacteriota*.

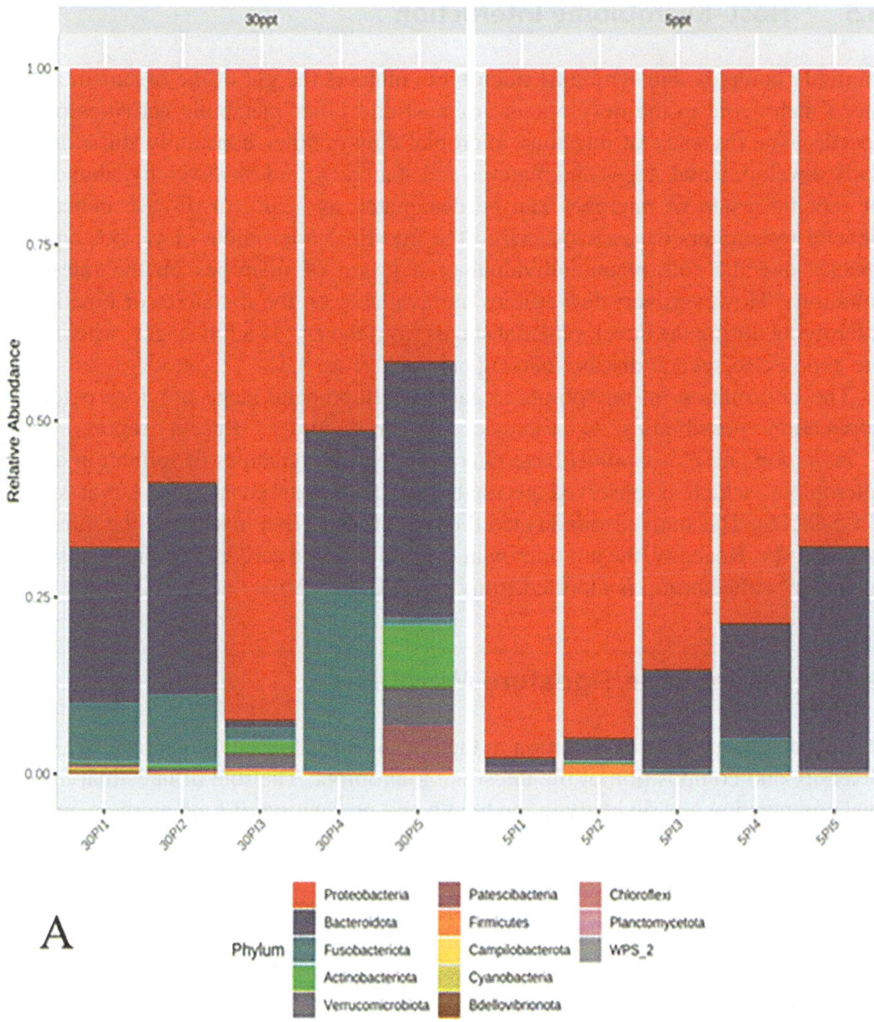

Fig. 8.2 Taxonomy at the phylum levels in the gut microbiota of *P. indicus* reared at 30 and 5 ppt salinity

However, as the salinity decreased abundance of *Proteobacteria* increased while richness of *Bacteroidota* and *Fusobacteriota* decreased significantly. At genus level, salinity reduction led to increase in the abundance of vibrio and *Photobacterium*, while it reduced the abundance of *Spongiimonas, Hypnocyclicus* (Liang et al. 2022).

8.5 Host–Microbiome Interaction

A highly complex and symbiotic microbiota inhabits the gut of the organism and it has a major role in the development of host and other metabolic activities of the species. The diversity of intestinal microbial communities constantly shifts during early developmental stages and matures and stabilizes, and the diversity, abundance of specific group of microbes can be designated as health or disease indicators. Several researchers have characterized the intestinal microbiota of various aquatic species and the role of an individual or a group of microbes on the intestinal immunity. However, very few studies are available on the dynamics or kinetics of microbiota during the developmental stages from the egg to adult stages, where they are exposed to several disease-causing elements (Fig. 8.3).

The information on microbiota at each developmental stage provides potential application possibilities to enhance the productivity. Recent reports from *P. monodon* and *P. vannamei* have characterized the dynamics of formation of gut microbiota, which is observed get stabilized and matured by post-larval stages (PL5-PL15). The study also indicated the role of larval diet in shaping the intestinal microbiome. Recently, the gut microbial dynamics of the early developmental stages of Indian white shrimp is reported from ICAR-CIBA (Vinay et al. 2022).

8.6 Pathobiome/Signature Microbiome

It is observed that the inflammation in the intestine due to EHP might lead to altered diversity and composition of the intestinal microbiota. The role of intestinal microbial composition and diversity in the health and production of food animals, including shrimp, has been extensively reviewed. Recently, variation or dysbiosis of gut microbiota in penaeid shrimp infected with *Vibrio harveyi, V. parahaemolyticus,* and White Spot Syndrome virus has been reported (Wahjuningrum et al. 2022). However, generating the "signature microbiome" of healthy gut or "pathobiome" for a particular disease for a specific host species is challenging and requires extensive studies. It is essential to generate data on gut microbial composition and diversity in different disease conditions to build a signature profile of biota for a particular disease in the given host species. Understanding the signature pathobiome might help design intervention strategies to overcome the diseases. Recently, the variation in gut microbial profile with an increase in the relative abundance of *Chloroflexi* and a decrease in the relative abundance of *Bacteroidetes* following EHP infection in China (Fig. 8.4).

8.7 Probiotics Role for Healthy Gut Microbiota

Application of probiotics in aquaculture is a regular practice to improve health and production. Probiotics are live beneficial microbes, which are known to enhance the health and production of aquatic organisms by various modes of action. Some of the

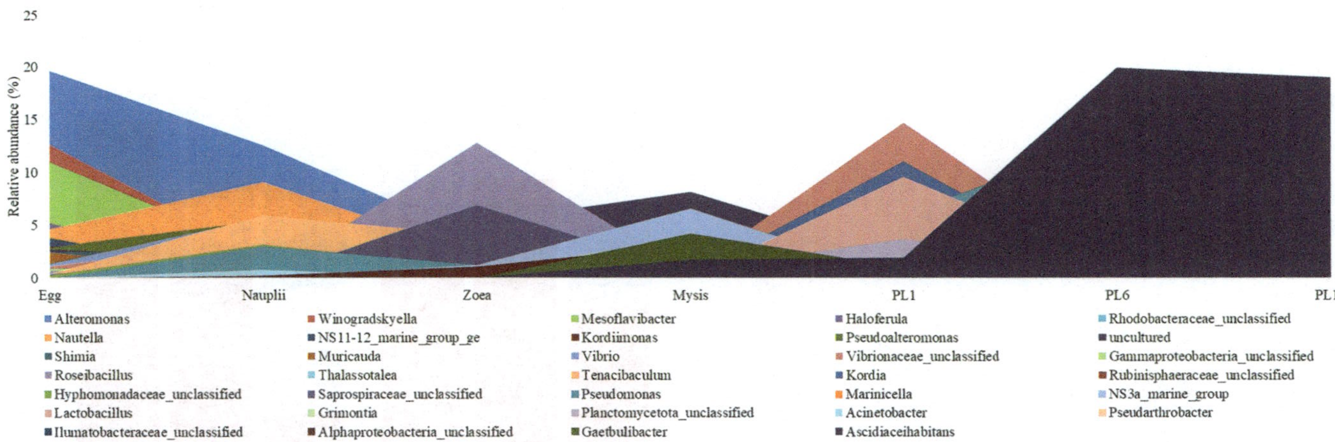

Fig. 8.3 Microbial taxonomy in the gut of various developmental stages of *P. indicus* at genus level (Vinay et al. 2022)

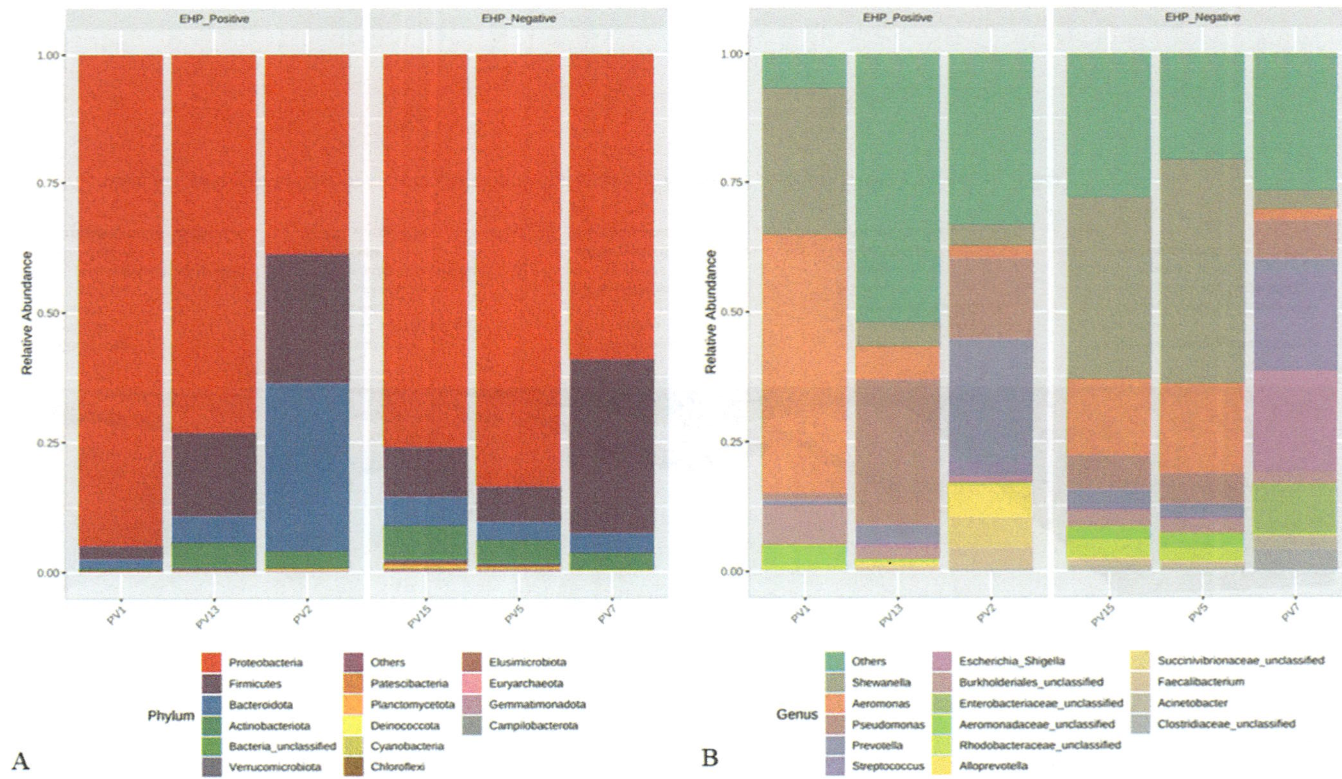

Fig. 8.4 Taxonomy at phylum (**a**) and genus (**b**) levels in the gut microbiota of EHP infected and uninfected shrimp

important modes of action of probiotics are: antimicrobial activity, competitive exclusion and improve digestion capabilities of the host and enhance water quality. Several strains of bacteria capable of producing enzymes are isolated and are promising as probiotic bacteria for commercial shrimp aquaculture. These enzyme-producing bacteria are having inhibitory activity against *Vibrio* to reduce biofilm formation. Many probiotic studies are mainly focused on understanding its effects on growth, immunity, and disease resistance in aquatic animals (Tseng et al. 2009). However, the understanding the mechanism of these beneficial microbes on the dynamics of gut microbiota will provide more information on the mode of action of probiotics (Foysal et al. 2020). Some studies have demonstrated the effect of applications of probiotics on the gut microbiota of shrimps (*P. vannamei*, *P. indicus*) without establishing in the gut.

The influence gut microbiota on factors such as survival, growth, nutrient absorption, immune response, and disease resistance is known from several terrestrial and aquatic animals. The microbiota, which colonizes the intestine in the early stages of the life, will influence the health status of the organisms. Probiotics are the microorganisms which, in isolated form in consortia, will help the host to develop and maintain a balanced microbiome in the system. Furthermore, probiotics are known to create an unfavorable condition for the proliferation of pathogenic microbes, thereby not allowing pathogens to establish and spread the disease (Sumon et al. 2022). The early exposure of the host to various non-pathogenic microbes leads to the development of necessary immune responses. It is also known that the gut microbiota will mature and stabilize by post-larval stages (Kewcharoen and Srisapoome 2019), the exposure of shrimp to beneficial microbes/probiotics during early developmental stages will help in the formation of healthy microbiota.

8.8 Heterotrophic Production System

Biofloc based intensive nursery/grow-out systems helps in increased fish/shrimp survival and enhanced growth performance in grow-out ponds by limiting the culture period. The standardization of innovative biofloc based culture practice improves the healthy production system. Standardization of carbon sources, C: N ratio, stocking density, etc. are essential factors in microbial management, and it was successfully demonstrated in shrimp and fishes such as catfish, tilapia, milkfish, etc. (Panigrahi et al. 2020). Also, the supplementation of probiotic with biofloc is an effective source of generating biofloc with beneficial microbes and proved in the shrimp larval rearing system (Panigrahi et al. 2022).

8.9 Growth Enhancement and Microbial Colonization

Microbial colonization associated with the growth improvement in shrimp could contribute to the understanding of mechanism in a BFT system. In BFT system due to the organic nutrient richness, the bacterial diversity and communities are found to be much higher when compared to the clean water system (Cardona et al. 2016). Vital functions like reducing of pathogen invasion, nutrient digestion and enhanced immune system are influenced by the colonization of beneficial microbes in the intestine (Nicholson et al. 2012). Based on the production of antimicrobial molecules against invasive pathogens and the production of extracellular enzymes (protease, amylase, lipase, cellulase, etc.) beneficiary of microbes was categorized. Enzyme synthesis by the potential isolates and their probiotic effect could possibly be useful for enhanced production of healthy shrimps (Tzuc et al. 2014).

Studies on beneficial bacterial colonization in shrimp *Penaeus vannamei* show significant results in the experiment conducted on different sources of biofloc and clear water. While screening the heterotrophic bacterial population in the intestine, 94 extracellular enzyme (amylase, protease, lipase, cellulase, xylanase, and pectinase) producing bacteria were isolated, among them, 36 strain produces amylase, 20 isolates produce protease, 27 produces lipase, 6 strains produce cellulase and 8 strains produces xylanase. The selected isolates were further identified as multiple species of *Cobetia*, *Exiguobacterium*, *Bacillus*, *Marinilactibacillus*, *Staphyllococcus*, and *Novosphingobium* genera were seen in biofloc system and *Bacillus*, *Exiguobacterium* genera were observed in control groups. In both the groups genus *Exiguobacterium* shows abundance. Results show that diversification of beneficial bacteria in gut is higher in biofloc system than control groups (Fig. 8.5).

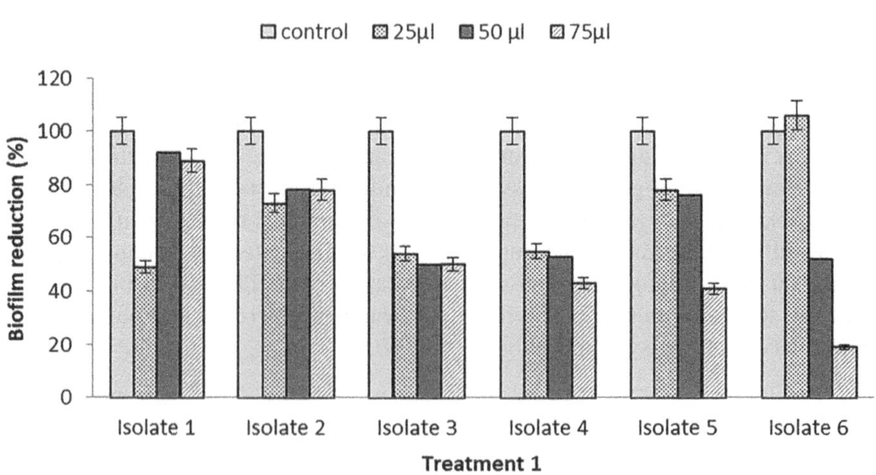

Fig. 8.5 Antibiofilm activity of culture supernatant of bacterial strains isolated from the intestine of biofloc reared shrimp (Panigrahi et al. 2019a, b, c)

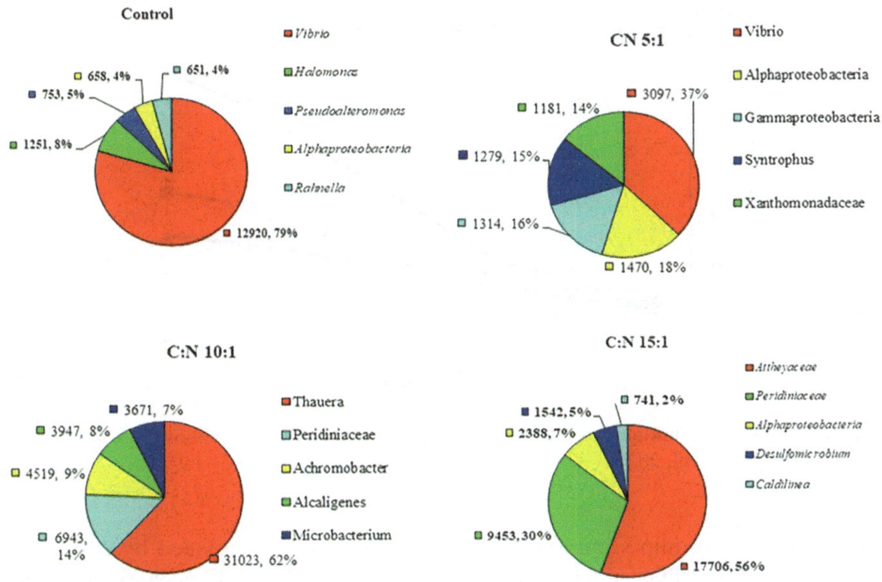

Fig. 8.6 Relative abundance of the most frequently identified bacterial phyla in water from different rearing conditions: clear water and biofloc (BFT) with different CN ratio (Panigrahi et al. 2018)

Following the experimental results, beneficial bacterial colonies in shrimp *Penaeus vannamei* were characterized as their colonization pattern is associated with different nutrient sources in biofloc systems.

In comparison to the bacterial community in water with different C:N ratio (BFT), the control groups showed the dominance of bacterial flora, potential pathogenic Vibrio colonies showed abundance (79%) in Control which reduced to 37% in C:N5 group. However, genus *Thaurea* (62%) was most abundant in C:N10, similarly *Attheyaceae* (56%) and *Peridiniaceae* (30%) shows abundance in C:N15 groups, and in C:N20 the groups *Psychrobacter* (26%), Proteobacteria (25%) and *Peridiniaceae* (20%) diversity of bacterial flora were found to be higher. From the results its clearly seen that *Vibrio* dominance was replaced by heterotrophic bacteria on increasing the C:N ratios (Fig. 8.6).

8.10 Microbial Community and Health Management

The complex microbial consortia developed in biofloc system leads to competitive strategies, like competitive exclusion, damage the other microbes and stop the proliferation of pathogens in the system. Additionally, the microbial by-products in biofloc like their cell wall components, metabolites act as immunostimulants and enhance the nonspecific immunity of fish/ shellfish. In shrimp, the immune system lacks specific immunity and relies entirely on nonspecific immunity to fight against

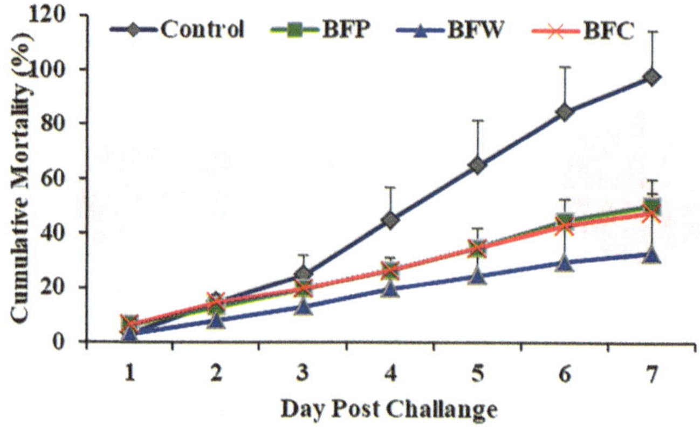

Fig. 8.7 Cumulative mortality of the shrimps grown in biofloc and conventional nursery rearing systems (Panigrahi et al. 2020)

pathogens. The shrimp's immune system is known to get activated by coming in contact with microbial cell wall components such as lipopolysaccharides (LPS), peptidoglycan (PG), and β-1, 3-glucans (BG). The biofloc has an abundance of microbes and acts as an immune stimulator and keeps the immune system active. ICAR-CIBA has developed "CIBAFLOC" as a consortium for biofloc generation. A mixture of extracellular enzyme-producing antibiofilm bacteria having natural probiotics effect. BFT system uses fermentation of carbon sources and helps in the development of biofloc efficiently. It inhibits the growth of pathogenic bacteria in biofloc system and improves growth and immunity of cultured species. Similarly, another immunostimulant developed is CIBASTIM as one consortium of microbes as a feed additive for promoting immunity and growth in shrimp. Shrimps reared in biofloc system able to tolerate pathogen load. In vitro pathogen tolerance test of the shrimps reared in biofloc treatments showed the protective effect when challenged against the pathogenic strain *Vibrio parahaemolyticus* (Fig. 8.7).

The antagonistic activity between pathogens and other heterotrophic microbes in the biofloc will limit the growth of pathogens. The opportunistic bacterium *vibrio* may grow in higher pace during the initial stages, however, as the biofloc matures, the system will develop healthy and diverse group of microbes, which act against the pathogenic bacteria. Furthermore, the animals reared in biofloc system will develop immunity to fight against various bacterial and viral pathogens. It is observed in several studies that the immune system of animals reseed in biofloc system is usually robust and the growth is improved. Furthermore, the shrimp intestine is exposed to various beneficial microbes in the biofloc system, which will enhance the gut immunity. Hence, this will improve the overall development of the shrimps in biofloc system. Additionally, the complex carbon sources supplemented biofloc reared shrimps also showed an upregulation of several digestive enzyme and metabolism-related genes (Chakrapani et al. 2021) (Fig. 8.8).

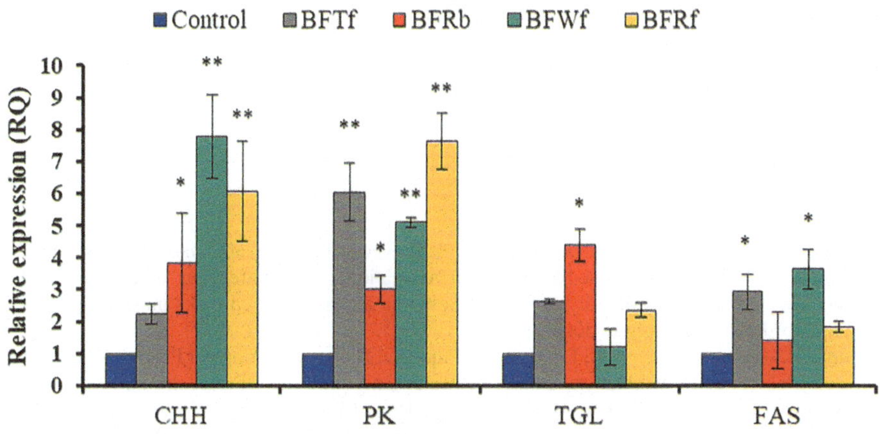

Fig. 8.8 Comparative mRNA expression levels of carbohydrate and Fatty Acid metabolism-related genes

8.11 Conclusion

Understanding gut microbiota of aquatic organisms has lot of application possibilities. Some of the important applications are as follows.

- Development of probiotics for healthy shrimp/ fish production.
- Understanding the pathobiome and healthy signature microbiome of an organism will help in health fecal transplant to reverse the disease conditions in aquatic animals. This will aid in the development of fecal microbiome therapy.
- Understanding the core microbiome of an organism will aid in the development of species-specific intervention strategies.
- Recent studies have shown that certain signature microbiome are associated with specific traits and are also heritable.
- Characterizing these microbiome signatures will help in developing non-invasive methods to study genetic parameters in selective breeding programs.

More and more studies focusing on microbiome of shellfish and their significance started appearing. Some innovative systems allow colonizing shrimp with more of enzyme-producing and biofilm inhibiting bacteria than the conventional system. Colonizing the gut with promising strains like enzyme producers, or with bioremediating bacteria or with active antibiofilm activities will be helpful to improve immunity, health, reduce the organic discharge and achieving higher productivity sustainably.

Acknowledgements We are thankful to the director, CIBA for all his support for conducting this research activity.

Disclosure of Potential Conflicts of Interest N/A.

Funding This work was financially supported by Department of Biotechnology, Ministry of Science and Technology, New Delhi, India.

References

Alavandi SV, Muralidhar M, Dayal JS, Rajan JS, Praveena PE, Bhuvaneswari T, Saraswathy R, Chitra V, Vijayan KK, Otta SK (2019) Investigation on the infectious nature of running mortality syndrome (RMS) of farmed Pacific white leg shrimp, Penaeus vannamei in shrimp farms of India. Aquaculture 500:278–289

Cardona E, Gueguen Y, Magré K, Lorgeoux B, Piquemal D, Pierrat F, Noguier F, Saulnier D (2016) Bacterial community characterization of water and intestine of the shrimp *Litopenaeus stylirostris* in a biofloc system. BMC Microbiol 16(1):157

Chaiyapechara S, Uengwetwanit T, Arayamethakorn S, Bunphimpapha P, Phromson M, Jangsutthivorawat W, Tala S, Karoonuthaisiri N, Rungrassamee W (2022) Understanding the host-microbe-environment interactions: intestinal microbiota and transcriptomes of black tiger shrimp Penaeus monodon at different salinity levels. Aquaculture 546:737371

Chakrapani S, Panigra A, Jayashree S, Sundaram M, Esakkiraj P, Rameshbabu VS, Ambiganandham K (2021) Utilization of complex carbon sources on biofloc system and its influence on the microbial composition, growth, digestive enzyme activity of pacific white shrimp, Penaeus vannamei culture. Turk J Fish Aquat Sci 22:TRJFAS18813

Diwan AD, Harke SN, Panche AN (2022) Aquaculture industry prospective from gut microbiome of fish and shellfish: an overview. J Anim Physiol Anim Nutr 106(2):441–469

FAO (2020) The state of world fisheries and aquaculture. Sustainability in action, Rome

Foysal MJ, Fotedar R, Siddik MA, Tay A (2020) Lactobacillus acidophilus and L. plantarum improve health status, modulate gut microbiota and innate immune response of marron (Cherax cainii). Sci Rep 10(1):1–3

Frémont M, Coomans D, Massart S, De Meirleir K (2013) High-throughput 16S rRNA gene sequencing reveals alterations of intestinal microbiota in myalgic encephalomyelitis/chronic fatigue syndrome patients. Anaerobe 22:50–56

Kewcharoen W, Srisapoome P (2019) Probiotic effects of bacillus spp. from Pacific white shrimp (Litopenaeus vannamei) on water quality and shrimp growth, immune responses, and resistance to Vibrio parahaemolyticus (AHPND strains). Fish Shellfish Immunol 94:175–189

Khanjani MH, Mohammadi A, Emerenciano MG (2022) Microorganisms in biofloc aquaculture system. Aquac Rep 26:101300

Kostic AD, Howitt MR, Garrett WS (2013) Exploring host–microbiota interactions in animal models and humans. Genes Dev 27(7):701–718

Liang X, Luo X, Lin H, Han F, Qin JG, Chen L, Xu C, Li E (2022) Growth, health, and gut microbiota of female Pacific white shrimp, Litopenaeus vannamei broodstock fed different phospholipid sources. Antioxidants 11(6):1143

Nicholson JK, Holmes E, Kinross J, Burcelin R, Gibson G, Jia W, Pettersson S (2012) Host-gut microbiota metabolic interactions. Science 336(6086):1262–1267

Otta SK, Bhuvaneswari T, Praveena E, Joseph SR (2014) Protocols for detection of shrimp EMS/AHPND. Training manual (CIBA), p 115

Panigrahi A, Saranya C, Sundaram M, Kannan SV, Das RR, Kumar RS, Rajesh P, Otta SK (2018) Carbon: nitrogen (C: N) ratio level variation influences microbial community of the system and growth as well as immunity of shrimp (*Litopenaeus vannamei*) in biofloc based culture system. Fish Shellfish Immunol 81:329–337

Panigrahi A, Esakkiraj P, Jayashree S, Saranya C, Das RR, Sundaram M (2019a) Colonization of enzymatic bacterial flora in biofloc grown shrimp *Penaeus vannamei* and evaluation of their beneficial effect. Aquac Int 27(6):1835–1846

Panigrahi A, Sundaram M, Saranya C, Kumar RS, Dayal JS, Saraswathy R, Otta SK, Anand PS, Rekha PN, Gopal C (2019b) Influence of differential protein levels of feed on production performance and immune response of pacific white leg shrimp in a biofloc–based system. Aquaculture 503:118–127

Panigrahi A, Sundaram M, Saranya C, Swain S, Das RR, Dayal JS (2019c) Carbohydrate sources deferentially influence growth performances, microbial dynamics and immunomodulation in Pacific white shrimp (*Litopenaeus vannamei*) under biofloc system. Fish Shellfish Immunol 86: 1207–1216

Panigrahi A, Saranya C, Ambiganandam K, Sundaram M, Sivakumar MR (2020) Evaluation of biofloc generation protocols to adopt high density nursery rearing of Penaeus vannamei for better growth performances, protective responses and immuno modulation in biofloc based technology. Aquaculture 522:735095

Panigrahi A, Esakkiraj P, Saranya C, Das RR, Sundaram M, Sudheer NS, Biju IF, Jayanthi M (2022) A biofloc-based aquaculture system bio-augmented with probiotic Bacteria Bacillus tequilensis AP BFT3 improves culture environment, production performances, and proteomic changes in Penaeus vannamei. Probiotics Antimicro Prot 14:277–287

Patil PK, Vinay TN, Ghate SD, Baskaran V, Avunje S (2021) 16 S rRNA gene diversity and gut microbial composition of the Indian white shrimp (Penaeus indicus). Antonie Van Leeuwenhoek 114(12):2019–2031

Rajendran KV, Shivam S, Praveena PE, Rajan JJS, Kumar TS, Avunje S, Jagadeesan V, Babu SVANVP, Pande A, Krishnan AN, Alavandi SV, Vijayan KK (2016) Emergence of Enterocytozoon hepatopenaei (EHP) in farmed Penaeus (Litopenaeus) vannamei in India. Aquaculture 454:272–280

Saeed M, Arain MA, Ali Fazlani S, Marghazani IB, Umar M, Soomro J, Bhutto ZA, Soomro F, Noreldin AE, Abd El-Hack ME, Elnesr SS (2021) A comprehensive review on the health benefits and nutritional significance of fucoidan polysaccharide derived from brown seaweeds in human, animals and aquatic organisms. Aquac Nutr 27(3):633–654

Sumon TA, Hussain MA, Sumon MA, Jang WJ, Abellan FG, Sharifuzzaman SM, Brown CL, Lee EW, Kim CH, Hasan MT (2022) Functionality and prophylactic role of probiotics in shellfish aquaculture. Aquac Rep 25:101220

Tepaamorndech S, Nookaew I, Higdon SM, Santiyanont P, Phromson M, Chantarasakha K, Mhuantong W, Plengvidhya V, Visessanguan W (2020) Metagenomics in bioflocs and their effects on gut microbiome and immune responses in Pacific white shrimp. Fish Shellfish Immunol 106:733–741

Tseng DY, Ho PL, Huang SY, Cheng SC, Shiu YL, Chiu CS, Liu CH (2009) Enhancement of immunity and disease resistance in the white shrimp, Litopenaeus vannamei, by the probiotic, Bacillus subtilis E20. Fish Shellfish Immunol 26(2):339–344

Tzuc JT, Escalante DR, Herrera RR, Cortés GG, Ortiz MLA (2014) Microbiota from *Litopenaeus vannamei*: digestive tract microbial community of Pacific white shrimp (*Litopenaeus vannamei*). Springerplus 3(1):280

Vinay TN, Patil PK, Aravind R, Anand PS, Baskaran V, Balasubramanian CP (2022) Microbial community composition associated with early developmental stages of the Indian white shrimp, Penaeus indicus. Mol Genet Genomics 297(2):495–505

Wahjuningrum D, Hariati S, Yuhana M, Effendi I, Citarasu T, Utami DA, Tarman K (2022) Low dose of Nodulisporium sp. KT29 metabolite promotes production performance and innate immunity of Pacific white leg shrimp (Litopenaeus vannamei) against co-infection of white spot syndrome virus and Vibrio harveyi. Aquac Int 30(5):2611–2628

Zhou L, Li H, Qin JG, Wang X, Chen L, Xu C, Li E (2020) Dietary prebiotic inulin benefits on growth performance, antioxidant capacity, immune response and intestinal microbiota in Pacific white shrimp (Litopenaeus vannamei) at low salinity. Aquaculture 518:734847

Shrimp Health and Microbiome

9

Sanath H. Kumar, Jerusha Stephen, K. Jeena, Manjusha Lekshmi, K. Sreedharan, and Rajendran Kooloth Valappil

Abstract

Farmed shrimp are exposed to environmental conditions that are distinctly different from their natural habitat. Bacteria that colonize the shrimp surface and the intestine vary depending on the culture conditions as well as the health status of shrimp. From disease perspective, the gut microbiome of shrimp has been of intense focus in the recent times owing to its potential application in disease prediction and management. Gut being the focal point of various diseases in shrimp of known and unknown etiologies, microbiome studies using powerful next-generation sequencing (NGS) technologies can unravel the microbiome dynamics of shrimp gut under different pond conditions and the life stages of shrimp, and help understand their influence on the growth, physiology, immunity, and health status of shrimp. Disturbances in gut microbiome (dysbiosis), which predispose the shrimp, have been noticed in conditions such as early mortality syndrome (EMS)/acute hepatopancreatic necrosis disease (AHPND), slow/retarded growth, white feces syndrome (WFS), etc. As the composition of gut microbiome changes with the infection status of shrimp, identification of specific microbiota associated with a specific disease can serve as a useful diagnostic biomarker to predict the onset of disease. This can eventually help in disease

S. H. Kumar · J. Stephen · M. Lekshmi
Fisheries Resources, Harvest & Post-harvest Management Division, ICAR-Central Institute of Fisheries Education (CIFE), Mumbai, Maharashtra, India

K. Jeena · R. K. Valappil (✉)
Aquatic Environment and Health Management Division, ICAR-Central Institute of Fisheries Education (CIFE), Mumbai, Maharashtra, India
e-mail: kvrajendran@cife.edu.in

K. Sreedharan
ICAR-Central Institute of Fisheries Education (CIFE) Rohtak Centre, Rohtak, Haryana, India

A. Diwan et al. (eds.), *Microbiome of Finfish and Shellfish*,
https://doi.org/10.1007/978-981-99-0852-3_9

management and decision-making. However, a multitude of factors in shrimp aquaculture system influence the gut microbiome of shrimp and, therefore, healthy gut microbiome can vary depending on the culture system and the environmental conditions. Nevertheless, microbiome studies can provide deeper insights into the etiology of various diseases and help to delineate the interactions between various factors, both biotic and abiotic, that determine the health of farmed shrimp.

Keywords

Shrimp · Gut microbiome · Diseases · Biofloc · Metagenomics

9.1 Introduction

Farming of aquatic organisms and trade have been documented by various historians for centuries. The earliest Indian record of aquaculture dates back to 321–300 B.C (Salunke et al. 2020). Over time, farming of aquatic animals changed from artisanal to commercial aquaculture. The trade of aquatic food was facilitated by rising purchasing power, changing dietary preferences, and international trade. While the population grew at a pace of 1.6% between 1961 and the present, the consumption of aquatic food increased at an average yearly rate of 3% (FAO 2020). Aquatic food consumption per capita was 20.2 kg in 2020, and it is expected to increase by another 15% by 2030. On an average, 17% of animal protein consumed worldwide comes from aquatic sources, but in Asian and African nations, it is 50%. For developing nations in Asia and Latin America, including India, the export of shrimp is a crucial source of foreign exchange earnings. The culture of shrimp had a boom with the introduction of new cultivable species and methods. Consumption of crustaceans was previously high in advanced economies due to the greater unit cost of these animals. However, traditional and semi-intensive farming practices were insufficient to meet the need demand and, as a result, shrimp farming shifted to the intensive culture system. And, as a result of expansion and intensification of shrimp culture, shrimp availability increased by five times, from 0.4 kg in 1961 to 2.2 kg in 2019. About 8.2% of the overall aquaculture production is made up of crustaceans by live weight and in that *Penaeus vannamei*, known as the white leg shrimp, is responsible for 53% of all cultivated crustaceans. However, though intensification of shrimp farming increased productivity, the occurrence of new diseases and disease outbreaks has also become frequent and consequential. Despite a positive production trend, India and several other leading shrimp-producing countries of Asia are severely hit by the constant outbreak of diseases involving viruses, bacteria, and diseases of unknown etiology.

The intensification of farming practices causes changes in the aquaculture environment as well as the normal physiology of farmed animals resulting in disease emergence. The biosecurity measures and the farming practices were formulated with the primary goal of improving the microbiological and physiochemical components of the culture environment. There are two broad approaches for maintaining an optimal environment for shrimp growth and pathogen suppression. The first is a water exchange system, such as a flow-through system, and recirculating aquaculture system (RAS). The other approach is to use microbial-based systems such as biofloc and periphyton systems. The introduction and/or development of beneficial heterotrophic bacteria play an important role in the maintenance of water quality by converting inorganic toxicants into organic biomass, which is then consumed by shrimp. Another important role of beneficial bacteria is to suppress pathogens through competition, reduced pathogen colonization, and improve shrimp immunity (Emerenciano et al. 2022).

The microbial community of the gut or the gut microbiome is becoming increasingly recognized as an important driver of cultivation success (Holt et al. 2021). The deviations in the gut microbiome of healthy and diseased shrimps are closely correlated with disease severity. Several bacterial indicators can demonstrate disease initiation and severity of diseases in individuals. Increasing evidence suggests a tight link among the gut microbiota, host age, and health status (Xiong et al. 2017). The variation in microbiome at different life stages, in different species, culture conditions, healthy and diseased stages, during and after treatments, pond ecology, feed supplements, etc. are to be deeply investigated to develop interventions for shrimp health management. The identification of specific microbes as biomarkers in healthy and diseased shrimp can supplement the existing management practices. Better understanding of shrimp gut microbiota can facilitate approaches to the improvement of growth, reduce disease susceptibility in cultured shrimp, aid in disease prediction, and thus can ensure sustainability of the production system.

9.2 Shrimp Culture Environment: A Microbial Perspective

9.2.1 Shrimp Aquaculture and Diseases

Over the decades, shrimp aquaculture has emerged as a major industry globally, contributing significantly to the economy and social development of the farming communities. *Penaeus monodon*, the black tiger shrimp was the foremost-cultivated shrimp species until the introduction of *P. vannamei* with the specific pathogen-free (SPF) status, which prompted the aquaculture industry to adopt *P. vannamei* farming. Currently, ~53% of total global crustacean aquaculture production is contributed by *P. vannamei* (87% of the total shrimp production) with *P. monodon* comprising only ~8% (FAO 2020).

However, the intensification in culture practices and the associated stress, and the transboundary movements in trade have resulted in the emergence and spread of various diseases resulting in huge economic losses. Shrimp farming, concomitant with its expansion, diversification, and intensification is constantly being threatened by emerging and existing diseases (Kooloth Valappil et al. 2021). The global shrimp production losses due to diseases have been estimated to the tune of 4 billion USD annually (Shinn et al. 2018).

Disease expression in the aquaculture system is the result of interaction among host, pathogen, and environment. Furthermore, the interaction of pathogenic agents with the cohabiting microorganisms in a system also contributes significantly to the disease processes. These cohabiting microorganisms are referred to as microbiota. Many vital host functions are provided by the microbiota, as the gut-inhabiting microbes are recognized as key players in several metabolic processes in the host (Holt et al. 2021). The gut microbiome contributes to multiple aspects of physiological health status of the host, including nutritional regulation, immune modulation, etc. (Chen et al. 2022). The microbiome protects its host from pathogen colonization, stress conditions, and other environmental factors and several endogenous and exogenous factors that impact the shrimp microbiome. Microbiota of the shrimp intestine and hepatopancreas participate in disease resistance mechanisms owing to the constant interaction between the shrimp digestive system and its surrounding environment (water and sediment) (Cornejo-Granados et al. 2017). There is a correlation of high plasticity between the development of microbiota of shrimp intestine and changes in the environmental conditions and eventually the status of shrimp health. Furthermore, it has been reported that the composition of shrimp microbiota is impacted by the changes in aquaculture ecosystem across age, environment, diet, and diseases or exposure to new habitats (El-Saadony et al. 2022). Understanding what is "normal" in a wide range of situations such as during or predisposing to microbial infections/diseases is important and can serve as signature/markers contributing to the early prediction of the disease onsets. Better characterization of the microbiota in different compartments along the entire length of the gut, and in different growth and life stages will likely facilitate the improvement and development of shrimp gnotobiotics, antibiotics, pro-, pre- and synbiotics to help in improving growth and reducing disease susceptibility (Butt and Volkoff 2019), and will ultimately maximize the sustainability of the global shrimp industry.

9.2.2 Major Diseases of Shrimp

Even though numerous infectious diseases are known to impact the shrimp aquaculture production systems causing huge economic losses (Stentiford et al. 2012), major disease losses in shrimp culture (60%) are caused by viral pathogens followed by bacterial pathogens (20%); however, comparatively, losses due to fungi and parasites have been small. Globally, impact of diseases of viral etiology have been very significant as these led to huge economic losses in all shrimp farming countries. Other than the OIE-listed shrimp diseases such as white spot syndrome virus (WSSV), infectious myonecrosis virus (IMNV), infectious hypodermal and haematopoietic necrosis virus (IHHNV), and acute hepatopancreatic necrosis diseases (AHPND), the major diseases encountered in shrimp farming include the infection caused by *Enterocytozoon hepatopenaei*, slow/retarded growth syndrome, white feces/gut syndrome, running mortality syndrome, loose shell syndrome, white muscle disease, etc. (Kooloth Valappil et al. 2021; Patil et al. 2021). Furthermore, shrimp aquaculture is currently affected by several syndromic diseases of unknown etiology which needs to be elucidated to determine the pathogen(s) involved with the necessary pre-conditions for the detection and control of diseases.

9.2.2.1 Bacterial Diseases

The major bacterial diseases/pathogens affecting farmed shrimps include Acute Hepatopancreatic Necrosis Disease (AHPND) caused by toxins encoded by plasmid bearing genes of *Vibrio parahaemolyticus*, septic hepatopancreatic necrosis (SHPN) or Vibriosis caused by *V. harveyi*, *V. parahaemolyticus*, *V. alginolyticus*, and other *Vibrios* sp. such as *V. campbelli*, *V. anguillarum*, *V. splendidus*, and *V. vulnificus*, necrotizing hepatopancreatitis (NHP) caused by *Hepatobacter penaei* (an intracellular rickettsia- like bacterium) (Morales-Covarrubias et al. 2018).

Vibriosis, caused by several *Vibrio* spp. viz., *V. parahaemolyticus*, *V. harveyi*, is a major disease condition with different manifestations, exacerbated by compromised management practices and environmental triggers. Different manifestations of vibriosis include Zoea II syndrome, Sea gull syndrome, septic hepatopancreatic necrosis (SHPN), luminous bacterial disease, and Acute Hepatopancreatic Necrosis Disease (AHPND). AHPND is one of the major threats to the shrimp industry in many Asian countries (Tran et al. 2013).

9.2.2.2 Viral Diseases

The most significant economic losses in shrimp farming industry are caused by viral diseases. The major viruses infecting farmed shrimp include white spot syndrome virus (WSSV), yellow head virus (YHV), Taura syndrome virus (TSV), infectious myonecrosis virus (IMNV), and infectious hypodermal and hematopoietic necrosis virus (IHHNV) (Kooloth Valappil et al. 2021). Other viruses known to infect shrimp are Monodon baculovirus (MBV), Decapod iridescent virus (DIV1), and Laem-Singh virus (LSNV).

9.2.2.3 Other Microbial Diseases

The losses due to other microbial diseases were considered negligible until the emergence of the obligate intracellular microsporidian parasite, *Enterocytozoon hepatopenaei* (EHP) causing Hepatopancreatic microsporidiosis (HPM), which impacts the production due to severe growth retardation. EHP infection is a major problem in global shrimp farming (Shen et al. 2022). Other diseases reported in farmed shrimp are due to ectocommensals such as *Zoothamnium* sp., *Vorticella* sp., and *Epistylis* sp.

9.2.2.4 Diseases of Unknown Etiology

Shrimp aquaculture is affected by several syndromic diseases of unknown etiology (Kooloth Valappil et al. 2021). These include a series of emerging conditions such as "slow/retarded/stunted growth syndrome," "white gut/feces syndrome (WFS)," "zoea syndrome," "running mortality syndrome," and other infections of gut system with opportunistic microsporidians (Stentiford et al. 2019). Insights into the pathobiome concepts and the microbiome dysbiosis are unraveling the multifactorial perspective of disease etiologies (Sathish Kumar et al. 2022). The EHP/ATM-WFS,

EHP-WFS and its association with *Propionigenium* and *Vibrio* spp., but no association with ATM (Munkongwongsiri et al. 2022), association of EHP with septic hepatopancreatic necrosis (SHPN) (Tang et al. 2016), heavy infestation of EHP without exhibiting WFS (Tangprasittipap et al. 2013) are all the examples of the contributions of possible interactions of multiple agents including eukaryotes and bacterial species (Huang et al. 2020; Chaijarasphong et al. 2021).

9.3 Shrimp Microbiome and Its Role in Health and Disease

The gut microbiota represents a complex microbial ecosystem which plays significant roles in the development and health of organisms (Bäckhed et al. 2005; Lynch and Pedersen 2016; Adelman et al. 2020). The dominant factors which shape the diversity of gut microbiota in a host include immune system, developmental cycle intestinal pH and dietary composition (Donaldson et al. 2016; Kim and Jazwinski 2018). Similarly, microbiota influences various physiological processes by regulating multiple processes in the host. Furthermore, as the gut microbiota is influenced by complex synergistic interactions involving environment, gut microbiota in different habitats can be divided into core microbes (microorganisms that are prevalent among different hosts regardless of environmental factors) and habitat-specific microbes (Coyte et al. 2015; Kokou et al. 2019; Pacheco-Sandoval et al. 2019). Though significant information on the core microbiota of vertebrates is available, data on those of invertebrates, especially aquatic invertebrates, is very limited.

Microorganisms form indispensable components of eukaryotic life and pathogens represent only a tiny fraction of these microorganisms. "Dysbiosis," which represents the altered composition of microbes, is caused by the loss of microbial diversity, which will have cascading effect on the host immune system. Eventually, this favors the emergence and outbreak of disease conditions (Tamboli et al. 2004; Hooks and O'Malley 2017). Shrimp, being one of the most commercially important farmed aquatic organisms and intensive culture being widely employed, "dysbiosis" has a critical role in in the occurrence of disease outbreaks leading to large scale mortality.

9.3.1 Microbiome of Farmed Shrimp

The bacterial composition of shrimp gut is diverse, and generally resembles the composition of the surrounding environment. Diverse bacterial communities inhabit the natural habitat of shrimp, representing both beneficial and pathogenic to shrimp. Changes in the composition of bacterial communities in shrimp habitat can have profound influence on the gut microbial communities of shrimp. Furthermore, the abundance and the diversity of bacterial communities in aquatic invertebrates depend on the diet, age or life stage, host physiology, season, etc. (Harris 1993). The bacterial density of water and sediment in shrimp farms can be as high as 10^6 to

10^7 per mL or gram, and depending on the culture system, the composition of bacterial communities can vary significantly (Alfiansah et al. 2018). Temperature, salinity, rate of water exchange, feeding strategies, use of probiotics, sediment composition, etc. can remarkably influence the bacterial composition of shrimp farms and the shrimp digestive tracts (Alfiansah et al. 2018). As a consequence, it is practically impossible to define certain bacterial groups as resident and others as transient communities in shrimp gut. Studies, however, have shown that the shrimp gut microbiome can vary in diversity compared to the pond water. According to Zoqratt et al. (2018), although Proteobacteria, Actinobacteria, Bacteroidetes, and Fusobacteria were the dominant phyla in both pond water and shrimp gut, the relative abundance of these phyla significantly varied between them. For example, Actinobacteria and Bacteroidetes were the abundant phyla in water compared to the higher abundance of Proteobacteria in shrimp gut. No spatial variations in the microbiome of water were observed, as the microbial diversity and the abundance in samples collected from different locations in the same pond did not vary significantly (Zoqratt et al. 2018). Some of the dominant bacterial families commonly shared between the pond water and shrimp gut were Alcaligenaceae, Flavobacteriaceae, Microbacteriaceae, Acidimicrobiaceae, and Rhodobacteraceae. However, Vibrionaceae were found to be more dominant in shrimp gut. In contrast, a recent study revealed that sediment contributed more to shrimp gut microbiome compared to water, and also the gut microbiota of adult shrimp is largely derived from the juvenile shrimp (Zhang et al. 2021a). Furthermore, it is rational to assume that the gut microbiome of farmed shrimp would be different from that of wild shrimp; the latter living in its natural environment is expected to have a more stable microbiome composed of bacteria that play critical role in the development of shrimp.

From the larval to the adult stage, the gut microbiome of shrimp exhibits definitive changes that are predicted to influence the health and well-being of shrimp. A study comparing the gut microbiome of larval (14 days) and juvenile (1, 2, and 3 months old) *P. vannamei* identified members of Flavobacteriaceae and Rhodobacteraceae (Alphaproteobacteria) as core gut microbiome, as they were identified through all growth stages of shrimp although their relative abundances varied (Huang et al. 2016). The gut microbiome exhibited a definite change from Betaproteobacteria in larval and early juvenile (1 month) stages to Bacteroidetes in 2-month-old juvenile, and Gammaproteobacteria in 3-month-old juvenile. Members of Vibrionaceae (Gammaproteobacteria) became dominant in latter developmental stages of shrimp, which are also important opportunistic pathogens of shrimp. Gammaproteobacteria is the most abundant group in the intestine of healthy *P. vannamei*, and *P. monodon* in all growth stages suggesting that the changes in the microbiome composition rather than the abundance of a particular group can better represent the health status of shrimp (Rungrassamee et al. 2013). The association between shrimp gut and *Vibrio* spp. is evidenced by their higher abundance in shrimp gut compared to the pond water (Zoqratt et al. 2018). The fine balance between pathogenic and commensal bacteria is altered leading to the dominance of pathogenic groups, and various factors such as physical, chemical, and biological, contribute to this dysbiosis in intensive shrimp farms. For example, the gut microbiome of starved shrimp is more abundant with opportunistic pathogens which correlated with impaired physiological and immune functions in starved shrimp (Dai et al. 2018).

Oxley et al. (2002) compared the gut microbiota of wild and farmed *Penaeus merguiensis* and found similar bacterial composition. *Vibrio* spp. were quantitatively dominant, followed by *Aeromonas*, *Plesiomonas*, *Photobacterium*, *Pseudoalteromonas*, and *Pseudomonas* suggesting that the host factors might play important role in selecting gut bacterial communities. Interestingly, some *Vibrio* species such as *Vibrio gazogenes* were isolated only from farmed shrimp gut. Another study similarly reported the predominance of *Vibrio* spp. in the digestive tract of juvenile *P. monodon* from aquaculture ponds, and also observed that *Shewanella* was more abundant in shrimp gut than in the pond water and the sediment (Shakibazadeh et al. 2009). Differences in gut bacterial communities with respect to the type of aquaculture system have been reported. For example, in contrast to the above studies, *P. vannamei* reared in recirculatory aquaculture system (RAS) was dominated by *Roseobacter* spp. (Li et al. 2007). It has also been recorded that application of prebiotics and probiotics can significantly influence the gut bacterial composition (Farzanfar 2006; Li et al. 2007).

9.3.2 *Vibrio* Species in Health and Diseases of Shrimp

Vibrio spp. are commonly associated with the gut of farmed shrimp and their numbers, and the species diversity may vary with the age and life stage of shrimp. Some *Vibrio* spp., such as the luminous *V. harveyi* is associated with all life stages of shrimp, and is also an important opportunistic pathogen, particularly of shrimp larvae, causing mass mortalities (Karunasagar et al. 1994). Their numbers increase from egg stage to postlarvae and can reach as high as 5 log CFU/g of larval shrimp, and 7 log CFU/g in the gut contents of adult shrimp (Abraham and Palaniappan 2004). Although luminous vibrios have been generally regarded as important opportunistic pathogens of all life stages of shrimp, the role of *V. parahaemolyticus* as a shrimp pathogen has come to light only recently with the association of *V. parahaemolyticus* with the early mortality syndrome (EMS), also known as acute hepatopancreatic necrosis disease (AHPND) which was first detected in China in 2009 and later in several Southeast Asian countries. Unique clones of *V. parahemolyticus* harboring *pirA* and *pirB* toxin-encoding genes on a large plasmid were identified as the causative agents of AHPND. These strains produce toxins that erode the gut lining and severely affect the feed assimilation and the consequent growth of shrimp resulting in retarded growth and slow mortality (Tran et al. 2013; Kondo et al. 2014).

According to a study employing 16S rDNA PCR-DGGE, bacteria belonging to the γ-proteobacteria class were commonly found in the gut of juvenile *P. monodon* represented predominantly by the genera such as *Vibrio*, *Photobacterium*, *Aeromonas*, *Pseudomonas*, *Mycoplasma*, *Shewanella*, and *Actinomycetes*

(Chaiyapechara et al. 2012). Using the same technique, Liu et al. (2010) reported that the gut of kuruma shrimp (*Marsupenaeus japonicus*) was dominated by *Vibrio* spp., and uncultured γ-proteobacteria.

9.3.3 Sequence-Based Microbiome of Shrimp

Most studies have focussed on the microbiome of the entire shrimp, although some studies have reported significant differences in the microbiome of different gut portions. Shrimp gut or the intestine is continuous in nature and is divided into foregut, midgut, and hindgut. Hepatopancreas constitutes the midgut and performs the important functions of food digestion and nutrient absorption and is also responsible for the metabolism of lipids akin to the liver of higher animals. Hepatopancreas is the site of secretion of several hormones and plays a major role in the growth, immunity, and reproductive development of shrimp. At the same time, hepatopancreas is a target for a majority of infectious agents of bacterial, viral, and fungal groups. Considering the physiological importance of hepatopancreatic region, a few studies have investigated its microbiome composition to understand the contribution of bacteria to nutrient assimilation, enzyme secretion, and immunity. Cornejo-Granados et al. (2017) compared the microbiomes of intestine and the hepatopancreas of healthy and AHPND/EMS affected shrimp and found differences in the microbiota of intestines of wild and farmed shrimp. The appearance of AHPND/EMS prompted a shift toward disease-specific microbiome enriched with *Aeromonas taiwanensis*, *Simiduia agarivorans,* and *Photobacterium angustum* (Cornejo-Granados et al. 2017).

Garibay-Valdez et al. (2021) found differences in the microbiome compositions of foregut, midgut, and the hindgut of shrimp, *L. vannamei*. The foregut was abundant with Rhodobacteraceae and Rubritalaceae, midgut with Aeromonadaceae, Beijerinckiaceae, and Propionibacteriaceae, and hindgut with Cellulomonadaceae. Functionally, foregut microbiota was involved in carbohydrate and amino acid metabolism. The midgut microbiome was enriched with lipid, terpenoid, and polyketide metabolism, and the hindgut microbiome was associated largely with co-factor and vitamin metabolism. Wang et al. (2019) compared the intestinal microbiome of healthy and WSSV-infected *L. vannamei* and found no difference in terms of diversity and species richness, although the abundance of certain genera such *as Photobacterium, Propionigenium,* and *Arcobacter* significantly increased while others such as *Candidatus bacilloplasma* and *Flavobacterium* decreased in WSSV-infected shrimps. In *P. japonicus*, infection with WSSV decreased the bacterial diversity in the cardiac and pyloric stomachs but did not affect microbial diversity in the intestine (Niu et al. 2022). Following WSSV infection, the abundance of *Vibrio* increased in cardiac stomach, while the abundance of *Aliivibrio* increased in the pyloric stomach along with other genera such as *Trichococcus* and *Photobacterium.* Virus infection significantly altered the microbiota functions which might negatively impact the physiology and the immune response in shrimp. However, the stomach microbiome of *L. vannamei* from two different ponds in the same study varied significantly suggesting the influence of culture conditions on the gut microbiome (Imaizumi et al. 2022). Zhang and Sun (2022) reported significant variations in the core gut microbiota among the four different species of shrimp. Challenge with WSSV resulted in alterations in core microbial structures; however, variation of core microbiota followed the same trend among different species and displayed immune-related function. *Candidatus bacilloplasma* of class Mollicutes was the dominant species in the stomach of shrimp from one pond, and this species might have a role in certain disease conditions as it has been associated with AHPND-infected *L. vannamei* (Chen et al. 2017) and is reportedly abundant in the intestine of *L. vannamei* with white feces syndrome (WFS) (Hou et al. 2018; Restrepo et al. 2021). Similar to other studies, *Vibrio* and *Photobacterium* were also abundant in the midgut, but were relatively less abundant in the stomach.

Comparison of the intestinal bacterial communities of healthy and WFS-affected shrimp has revealed clear distinctions in the microbiomes, with less diverse but more heterogenous bacterial communities associated with WFS shrimp compared to healthy shrimp (Hou et al. 2018). The abundance of *Candidatus bacilloplasma* and *Phascolarctobacterium* increased significantly in WFS-affected shrimp, while the beneficial bacteria such as *Paracoccus* and *Lactococcus* decreased in abundance. Although the exact role of these bacteria in WFS is unclear, the dynamic shift in the abundance toward certain species of bacteria under WFS conditions might indicate significant change in the intestinal physiology of shrimp. In WFS-affected shrimp, *Bacteroides*, *Subdoligranulum*, *Clostridium*, *Blautia,* and *Phascolarctobacterium* were more common compared to the healthy shrimp. Pattarayingsakul et al. (2022) investigated the WFS associated with EHP and found that the gut microbiomes in white gut shrimp displayed less diversity and had higher abundance of *Vibrio* sp. and *Propionigenium* sp. than the normal gut. The authors noticed that even though EHP was confirmed in the hepatopancreas and midgut of both white gut and normal gut, the EHP load, and tissue damage were severe in white gut-affected shrimp. Similarly, Kurniawinata et al. (2021) reported less diversity of bacterial flora and abundance of pathogenic bacteria including *Vibrio* sp. in shrimp gut and rearing water during a WFS outbreak. The dominance of *Rubritalea* sp. has been suggested as the trigger for the reduced bacterial diversity in the gut. Similar to WFS, a decrease in the bacterial diversity was observed in the intestine of AHPND-infected shrimp, which was dominated by Vibrionales (Restrepo et al. 2021). A similar decrease in diversity and increase in the populations of *Vibrio* (Vibrionales) and *Candidatus bacilloplasma* (Mycoplasmatales) in AHPND-infected shrimp has been reported by (Chen et al. 2017). In contrast, populations of *Candidatus bacilloplasma* were higher in healthy shrimp, which decreased significantly upon artificial infection with *Vibrio* species (*V. sinaloensis* and *V. parahaemolyticus*) causing white feces syndrome (Wang et al. 2020). A similar study comparing the gut microbiome of healthy and AHPND-affected subadult *L. vannamei* found a decrease in the community diversity in AHPND shrimp, with many dysbiosis-related biomarkers associated with *Vibrio* spp. (Dong et al. 2021). The genus *Vibrio* was found highly enriched in diseased shrimp (61.69%) compared to the healthy shrimp (4.99%). Reyes et al. (2022) compared the microbiome of *P. vannamei* larvae infected with AHPND-causing bacteria and studied the biomarkers associated with high and low survival in hatchery rearing tanks. The genera *Bacillus*, *Vibrio*, *Yangia*, *Roseobacter*, *Tenacibaculum*, *Bdellovibrio*, *Mameliella*, and *Cognatishimia* were predominant in the tanks with high-survival. However, low survival tanks showed the enrichment of *Gilvibacter*, *Marinibacterium*, *Spongiimonas*, *Catenococcus*, and *Sneathiella*.

Along similar lines, EHP infection altered gut bacterial communities resulting in massive changes in digestive enzyme activities that negatively affected the growth rate of shrimp (Shen et al. 2022). Some of the taxa that got enriched in the gut of EHP-infected shrimp belonged to Piscirickettsiaceae and Xanthomonadaceae, while the abundance of Rhodobacteraceae, *Rubritalea*, *Amaricoccus*, and *Frankiaceae* decreased significantly. The dysbiosis of gut microbiota was also reflected in reduced protein, lipid, and carbohydrate metabolism, suppressed immune system, while pathways representing bacterial infections increased significantly (Shen et al. 2022). Table 9.1 summarizes some important works on shrimp microbiome.

9.3.4 Biofloc-Associated Microbiome

Biofloc-based shrimp culture system is gaining prominence. Biofloc is a conglomeration of microbial communities on floating particles, which are held together by extracellular polymeric substances. Bioflocs are a ready source of protein-rich diet for shrimp, in addition to being good sources of lipids and vitamins. Furthermore, biofloc bacteria contribute profoundly to nutrient cycling leading to remediation of organic waste generated in the shrimp ponds. Biofloc technology is rapidly becoming popular owing to a multitude of benefits in terms of providing a healthy and nutritionally rich diet to shrimp, better feed conversion, cleaning up of pond water, prevention of diseases and is considered as an environment-friendly shrimp culture technique. The positive effect of bioflocs on shrimp health and physiology have been documented.

The biofloc bacterial communities vary depending on the size of biofloc, as reported by a recent study (Huang et al. 2020). A significant difference was observed in the bacterial communities between small (<20 μm) and large (>20 μm) bioflocs. The large-sized bioflocs predominantly harbored Rhodobacteraceae and Flavobacteriaceae, which resembled the gut microbiota of shrimp, and these positively contributed to shrimp health and pond water quality, while the small-sized biofloc. On the other hand, small-sized bioflocs carried shrimp pathogenic groups such as Vibrionaceae, Mycoplasmataceae, and Pseudoalteromonadaceae, the presence of which negatively correlated diversity in cardiac and pyloric stomachs.

9.3.5 Microbiome of Shrimps Farmed in Inland Saline Water

In some of the arid regions of India, the soil is affected by salinization rendering it unsuitable for agriculture. The inland saline soils have been found to be suitable for aquaculture as the soil and water salinity in majority of areas range from 5 to 15 ppt,

Table 9.1 Composition of shrimp microbiome, its dynamics and the associated disease conditions

Shrimp species	Sample	Health status	Dominant bacterial groups	Reference
L. vannamei	Hepatopancreas	Healthy	*Pantoea agglomerans*	Cornejo-Granados et al. (2017)
		AHPND-infected	*Photobacterium angustum*	
L. vannamei	Intestine	Healthy	*Vibrio shilonii* and *Faecalibacterium prausnitzii*	
		AHPND-infected	*Aeromonas taiwanensis*, *Microbispora rosea*, and *Simiduia agarivorans*	
L. vannamei	Intestine	WSSV-infected	*Proteobacterium*, *Propionigenium*, and *Arcobacter*	Wang et al. (2019)
L. vannamei	Gut	Experimentally infected with *V. parahaemolyticus*	*Oceanospirillum*, *Lewinella*, and *Saprospira* (pathogenic) *Agarivorans*, *Ferrimonas*, and *Shewanella*	Zhang et al. (2021b)
L. vannamei	Gut	WFS-affected	*Vibrio* spp.	Wang et al. (2020)
L. vannamei	Foregut	Healthy	Rhodobacteraceae and Rubritalaceae	Garibay-Valdez et al. (2021)
	Midgut	Healthy	Aeromonadaceae, Beijerinckiaceae, and Propionibacteriaceae	
	Hindgut	Healthy	Cellulomonadaceae	
L. vannamei	Stomach and the midgut	Healthy	Mollicutes (Candidatus Bacilloplasma)	Imaizumi et al. (2022)
L. vannamei	Intestine	WFS	*Tenericutes* *Firmicutes* Candidatus Bacilloplasma and Phascolarctobacterium	Hou et al. (2018)
L. vannamei	Hepatopancrease and stomach	AHPND	Vibrionales	Restrepo et al. (2021)
L. vannamei	Stomach	AHPND	Vibrionales Mycoplasmatales	(Chen et al. 2017)
Marsupenaeus japonicus	Cardiac stomach, pyloric stomach, and intestines	WSSV-infected	Firmicutes in stomachs. No change in the intestine	Niu et al. (2022)

L. vannamei	Intestine	Healthy	Proteobacteria, Bacteroidetes, and Actinobacteria	Huang et al. (2016)
P. monodon	Different growth stages (PL, juveniles)	Healthy	Gamma proteobacteria	Rungrassamee et al. (2013)
L vannamei	Intestine	Healthy	(Alcaligenaceae, Flavobacteriaceae, Microbacteriaceae, Acidimicrobiaceae, and Rhodobacteraceae)	Zoqratt et al. (2018)
L. vannamei	Intestine	Healthy and AHPND-affected	*Vibrio* spp.	Dong et al. (2021)
L. vannamei	Gut	EHP-infected	Piscirickettsiaceae and Xanthomonadaceae	Shen et al. (2022)
L vannamei	Stomach and hepatopancreas	Healthy and starved	Mollicutes and Epsilonproteobacteria abundant in healthy shrimp; Gammaproteobacteria in starved shrimp	

which is ideal for shrimp farming. Inland saline farming is gaining prominence in countries such as USA, Israel, Australia, and India. Inland saline shrimp farming represents a new production system involving utilization of degraded land regions for food production which can significantly contribute to global food security in future. Inland saline water shrimp farming in India makes use of water extracted from aquifers which have a chemical composition different from the brackishwater. Successive shrimp farming attempts in inland saline water established that shrimps can adapt and grow in this unique ecosystem.

Inland saline water has unique geological origin, and has a completely different chemical and microbial composition compared to the brackishwater. The latter has a microbial composition representative of marine and freshwater environments. The microbial composition of inland saline water is predominantly acquired from the inland saline soil. Our recent study of the bacterial composition of inland saline shrimp farms revealed the presence of diverse *Vibrio* species in farm water and sediment. *Vibrio* spp., which are restricted to coastal-marine environments, can survive and persist in inland saline water and also colonize shrimps. The halophilic marine bacterium *V. parahaemolyticus* was detected in shrimp (100%), post-stocking pond water (31.25%), and sediment (28.57%) samples analyzed from inland saline farms in India. However, *Vibrio* spp. were not detected in inland saline water from aquifers or tube wells suggesting that *Vibrio* spp. were introduced into the inland saline farms through shrimp larvae. Although *Vibrio* spp. were not naturally found in inland saline water, once introduced, they could survive and persist in inland saline water throughout the shrimp culture cycle, and also colonize shrimp gut.

We compared the gut microbiomes of inland saline water- and brackishwater-reared shrimp using metagenomic sequencing which revealed significant differences in the composition and diversity of prokaryotic sequences. Bacterial sequences were found to be more abundant in brackishwater shrimp compared to shrimp reared in inland saline water. However, *Vibrio* composition of brackishwater and inland saline water shrimp was similar and comparable, and *Vibrio parahaemolyticus* was found to be predominant in both the ecosystems (Fig. 9.1). Diverse viral sequences were detected in both inland saline and brackishwater shrimp.

9.4 Microbiome and Disease Prediction

A balanced microbiome is essential for the normal physiology and health of shrimp. Changes in the microbiome can have a significant impact on the shrimp health. Recent findings involving humans and higher animals suggest the potential application of microbiomes in predicting health and disease status. Health of farmed shrimp is a consequence of population density, physico-chemical stress, presence of pathogens and the immune status of shrimp. It is practically impossible to control all of the above factors to achieve the desired health condition, but better management interventions can positively impact the farm and animal health. Among all the factors that influence the shrimp health, gut microbiome is of particular importance

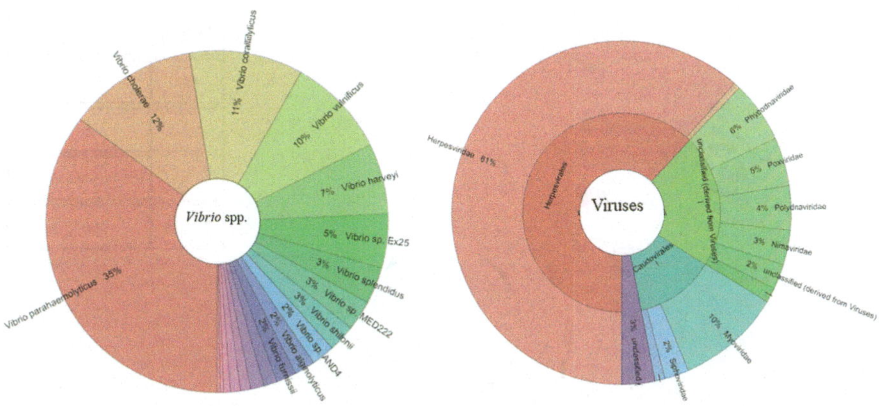

Fig. 9.1 Gut metagenome of shrimp (*L. vannamei*) farmed in inland saline water showing the abundance of *Vibrio* (**a**) and viral sequences (**b**) by metagenomic sequencing. About 35% of *Vibrio* sequences belonged to *V. parahaemolyticus*. Other species identified were *V. harveyi*, *V. vulnificus*, *V. alginolyticus*, *V. furnissi,* etc. Among viral sequences, Herpesviridae was predominant, followed by Phycodnaviridae, Poxviridae, Polydnaviridae, and Nimaviridae

as this transient population is subjected to alterations by extraneous as well as the factors intrinsic to shrimp. The extraneous factors include feed, water, and sediment microbial composition, application of chemicals and probiotics, while the intrinsic factors include the ontogeny, immune, and physiological conditions (Fig. 9.2). For example, the gut microbiome diversity is significantly reduced in starved shrimp with the consequent reduction in the metabolism of proteins, carbohydrates, and lipids, reduced intestinal enzyme activities and weakened immune response (Dai et al. 2018). Similarly, feeding shrimp with a diet containing inulin increased the relative abundance of the phylum Firmicutes and the genus *Bacillus*, resulting in better immune response, growth, and tolerance to low salinity stress (Zhou et al. 2020). The role of gut microbiome in certain disease conditions such as the white feces syndrome, AHPND, and slow growth syndromes has been under intense focus recently. Understanding healthy microbiome versus diseased microbiome is speculated to help eventually in maintaining a balanced microbial composition of shrimp gut for better health. Another potential application of gut microbiome is disease prediction, as changes in microbiome from "healthy" to "diseased," representing evolving dysbiosis, is presumed to provide important cues on the possible outbreak so that preventive measures can be taken well in advance, although this method has not yet been well established (Holt et al. 2021). Nevertheless, mounting evidence from shrimp gut microbiome-based studies suggest its potential application in disease prediction in near future.

The plasticity of shrimp gut microbiome offers numerous challenges. The gut microbiome changes with developmental stages and several extraneous factors listed above, and therefore, the signature microbiome representative of good health or diseased status can vary. Based on this hypothesis, a study identified 13 bacterial

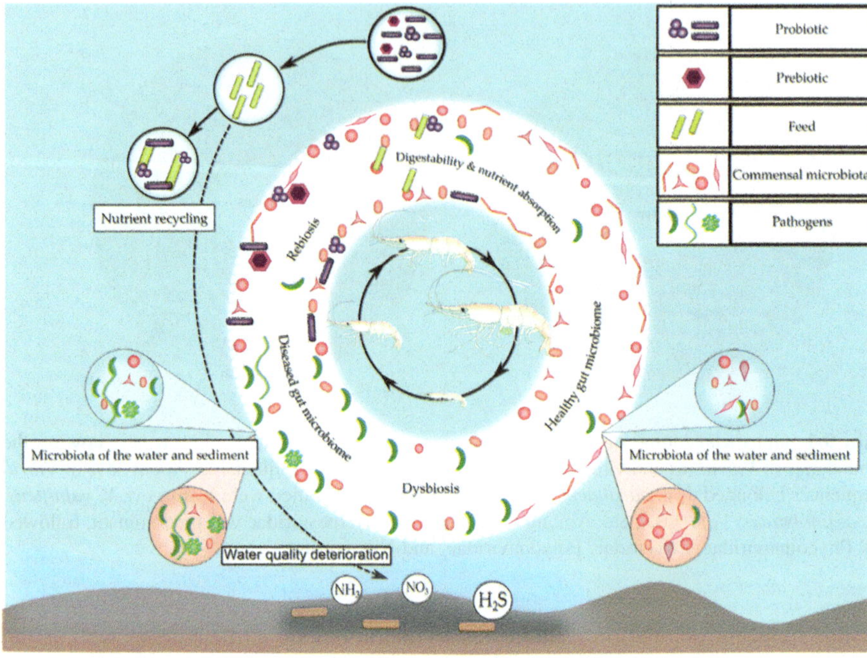

Fig. 9.2 A healthy gut microbiome is characterized by the presence of various groups of commensal bacteria, largely representing the microbiome of water and sediment. Water quality deterioration, as well as abiotic and biotic stresses can weaken the immune system of shrimp resulting in dysbiosis, a condition in which opportunistic pathogens begin to dominate and colonize the gut. External interventions such as probiotics, prebiotics, symbiotics, immunostimulants, and antimicrobial compounds can assist shrimp in rebiosis. The health of the gut microbiome is critical because it not only aids in disease resistance but also improves digestibility and nutrient availability to the shrimp. Beneficial microorganisms in aquaculture environment not only aid in good gut health, but also play important roles in water quality, nutrient recycling, and the conversion of unconsumed feed into useful biomass (e.g., biofloc)

families as indicators of shrimp health, with seven as indicators of good health and six as indicators of disease (Xiong et al. 2014). A logistic regression model based on the indicator families could predict the disease outbreak with more than 79% accuracy. A separate study showed changes in shrimp gut microbiome with developmental stages and the progression of disease, and using age- and disease-discriminatory bacterial species, a model was developed to predict disease outbreaks with 91.5% accuracy (Xiong et al. 2017). This model was based on the relative abundance of bacterial indicators of shrimp disease. A recent metanalysis study re-analyzed ten different studies involving 261 samples, four different life stages of shrimp and six disease conditions, and developed a diagnostic model for disease prediction (Sha et al. 2022). The study identified 18 taxa that were strongly associated with shrimp disease irrespective of disease type or condition, and a model involving these taxa was used to predict diseased health status of shrimp with 90% accuracy. This model could predict the general disease status of shrimp, and was not specific to any disease or the causative agent. A different approach adopted qPCR method for quantitatively determining the abundance of top ten disease-discriminatory phyla. These ten phyla contributed to 93.2% accuracy of disease diagnosis in shrimp with apparent disease symptoms, and 70% accuracy at disease onset stage in asymptomatic individuals (Yu et al. 2018).

According to another study, the abundance of *Vibrio* species and the *pirA* gene positively correlated with the occurrence of AHPND, which in turn correlated with the dysbiosis of gut microbiome (Dong et al. 2021). Shen et al. (2022) identified 20 EHP-discriminatory taxa that could accurately represent the onset of EHP in shrimp, suggesting that these taxa could be used as biomarkers for identifying EHP infection in *L. vannamei*. A study investigated dynamics of microbiome during white feces syndrome (WFS) progression in *L. vannamei* which revealed that although the gut microbiome of both healthy and diseased shrimp predominantly contained *Vibrio fluvialis*, *V. coralliilyticus*, and *V. tubiashii*, their abundance increased significantly with the progression of WF disease (Lu et al. 2020). This strong association enabled the prediction of WFS onset with 91.4% accuracy using these three signature species. This study also showed that these three species were important "drivers" of WFS in shrimp that contribute to shift from healthy to diseased shrimp.

9.5 Conclusions

In the present context of disease outbreaks affecting global shrimp farming, scientific approaches to management of disease need emphasis. Study of gut microbiome offers potential avenues to understand the disease etiology, interactions between the host and the pathogen, and useful markers for disease diagnosis. Microbiome-based studies have immensely contributed to our understanding of shrimp diseases of complex etiologies such as the white feces and slow growth syndromes, and also the microbial dynamics that precipitate infections by viral pathogens such as WSSV. The recent modeling approaches inolving disease prediction based on signature microbiota have positively impacted the research efforts towards disease management in shrimp aquaculture.

References

Abraham TJ, Palaniappan R (2004) Distribution of luminous bacteria in semi-intensive penaeid shrimp hatcheries of Tamil Nadu, India. Aquaculture 232:81–90

Adelman MW, Woodworth MH, Langelier C, Busch LM, Kempker JA, Kraft CS, Martin GS (2020) The gut microbiome's role in the development, maintenance, and outcomes of sepsis. Crit Care 24:278

Alfiansah YR, Hassenrück C, Kunzmann A, Taslihan A, Harder J, Gärdes A (2018) Bacterial abundance and community composition in pond water from shrimp aquaculture systems with different stocking densities. Front Microbiol 9:2457

Bäckhed F, Ley RE, Sonnenburg JL, Peterson DA, Gordon JI (2005) Host-bacterial mutualism in the human intestine. Science 307:1915–1920

Butt RL, Volkoff H (2019) Gut microbiota and energy homeostasis in fish. Front Endocrinol (Lausanne) 10:9

Chaijarasphong T, Munkongwongsiri N, Stentiford GD, Aldama-Cano DJ, Thansa K, Flegel TW, Sritunyalucksana K, Itsathitphaisarn O (2021) The shrimp microsporidian *Enterocytozoon hepatopenaei* (EHP): biology, pathology, diagnostics and control. J Invertebr Pathol 186:107458

Chaiyapechara S, Rungrassamee W, Suriyachay I, Kuncharin Y, Klanchui A, Karoonuthaisiri N, Jiravanichpaisal P (2012) Bacterial community associated with the intestinal tract of *P. monodon* in commercial farms. Microb Ecol 63:938–953

Chen W-Y, Ng TH, Wu J-H, Chen J-W, Wang H-C (2017) Microbiome dynamics in a shrimp grow-out pond with possible outbreak of acute Hepatopancreatic necrosis disease. Sci Rep 7: 9395

Chen C-Z, Li P, Liu L, Li Z-H (2022) Exploring the interactions between the gut microbiome and the shifting surrounding aquatic environment in fisheries and aquaculture: a review. Environ Res 214:114202

Cornejo-Granados F, Lopez-Zavala AA, Gallardo-Becerra L, Mendoza-Vargas A, Sánchez F, Vichido R, Brieba LG, Viana MT, Sotelo-Mundo RR, Ochoa-Leyva A (2017) Microbiome of Pacific Whiteleg shrimp reveals differential bacterial community composition between wild, aquacultured and AHPND/EMS outbreak conditions. Sci Rep 7:11783

Coyte KZ, Schluter J, Foster KR (2015) The ecology of the microbiome: networks, competition, and stability. Science 350:663–666

Dai W-F, Zhang J-J, Qiu Q-F, Chen J, Yang W, Ni S, Xiong J-B (2018) Starvation stress affects the interplay among shrimp gut microbiota, digestion and immune activities. Fish Shellfish Immunol 80:191–199

Donaldson GP, Lee SM, Mazmanian SK (2016) Gut biogeography of the bacterial microbiota. Nat Rev Microbiol 14:20–32

Dong P, Guo H, Wang Y, Wang R, Chen H, Zhao Y, Wang K, Zhang D (2021) Gastrointestinal microbiota imbalance is triggered by the enrichment of *Vibrio* in subadult *Litopenaeus vannamei* with acute hepatopancreatic necrosis disease. Aquaculture 533:736199

El-Saadony MT, Swelum AA, Abo Ghanima MM, Shukry M, Omar AA, Taha AE, Salem HM, El-Tahan AM, El-Tarabily KA, Abd El-Hack ME (2022) Shrimp production, the most important diseases that threaten it, and the role of probiotics in confronting these diseases: a review. Res Vet Sci 144:126–140

Emerenciano MGC, Rombenso AN, Vieira FDN, Martins MA, Coman GJ, Truong HH, Noble TH, Simon CJ (2022) Intensification of Penaeid shrimp culture: an applied review of advances in production systems, nutrition and breeding. Animals (Basel) 12:236

FAO (2020) The state of world fisheries and aquaculture 2020. Sustainability in action. FAO, Rome

Farzanfar A (2006) The use of probiotics in shrimp aquaculture. FEMS Immunol Med Microbiol 48:149–158

Garibay-Valdez E, Cicala F, Martinez-Porchas M, Gómez-Reyes R, Vargas-Albores F, Gollas-Galván T, Martínez-Córdova LR, Calderón K (2021) Longitudinal variations in the gastrointestinal microbiome of the white shrimp, *Litopenaeus vannamei*. PeerJ 9:e11827

Harris JM (1993) The presence, nature, and role of gut microflora in aquatic invertebrates: a synthesis. Microb Ecol 25:195–231

Holt CC, Bass D, Stentiford GD, van der Giezen M (2021) Understanding the role of the shrimp gut microbiome in health and disease. J Invertebr Pathol 186:107387

Hooks KB, O'Malley MA (2017) Dysbiosis and its discontents. mBio 8:e01492–e01417

Hou D, Huang Z, Zeng S, Liu J, Wei D, Deng X, Weng S, Yan Q, He J (2018) Intestinal bacterial signatures of white feces syndrome in shrimp. Appl Microbiol Biotechnol 102:3701–3709

Huang Z, Li X, Wang L, Shao Z (2016) Changes in the intestinal bacterial community during the growth of white shrimp, *Litopenaeus vannamei*. Aquac Res 47:1737–1746

Huang L, Guo H, Chen C, Huang X, Chen W, Bao F, Liu W, Wang S, Zhang D (2020) The bacteria from large-sized bioflocs are more associated with the shrimp gut microbiota in culture system. Aquaculture 523:735159

Imaizumi K, Molex W, Jitnavee C, Direkbusarakom S, Kondo H, Hirono I (2022) Bacterial and eukaryotic communities in pond water of whiteleg shrimp *Litopenaeus vannamei* and the bacterial communities of their stomach and midgut. Aquaculture 554:738139

Karunasagar I, Pai R, Malathi GR, Karunasagar I (1994) Mass mortality of *Penaeus monodon* larvae due to antibiotic-resistant *Vibrio harveyi* infection. Aquaculture 128:203–209

Kim S, Jazwinski SM (2018) The gut microbiota and healthy aging: a mini-review. Gerontology 64: 513–520

Kokou F, Sasson G, Friedman J, Eyal S, Ovadia O, Harpaz S, Cnaani A, Mizrahi I (2019) Core gut microbial communities are maintained by beneficial interactions and strain variability in fish. Nat Microbiol 4:2456–2465

Kondo H, Tinwongger S, Proespraiwong P, Mavichak R, Unajak S, Nozaki R, Hirono I (2014) Draft genome sequences of six strains of *Vibrio parahaemolyticus* isolated from early mortality syndrome/acute Hepatopancreatic necrosis disease shrimp in Thailand. Genome Announc 2: e00221–e00214

Kooloth Valappil R, Stentiford GD, Bass D (2021) The rise of the syndrome – sub-optimal growth disorders in farmed shrimp. Rev Aquac 13:1888–1906

Kurniawinata MI, Sukenda S, Wahjuningrum D, Widanarni W, Hidayatullah D (2021) White faeces disease and abundance of bacteria and phytoplankton in intensive pacific white shrimp farming. Aquac Res 52:5730–5738

Li P, Burr GS, Gatlin DM, Hume ME, Patnaik S, Castille FL, Lawrence AL (2007) Dietary supplementation of short-chain fructooligosaccharides influences gastrointestinal microbiota composition and immunity characteristics of Pacific white shrimp, *Litopenaeus vannamei*, cultured in a recirculating system. J Nutr 137:2763–2768

Liu H, Liu M, Wang B, Jiang K, Jiang S, Sun S, Wang L (2010) PCR-DGGE analysis of intestinal bacteria and effect of *Bacillus* spp. on intestinal microbial diversity in kuruma shrimp (*Marsupenaeus japonicus*). Chin J Ocean Limnol 28:808–814

Lu J, Zhang X, Qiu Q, Chen J, Xiong J (2020) Identifying potential polymicrobial pathogens: moving beyond differential abundance to driver taxa. Microb Ecol 80:447–458

Lynch SV, Pedersen O (2016) The human intestinal microbiome in health and disease. N Engl J Med 375:2369–2379

Morales-Covarrubias MS, Cuéllar-Anjel J, Varela-Mejias A, Elizondo-Ovares C (2018) Shrimp bacterial infections in Latin America: a review. Asian Fish Sci 31:76–87

Munkongwongsiri N, Thepmanee O, Lertsiri K, Vanichviriyakit R, Itsathitphaisarn O, Sritunyalucksana K (2022) False mussels (*Mytilopsis leucophaeata*) can be mechanical carriers of the shrimp microsporidian *Enterocytozoon hepatopenaei* (EHP). J Invertebr Pathol 187:107690

Niu G-J, Yan M, Li C, Lu P, Yu Z, Wang J-X (2022) Infection with white spot syndrome virus affects the microbiota in the stomachs and intestines of kuruma shrimp. Sci Total Environ 839: 156233

Oxley APA, Shipton W, Owens L, McKay D (2002) Bacterial flora from the gut of the wild and cultured banana prawn, *Penaeus merguiensis*. J Appl Microbiol 93:214–223

Pacheco-Sandoval A, Schramm Y, Heckel G, Brassea-Pérez E, Martínez-Porchas M, Lago-Lestón A (2019) The Pacific harbor seal gut microbiota in Mexico: its relationship with diet and functional inferences. PLoS One 14:e0221770

Patil PK, Geetha R, Ravisankar T, Avunje S, Solanki HG, Abraham TJ, Vinoth SP, Jithendran KP, Alavandi SV, Vijayan KK (2021) Economic loss due to diseases in Indian shrimp farming with special reference to *Enterocytozoon hepatopenaei* (EHP) and white spot syndrome virus (WSSV). Aquaculture 533:736231

Pattarayingsakul W, Munkongwongsiri N, Thitamadee S, Sritunyalucksana K, Aldama-Cano DJ (2022) Shrimp microsporidian EHP spores in culture water lose activity in 10 days or can be inactivated quickly with chlorine. Aquaculture 548:737665

Restrepo L, Domínguez-Borbor C, Bajaña L, Betancourt I, Rodríguez J, Bayot B, Reyes A (2021) Microbial community characterization of shrimp survivors to AHPND challenge test treated with an effective shrimp probiotic (*Vibrio diabolicus*). Microbiome 9:88

Reyes G, Betancourt I, Andrade B, Panchana F, Román R, Sorroza L, Trujillo LE, Bayot B (2022) Microbiome of *Penaeus vannamei* larvae and potential biomarkers associated with high and low survival in shrimp hatchery tanks affected by acute Hepatopancreatic necrosis disease. Front Microbiol 13:838640

Rungrassamee W, Klanchui A, Chaiyapechara S, Maibunkaew S, Tangphatsornruang S, Jiravanichpaisal P, Karoonuthaisiri N (2013) Bacterial population in intestines of the black tiger shrimp (*Penaeus monodon*) under different growth stages. PLoS One 8:e60802

Salunke M, Kalyankar A, Khedkar CD, Shingare M, Khedkar GD (2020) A review on shrimp aquaculture in India: historical perspective, constraints, status and future implications for impacts on aquatic ecosystem and biodiversity. Rev Fish Sci Aquac 28:283–302

Sathish Kumar T, Makesh M, Alavandi SV, Vijayan KK (2022) Clinical manifestations of White feces syndrome (WFS), and its association with *Enterocytozoon hepatopenaei* in *Penaeus vannamei* grow-out farms: a pathobiological investigation. Aquaculture 547:737463

Sha H, Lu J, Chen J, Xiong J (2022) A meta-analysis study of the robustness and universality of gut microbiota–shrimp diseases relationship. Environ Microbiol 24:3924–3938

Shakibazadeh S, Saad CR, Christianus A, Kamarudin MS, Sijam K, Nor Shamsudin M, Neela VK (2009) Bacteria flora associated with different body parts of hatchery reared juvenile *Penaeus monodon*, tanks water and sediment. Ann Microbiol 59:425–430

Shen H, Zhang X, Qian D, Chen J, Xiong J (2022) Pathobiology of *Enterocytozoon hepatopenaei* (EHP) in shrimp: diagnosis and interpretation from the gut bacterial community. Aquaculture 554:738169

Shinn AP, Pratoomyot J, Griffiths D, Trong TQ, Vu NT, Jiravanichpaisal P, Briggs M (2018) Asian shrimp production and the economic costs of disease. Asian Fish Sci S 31:29–58

Stentiford GD, Neil DM, Peeler EJ, Shields JD, Small HJ, Flegel TW, Vlak JM, Jones B, Morado F, Moss S, Lotz J, Bartholomay L, Behringer DC, Hauton C, Lightner DV (2012) Disease will limit future food supply from the global crustacean fishery and aquaculture sectors. J Invertebr Pathol 110:141–157

Stentiford GD, Bass D, Williams BAP (2019) Ultimate opportunists—the emergent Enterocytozoon group microsporidia. PLoS Pathog 15:e1007668

Tamboli CP, Neut C, Desreumaux P, Colombel JF (2004) Dysbiosis in inflammatory bowel disease. Gut 53:1–4

Tang KFJ, Han JE, Aranguren LF, White-Noble B, Schmidt MM, Piamsomboon P, Risdiana E, Hanggono B (2016) Dense populations of the microsporidian *Enterocytozoon hepatopenaei* (EHP) in feces of *Penaeus vannamei* exhibiting white feces syndrome and pathways of their transmission to healthy shrimp. J Invertebr Pathol 140:1–7

Tangprasittipap A, Srisala J, Chouwdee S, Somboon M, Chuchird N, Limsuwan C, Srisuvan T, Flegel TW, Sritunyalucksana K (2013) The microsporidian *Enterocytozoon hepatopenaei* is not the cause of white feces syndrome in whiteleg shrimp *Penaeus (Litopenaeus) vannamei*. BMC Vet Res 9:139

Tran L, Nunan L, Redman RM, Mohney LL, Pantoja CR, Fitzsimmons K, Lightner DV (2013) Determination of the infectious nature of the agent of acute hepatopancreatic necrosis syndrome affecting penaeid shrimp. Dis Aquat Org 105:45–55

Wang J, Huang Y, Xu K, Zhang X, Sun H, Fan L, Yan M (2019) White spot syndrome virus (WSSV) infection impacts intestinal microbiota composition and function in *Litopenaeus vannamei*. Fish Shellfish Immunol 84:130–137

Wang H, Wan X, Xie G, Dong X, Wang X, Huang J (2020) Insights into the histopathology and microbiome of Pacific white shrimp, *Penaeus vannamei*, suffering from white feces syndrome. Aquaculture 527:735447

Xiong J, Zhu J, Zhang D (2014) The application of bacterial indicator phylotypes to predict shrimp health status. Appl Microbiol Biotechnol 98:8291–8299

Xiong J, Zhu J, Dai W, Dong C, Qiu Q, Li C (2017) Integrating gut microbiota immaturity and disease-discriminatory taxa to diagnose the initiation and severity of shrimp disease. Environ Microbiol 19:1490–1501

Yu W, Cao J, Dai W, Qiu Q, Xiong J (2018) Quantitative PCR analysis of gut disease-discriminatory phyla for determining shrimp disease incidence. Appl Environ Microbiol 84: e01387–e01318

Zhang S, Sun X (2022) Core gut microbiota of shrimp function as a regulator to maintain immune homeostasis in response to WSSV infection. Microbiol Spectr 10:e02465–e02421

Zhang X, Li X, Lu J, Qiu Q, Chen J, Xiong J (2021a) Quantifying the importance of external and internal sources to the gut microbiota in juvenile and adult shrimp. Aquaculture 531:735910

Zhang X, Sun J, Han Z, Chen F, Lv A, Hu X, Sun X, Qi H, Guo Y (2021b) *Vibrio parahaemolyticus* alters the community composition and function of intestinal microbiota in Pacific white shrimp, *Penaeus vannamei*. Aquaculture 544:737061

Zhou L, Li H, Qin JG, Wang X, Chen L, Xu C, Li E (2020) Dietary prebiotic inulin benefits on growth performance, antioxidant capacity, immune response and intestinal microbiota in Pacific white shrimp (*Litopenaeus vannamei*) at low salinity. Aquaculture 518:734847

Zoqratt MZHM, Eng WWH, Thai BT, Austin CM, Gan HM (2018) Microbiome analysis of Pacific white shrimp gut and rearing water from Malaysia and Vietnam: implications for aquaculture research and management. PeerJ 6:e5826

Microbial Metagenomics and the Shellfish Microbiome 10

Andrea M. Tarnecki

Abstract

Taxonomic studies on the shellfish microbiome indicate that these complex microbial communities are influenced by the host and environment, and that shifts in the microbiome structure are associated with disease. These methods can predict the potential function of the microbes in these assemblages, but metagenomic studies are needed to determine the metabolic consequences of community alterations. Sequencing the genomes of microbial community members provides insight into the role of a healthy microbiome in symbiotic relationships, but also describes beneficial or detrimental functional changes in the face of climate change, anthropogenic influences, and intensive aquaculture. Understanding the interaction between aquaculture practices and microbiome function allows for the development of improved protocols that maintain healthy communities to prevent disease, improve digestion and growth, and improve the efficiency and sustainability of shellfish production. This chapter provides a review of research on microbial metagenomics in shellfish species, including crustaceans and bivalves, with emphasis on the benefits of this technology for aquaculture.

Keywords

Shellfish · Microbiome · Metagenomics · Symbionts · Diseases · Community functions

---

A. M. Tarnecki (✉)
Auburn University Shellfish Lab, School of Fisheries, Aquaculture and Aquatic Sciences, Auburn University, Dauphin Island, AL, USA
e-mail: atarnecki@auburn.edu

A. Diwan et al. (eds.), *Microbiome of Finfish and Shellfish*,
https://doi.org/10.1007/978-981-99-0852-3_10

10.1 Introduction

Microbial metagenomics is the study of all the genomic DNA present in a microbial community. This field of study is increasing as sequencing technologies become more affordable and microbiome research moves toward addressing not only community structure but also function. Although the high throughput sequencing of targeted housekeeping genes, such as the 16S rRNA gene present in bacteria, has widely been referred to as metagenomics, this method of microbiome characterization is more accurately defined as metataxonomics as it describes only the taxonomic structure of the community. With the data, community composition is used to infer function through predictive bioinformatics (Aßhauer et al. 2015; Iwai et al. 2016; Langille et al. 2013). True metagenomics on the other hand uses shotgun sequencing approaches to sequence all the DNA in a sample, resulting in partial to complete assembled genomes that are compared to reference genomes to identify functional genes. Using this method, the potential metabolic capability of the microbiome is described, with functions mapped to specific members within the community. This method provides greater insight into the role of the microbiome and its interplay with the host. Using microbial metagenomics, we can determine how the microbiome and its functional capacity change in response to environmental conditions such as climate change, disease, and dietary shifts, elucidate causative agents of disease, and detail relationships between hosts and their symbionts.

There is enormous potential for metagenomics to enhance our understanding of host–microbiome interactions, but in shellfish, the topic is relatively underexplored. Shellfish (crustacean and molluscan) aquaculture, whose contribution to aquatic animal production has surpassed that of wild capture fisheries, was valued at US $111.3 billion in 2020 (FAO 2022). This represents 42% of overall aquaculture production by value worldwide (FAO 2022) with finfish taking the majority. In line with the differences in production value, finfish metagenomics literature is more abundant than that for shellfish species. However, shellfish are valued at approximately US$1300 more per tonne than finfish and as intensive shellfish aquaculture expands, so do reports of disease-associated mortality. Therefore, it is becoming increasingly important to decipher the interactions between these organisms and their microbes.

This chapter will review the literature on shellfish metagenomics to date, including the application of this technology to aquaculture. A summary of the studies incorporating microbial metagenomics in shellfish is given in Table 10.1. Research has explored genetic heterogeneity in shellfish symbionts, examined relationships between disease and the microbiome, and connected changing environments to microbiome structure and function.

Table 10.1 Summary of papers describing the use of metagenomics in shellfish

Species and common name	Sample Type	Major finding	Citation
Molluscs			
Thyasira cf. *gouldi* Northern hatchet shell clam	Gill	Symbionts have metagenomic variation that allows for adaptability	McCuaig et al. (2020)
Actinonaias pectorosa Pheasantshell mussel	Hemolymph	Clinch densovirus 1 is associated with disease	Richard et al. (2020)
Bathymodiolus Deep-sea mussel	Gill	Symbionts are horizontally transmitted, continuously colonizing, and functionally diverse, improving the resiliency of the microbiome	Ansorge et al. (2019)
Crassostrea gigas Pacific oyster	Water Hemolymph	Dysbiosis associated with disease occurs in bacterial, protistan, and viral microbiomes. The structure of these communities is related to host genotype, selection pressure, and the environment	Dupont et al. (2020)
Crassostrea virginica Eastern oyster	GI tract	*Mollicutes* found in the oyster are closely related but functionally distinct from *Mollicutes* isolated from other marine invertebrates	Pimentel et al. (2021)
Crassostrea virginica Eastern oyster	Water Larvae	*Listonella* and *Vibrio* phages are associated with low larval survival	Ramachandran et al. (2018)
Crassadoma gigantea Purple-hinge rock scallop	Gill GI tract	Ocean acidification and climate change alter bacterial community functions	Alma et al. (2020)
Pecten maximus King scallop	Gill	*Endozoicomonas*-like organisms (ELO) are associated with mass mortality events	Cano et al. (2018)
Babylonia areolata Maculated ivory whelk	GI tract	Diseased whelk are associated with lower bacterial diversity and a greater number of virulence factors	Zhao and Gu (2021)
Crustaceans			
Unspecified crayfish and crabs	Water Sediment Crab GI contents Crayfish GI contents	Rearing crayfish with rice leads to a stable microbial community with greater connectivity that includes probiotic groups	Zhu et al. (2022)

(continued)

Table 10.1 (continued)

Species and common name	Sample Type	Major finding	Citation
Scylla olivacea Orange mud crab	GI tract	Microbiome structure and function differs with environment and diet, with artificial diets leading to increased diversity and metabolic potential	Lau et al. (2022)
Procambarus clarkii Red swamp crayfish	GI contents	Crayfish grown in aquaculture ponds have different bacterial and archaeal communities and functions than those reared in rice paddy fields	Chen et al. (2021)
Crangon crangon Common shrimp	Hepatopancreas Muscle	Viral community structure differs by both tissue and rearing environment	van Eynde et al. (2020)
Litopenaeus vannamei Pacific white shrimp	GI tract	Lower diversity, virulence genes, and enhanced host antimicrobial production are linked to disease	Zeng et al. (2020)
Litopenaeus vannamei Pacific white shrimp	GI contents	The microbiome of shrimp varies with age and can produce digestive enzymes and short-chain fatty acids	Gao et al. (2019)
Penaeus monodon Giant tiger shrimp	Water Biofloc GI tract	Biofloc, water, and shrimp intestine harbor differing microbiome structure and function. Biofloc harbored the lowest relative abundance of antibiotic resistance genes (ARGs)	Chen et al. (2022)
Rimicaris exoculata The "blind shrimp"	Appendages	Variability in the metagenomes of *Zetaproteobacteria* indicates differences in functions that may reduce competition and provide the potential for environmental adaptation	Cambon-Bonavita et al. (2021)

10.2 Shellfish Symbionts

The use of metagenomics to investigate microbial communities of shellfish has provided insight into specific bacterial strains that play a role in nutrient cycling and host adaptability in the face of changing conditions. These relationships have best been studied between hydrothermal vent organisms and their intra- and extra-cellular symbionts. In these locations, seawater is enriched with reduced inorganic compounds such as hydrogen sulfide, which is toxic to most life. However, microorganisms in these environments have evolved to use sulfide and/or other inorganic compounds as energy to generate carbon compounds from CO_2 in a process known as chemoautotrophy. This process may use the reverse tricarboxylic acid cycle (rTCA) whose enzymes are sensitive to oxygen or the Calvin-Benson-Bassham cycle (CBB) catalyzed by the enzyme RuBisCO which is not sensitive to oxygen but requires greater energy input (Sheik et al. 2020). Symbiotic relationships with these chemoautotrophs allow multicellular organisms to survive in these toxic environments, with the microbe removing harmful inorganics while providing organic compounds to their host organism (Nakagawa and Takai 2008). Three of these types of symbioses have been explored with metagenomics: appendage-associated *Zetaproteobacteria* of the deep-sea blind shrimp *Rimicaris exoculata*, intracellular sulfur-oxidizing (SOX) symbionts associated with deep-sea *Bathymodiolus* mussels, and extracellular gill-associated symbionts of Northern hatchet shell clams *Thyasira* cf. *gouldi*. The genomes of closely related symbionts within individuals and among sites are diverse, indicating functional heterogeneity which provides a benefit to the host in varying environmental conditions.

Cambon-Bonavita et al. (2021) assembled near complete genomes of two strains of *Zetaproteobacteria* associated with the cephalothoracic appendages of the blind shrimp *R. exoculata* from hydrothermal vents within the Mid-Atlantic Ridge (MAR). Upon phylogenetic analysis, these strains were dissimilar enough from each other and their nearest neighbors to represent potentially new species. As such, the authors recommended the names *Candidatus* Ghiorsea rimicarensis and *Candidatus* G. crypta for these two strains. CBB genes indicated full chemoautotrophic ability; however, the symbionts also carried genes associated with heterotrophy including various carbon compound transporters and genes associated with glycolysis and the TCA cycle. The potential for mixotrophy was also reported in extracellular gill symbionts associated with the clam *Thyasira* cf. *gouldi* (McCuaig et al. 2020). This metabolic diversification may be a result of a free-living stage, indicated by genes involved in flagella biosynthesis and chemotaxis (Cambon-Bonavita et al. 2021; McCuaig et al. 2020). These strains are therefore attracted to their host via chemical signals which improves colonization but can find food and energy sources in the external environment when necessary. Therefore, it is likely these symbiotic bacteria are transferred through horizontal transmission. Interestingly, metagenomics analysis indicated horizontal transmission of intracellular symbionts within the gill of *Bathymodiolus* mussels. This group of deep-sea mussels associates with chemoautotrophic intracellular symbionts that oxidize sulfur (SOX symbionts) and are sequestered within the gill in cells termed bacteriocytes. Despite previous studies focused on variability in 16S rRNA genes indicating low diversity, symbionts within the same mussel individual were genetically and functionally diverse when analyzed using metagenomics (Ansorge et al. 2019). This variability, resulting in 4–16 strains of SOX symbionts within an individual mussel, suggests continuous horizontal transmission of symbionts from the environment, allowing for establishment of symbiotic bacteria that are influenced by local environmental conditions.

Horizontal transmission of symbionts indicates these bacteria are subject to changing habitat parameters and therefore vary by site and over time, leading to shifts in the metabolic potential of the host microbiome. Indeed, studies show differences in symbionts based on sampling site. The two vent sites from which shrimp-associated *Zetaproteobacteria* were detected differed in conditions, with the TAG vent being characterized by high sulfur content and located 1000 m deeper than the Rainbow vent which had low concentrations of sulfur but high levels of hydrogen, methane, and iron. Although both *Ca.* Ghiorsea strains were detected at each vent, *Ca.* G. crypta was significantly more abundant at the TAG vent. In addition, only *Ca.* G. rimicarensis had genes associated with hydrogen oxidation and nitrate transport as well as some enzymes within the TCA cycle (i.e., citrate synthase, aconitate hydratase), whereas *Ca.* G. crypta had genes attributed to the reduction of nitric oxide and arsenate. It was suggested that the functional diversity of these symbionts may reduce competition between strains and allow for greater environmental adaptation of the epibiont community. Habitat partitioning is also likely in intracellular SOX symbionts as the potential microbiome functions of individual mussels collected from the same vent field were more similar than those across different fields (Ansorge et al. 2019). Some functions were missing from vents altogether, as all SOX strains from two vents had genes associated with phosphate metabolism while few to zero strains from the other two vents had these genes. These characteristics may reflect differing levels of phosphorus at the vent sites. Host-level partitioning was also reported in SOX symbionts, as those capable of hydrogen oxidation were found together in host bacteriocytes, while those without hydrogenase enzymes were found in separate bacteriocytes. Genes associated with nitrate reduction differed not only among sites, but also among individuals within a site. This data supports the influence of the local environment and the host on the microbiome at the strain level.

Upon colonization, metagenomic studies indicate that shellfish symbionts provide numerous benefits to their host. The clam symbionts have denitrification genes that the authors suggest enable them to occupy anaerobic niches to reduce competition for oxygen while removing harmful nitrogenous compounds from the environment to benefit the clam (McCuaig et al. 2020). The *Zetaproteobacteria* strains have numerous genes that suggest a beneficial relationship with their shrimp hosts including sulfide and iron oxidation, dissimilatory nitrate reduction to ammonium (DNRA), arsenic detoxification, generation of amino acids from ammonia, and production of vitamins (Cambon-Bonavita et al. 2021). These processes remove potentially harmful compounds from the water while providing the host with organic compounds and nutrients that benefit shrimp health and nutrition (Cambon-Bonavita et al. 2021).

The above studies demonstrate similar trends among intra- and extracellular shellfish symbionts. Many of the genes that are unique to specific strains are also expressed (Cambon-Bonavita et al. 2021), suggesting this genetic potential is realized and can influence host health. High population diversity is indicative of resiliency in the face of stressors and changing environmental conditions (Pelletier et al. 2020). The functional diversity within these symbiotic bacteria allows for reduced competition among related strains and encourages strain diversity within the population. Colonization by symbionts with varying metabolic capabilities establishes a diverse microbiome that is and can continue to be adapted to local environmental conditions. Once established, these microbes provide benefits to their hosts such as detoxification and amino acid biosynthesis and improve host fitness in these harsh environments. Functional redundancy that occurs within the microbiome, such as multiple organisms being capable of chemoautotrophy, provides continued benefits for the host if members of the community disappear. Although these studies focused on a subset of symbionts associated with particular shellfish species in hydrothermal vents, it can be assumed that the diverse functional potential of the whole microbiome provides numerous benefits for nutrient cycling and chemical crosstalk between microbes and their host, influencing host health and disease in all natural and culture environments.

10.3 Shellfish Disease

It is clear from studies focusing on the microbiome that deviations from a healthy microflora, a condition known as dysbiosis, is associated with disease (Infante-Villamil et al. 2021). As discussed in other chapters within this book, the diverse microbial community associated with a host organism exerts influence on host physiology through interactions with the digestive and immune systems. However, the community is also in direct contact with other members of the local microbiome, with space and resource competition, antimicrobial production, and secretion of growth factors mediating the ability of a microbe to establish within the microflora. To top it off, both the host and its microbiome are influenced by environmental conditions. These revelations are leading to a shift from the "one-pathogen-one-disease" paradigm to a more complex "pathobiome" perspective. In this community-based approach, the disease is not simply the result of the proliferation of a primary pathogen; it recognizes the importance of abiotic and biotic interactions between the pathogen, the host, and their environment (Bass et al. 2019). This paradigm shift has led to a multitude of studies relating the microbiome with the health and disease of the host organism. The original Koch's postulates set forth a stepwise procedure for demonstrating that a specific microbe causes a specific disease. The recognition that not all pathogens can be grown in typical culture conditions led to Molecular Koch's Postulates that relate virulence genes to disease. With the increasing evidence of disease states that cannot be attributed to a single pathogen, there is a new framework set forth that demonstrates that the pathobiome/dysbiosis is responsible for a disease state, with the procedure for testing this known as Ecological Koch's Postulates (Vonaesch et al. 2018).

Metagenomics provides a deeper understanding of the influence of disease on the microbiome than metataxonomics alone. The link between dysbiosis and disease is well established using targeted sequencing, but the impacts on the function of the microbiome and therefore its ability to influence host physiology is less understood. As functional redundancy exists within a healthy microbial community (Infante-

Villamil et al. 2021), alteration of microflora members will not necessarily result in altered function, and these changes can only be predicted when using metataxonomic strategies. However, numerous studies have demonstrated that a pathobiome contains different genetic potential than a healthy microbiome, and these changes in function have consequences for the shellfish host.

Metagenomic studies confirm conclusions from metataxonomic investigations that dysbiosis is associated with shellfish disease. This has been demonstrated in Pacific white shrimp *Litopenaeus vannamei* with white feces syndrome (WFS) (Zeng et al. 2020), Pacific oysters *Crassostrea gigas* exposed to Pacific oyster mortality syndrome (POMS) (Dupont et al. 2020), maculated ivory whelk *Babylonia areolata* affected by an unknown disease (Zhao and Gu 2021), and Eastern oyster *Crassostrea virginica* larvae exhibiting unexplained mortalities (Ramachandran et al. 2018). Decreased diversity and greater abundances of opportunists characterize the pathobiome of diseased shellfish (Zeng et al. 2020; Zhao and Gu 2021). The loss of taxa includes some that are considered keystone species in the healthy community, leading to an altered microbial network where other members of the population that rely on keystone species disappear (Dupont et al. 2020). This loss of diversity likely reduces functional redundancy, and the community may lose characteristics that benefit the host. At the same time, less species means reduced heterogeneity and diminished resilience to changing conditions. Thus, it is reasonable to hypothesize that reduced diversity leaves the host more susceptible to pathogenic infections which may advance to disease.

POMS, a disease of Pacific oysters, has been causing mortality issues for many years worldwide. This condition occurs in the spring when water temperatures increase and has been associated with *Ostreid herpesvirus* variants (OsHV-1 μVar) as well as opportunistic bacteria within the genus *Vibrio* (Dupont et al. 2020). When oysters reared in the hatchery were transplanted to the natural environment either before (March) or during the infectious period for POMS (May), significant shifts in the protistan, bacterial, and viral microbiomes occurred. Some of the indicator taxa for the infectious period included potential aquatic invertebrate pathogens such as ciliates *Acineta* (Hudson and Lester 1994; Kakoolaki and Afsharnasab 2016; Romero and Jimønez 2002) and *Ephelota* (Gacutan et al. 1977; Gómez-Gutiérrez and Morales-Ávila 2016), a dinoflagellate in the order *Syndiniales* (Guillou et al. 2008; Small et al. 2012), and two diatoms responsible for harmful algal blooms, *Chaetoceros* and *Pseudo-nitzschia* (Anderson et al. 2021). The viral communities associated with susceptible oysters had high abundances of OsHV-1 μVar which was confirmed by flow cytometry. There were also 60 bacterial operational taxonomic units (OTUs, a genetic designation roughly equivalent to species) indicative of oysters collected from the infectious period, with some of these having members known to cause disease in shellfish (i.e., *Francisella*, *Vibrio*, *Mycoplasma*, and *Rickettsiales*) (Birkbeck et al. 2011; de Souza Valente and Wan 2021; Destoumieux-Garzón et al. 2020; Ghadersohi and Owens 1999; Paillard et al. 2004). Thus, when environmental conditions were favorable for POMS, dysbiosis occurred in all microbial groups selecting for potential pathogens during the infectious period. This alteration of community structure would influence microbial interactions, perhaps allowing for proliferation of the pathogen(s) that lead to POMS mortalities; however, the only microbe related specifically to diseased individuals was OsHV-1 μVar. The authors suggest this may be due to the stability of the microbiome within the oyster hemolymph or that the sampling, which occurred 5 days following transplant, may not have been timed correctly to detect dysbiosis in the bacterial or protistan assemblages of diseased individuals. In another study relating microbiome structure to mortality events, Ramachandran et al. (2018) investigated bacterial and viral communities in hatchery-reared Eastern oysters with different survival rates. Low survival tanks were characterized by the presence of a *Roseovarius* spp. and phages that infect the bacterial genera *Listonella* and *Vibrio*. All three of these bacterial genera include species that have been identified as oyster pathogens, suggesting they may have played a role in mortality events (Boardman et al. 2008; Dubert et al. 2017; Hickey and Lee 2018). The microbiome of maculated ivory whelk affected by an unknown disease was characterized by a shift from the dominance of the *Proteobacteria* in healthy individuals to a community dominated by *Firmicutes* at the phylum level, with *Clostridiales*, *Lachnospiraceae*, *Holosporaceae*, and *Bradymonas* acting as taxonomic indicators of a disease state (Zhao and Gu 2021). The presence of different disease indicators in each study discussed above can be explained by the complexity of host–microbe interactions, i.e., species-specificity related to host animals, environmental influences (wild versus hatchery, varying geographical locations and seasons), and/or microbial interactions with the pathogen(s) responsible for the mortalities. Even within a host species, the influence of disease on the microbiome varies by tissue type (Pimentel et al. 2021). In Eastern oyster individuals, the abundance of a member of the *Mollicutes* was reduced in the gastrointestinal tract but enriched in the inner shell of individuals infected by the parasite Dermo *Perkinsus marinus*. However, a different Mollicute was decreased in the pallial fluid of infected individuals along with a bacterium within the *Bacteroidia*. Similarly, a sequenced member of the *Chlamydiae* was enriched in gill, mantle, and inner shell of infected oysters as compared to their uninfected counterparts. This study demonstrated a tissue-specific influence of this protozoan parasite that differed at the strain level within a given taxon (i.e., *Mollicutes*). This variability may be explained by diverse genetic potential or gene expression between the bacterial strains as well as different cell types, immune function, and biochemical signatures of each tissue within the host. This comparison only included a subset of the metagenomic data, and therefore it is likely the disease exerts more influence on the microbiome than was reported in this study.

Dysbiotic microbiomes often harbor genes coding for virulence factors. These include flagella biosynthesis and chemotaxis (Zeng et al. 2020; Zhao and Gu 2021), which indicates a greater ability for bacteria to colonize their hosts. However, these genes do not always correlate with pathogenesis, as it was stated earlier in this chapter that these genes are encoded in intra- and extracellular shellfish symbionts. Within the maculated ivory whelk, virulence factors in the diseased microbiome also included Type III secretion systems (T3SS), a series of proteins found in pathogenic Gram-negative bacteria that allow them to adhere to, infect, and alter function of host cells, often resulting in inflammation and tissue damage (Coburn et al. 2007). T3SS were reported in potentially pathogenic bacteria infecting scallops (Cano et al. 2018). The pathobiome associated with unexplained whelk mortalities also had enrichment of genes encoding lipopolysaccharide, a component of Gram-negative bacterial cell walls that acts as an endotoxin, and HSI-I (Hcp1 Secretion Island) which is associated with chronic disease (Zhao and Gu 2021).

To determine the influence of WFS on the microbiome and its Pacific white shrimp hosts, Zeng et al. (2020) combined metagenomic analysis with transcriptomics to decipher differences in gene expression within the infected shellfish. WFS-associated microflora had an increased capacity to break down antibiotic compounds, indicating greater resistance to antimicrobials (Zeng et al. 2020). This was an interesting finding as WFS led to an increase in the expression of genes associated with antimicrobial production in the shrimp, demonstrating an upregulated immune response in infected individuals (Zeng et al. 2020). Thus, the host was producing more antimicrobials in response to an altered microbial community structure while the bacteria were improving their ability to avoid them. Indeed, shellfish hosts exert a level of influence over their microbial community structures that may be directly tied to pathogen susceptibility. Dupont et al. (2020) reared oysters from five different families, with parents collected from different locations. From these families, 3 were assumed to be subjected to greater selection pressure for POMS resistance as they were collected from areas near aquaculture farms and faced mass mortality events each year. This selection pressure influenced the microbial community of bacteria and viruses, explaining 16% and 10% of the variability in these communities, respectively. This coincided with ~10 times higher survival in animals collected from aquaculture areas during the POMS infectious period. It may be these host-impacted members of the microbiome play a role in disease resistance, but it is also likely that continued disease pressure selected for host genotypes that are more resistant to POMS, and that the expressed phenotypes of the resistant animals alter niche availability and therefore bacterial colonization.

Unfortunately, mass mortality events are increasingly reported in shellfish, both in natural environments and in aquaculture. Metagenomics has been used in attempts to identify causative agents of mortalities that are suspected to be microbial in nature. In 2013–2014, mass mortality events occurred in English populations of the King scallop *Pecten maximus* (Cano et al. 2018). These events were characterized by intracellular rod-shaped Gram-negative bacteria within the gills of infected animals. Using a combination of 16S rRNA gene sequencing and metagenomics, these organisms were classified within the *Gammaproteobacteria* as *Endozoicomonas*-like organisms (ELO). Within the genome, genes related to pathogenicity were detected including virulence factors such as proteins associated with an enterotoxin and T3SS proteins. In addition, the genome encoded effector proteins that cause ruffling of the plasma membrane and altered cytoskeleton function, allowing the pathogen to enter a host cell, multiply, and avoid the immune system. These characteristics, alongside tissue damage in individuals with severe infection, suggest ELO played a role in these mass mortality events.

Between 2016 and 2019, mass mortality events resulted in a loss of over 85% of the population of pheasantshell mussel *Actinonaias pectorosa* in the Clinch River watershed (Richard et al. 2020). Based on the spread of the mortality, the authors hypothesized a viral pathogen, and viral metagenomes were assembled from hemolymph collected from 26 affected ("cases") and 32 unaffected ("control") mussels. Infection status was determined based on behaviors including burying, gaping, stimulus-response, and shell closure. Viral load was higher in affected individuals and there were five viruses that had significantly higher prevalence and/or load in cases as compared to controls. Using general linear models, one virus (Clinch densovirus 1) was significantly related to disease cases. Phylogenetic comparison suggests this is a new virus within the subfamily *Densovirinae*. These viruses are non-enveloped, icosahedral, and contain single-stranded DNA genomes (Baltimore Group II). Other viruses within this subfamily cause lethal infections in aquatic invertebrates including shrimp, crayfish, and sea stars (Richard et al. 2020) and it is possible the Clinch densovirus 1 may have played a role in the mortality events reported in pheasantshell.

It is important to point out that the association of metagenomes with disease is not a conclusive way to diagnose an etiological agent. As stated, environmental factors, host physiology, and other members of the microbiome play roles in infection and disease. In addition, limiting the analysis to a specific group of organisms (i.e., viruses or bacteria) may miss pathogens in other domains, and analyzing limited tissue types does not give a complete picture of the condition. However, these methods provide information on virulence genes and describe the pathobiome during mortality events. Future investigations may harvest the dysbiotic microbiome from affected individuals to fulfill Ecological Koch's Postulates as well as attempt to isolate the suspected pathogen(s). In any case, investigations into mortality-associated microbiomes can provide insight into potential pathogens and increase our understanding of healthy and dysbiotic microbial communities.

10.4 Environmental Influence on Community Function

It is often reported that the environment exerts a significant influence on the microbiome of shellfish. Aquatic organisms are in constant contact with surrounding microbes, are exposed to different strains that are adapted to various habitat conditions, and therefore can be colonized by the changing environmental microbiome. Being colonized by microbes that are best adapted to the surrounding conditions may provide a benefit for the host and it is possible the host is continuously "sampling" from the external environment to select a microbial community that is resilient in changing conditions. This could prove to be important for shellfish adaptation in areas facing temperature and pH changes due to climate change and ocean acidification as well as anthropogenic stressors including increased nutrient and pollutant input from a growing human population. However, it also provides a fitness strategy simply to allow populations of shellfish to thrive in sites that vary geographically and are influenced by differing water quality parameters. To understand the adaptability of the microbiome, and how that translates to host fitness, metagenomics has been used to describe the potential function of shellfish-associated microbiomes in varying environmental conditions including between different sites, comparing wild versus captive rearing, and during exposure to ocean acidification and climate change conditions.

Cephalothoracic appendages were sampled from blind shrimp collected from three hydrothermal vents within the Mid-Atlantic Ridge with different chemical signatures (Cambon-Bonavita et al. 2021). The vent Rainbow is characterized by high methane, hydrogen, and ferrous iron versus low sulfur and is approximately 1000 m shallower than the TAG which is characterized by higher sulfur content. Microbiomes from two shrimp per site were sequenced using metagenomics and structure and function were compared among sites. Genome assembly resulted in 49 MAGs (metagenome-assembled genomes). When site-related differences between differentially abundant MAGs and their functions are examined, interesting patterns arise. Out of five MAGs identified as *Flavobacteriaceae*, only one had the gene encoding β-glucosidase which plays a role in cellulase catabolism. Interestingly, this was the only MAG within the *Flavobacteriaceae* that was present in equal abundances at each site, with four others significantly enriched at the TAG vent. Similarly, the complete gene for glucoamylase was present in 1 of 6 *Gracilibacteria* MAGs which was equally abundant at both sites, whereas 3 of the others were increased at Rainbow and the remaining 2 were higher at TAG. Sulfur dioxygenase was present in one MAG within the *Zetaproteobacteria* which was equally abundant at each site versus the other MAG which was increased at TAG. The gene encoding Biofilm PGA Synthesis protein was present in 1 of 3 MAGs identified as *Desulfocapsaceae* and was equally abundant at both sites, whereas the other two MAGs in this family were enriched at Rainbow. Thus, although there were clear differences between sites in terms of the abundance of certain MAGs, when MAGs within a taxon had unique functions, these were represented equally at both sites, indicating that the microbiome may select for core functions in the community. However, genes for the rTCA cycle were only present in *Campylobacteria* enriched at Rainbow, but the authors suggest that the presence of multiple *Campylobacter* bacteria, as seen based on 16S rRNA genes, may have prevented assembly and therefore detection of the genomes. Overall, the results from this study suggest that the microbiome associated with blind shrimp differs based on the sampling site, and this is particularly true when the strains are functionally redundant. Strains encoding unique metabolic abilities are seemingly less influenced by environmental changes and may represent a true core microbiome in shellfish.

In an experiment investigating warming temperatures and ocean acidification, investigators performed metagenomics on purple-hinge rock scallop *Crassostrea gigantea* gill and gastrointestinal tract (Alma et al. 2020). In high (21.5 °C) temperatures, *Vibrionaceae* became dominant, whereas *Propionibacteraceae* was more abundant at lower (14 °C) temperatures. As previously mentioned, *Vibrionaceae* contains numerous species associated with bivalve disease (Destoumieux-Garzón et al. 2020) and this shift may indicate dysbiosis as waters warm. Analysis of the overall community, however, indicated no significant difference between tissues or between treatments. This result is different from that of previous studies that show an impact of environmental conditions as well as tissue type which have been discussed in this chapter. The authors state that OTUs were classified at the family and species level, but the presented analyses are all at the family level. If comparisons were only made among bacterial families, it is possible that differences were apparent at lower taxonomic levels. However, there were also very few differences in functions, with the only alterations seen within gill tissue and associated with increased "Cofactors Vitamins and Prosthetic Groups Pigments" in ambient partial pressure of CO_2 (365 μatm) treatments and "Regulation and Cell Signaling" in the high temperature, acidification (1050 μatm) treatment. This suggests that the ability to communicate within the microbiome is altered by climate change conditions, with an increase in bacterial cell-to-cell communication often associated with virulence (Leitão 2020). Compounding this issue is the reduced production of beneficial compounds including vitamins and pigments. Pigments often have antimicrobial properties (Numan et al. 2018), and the decrease in these products in climate-altered communities may indicate a reduction in beneficial bacteria. Together, these results suggest a shift toward a potentially harmful bacterial community structure and function, although, the minor shifts in community function suggest a level of functional resiliency in the scallop microbiome.

Studies have shown that the shellfish microbiome differs between captive and wild individuals, although many did not focus on the functional potential of these communities. van Eynde et al. (2020) compared the viromes of muscle and hepatopancreas in captively reared European brown shrimp *Crangon crangon*, as well as the viromes within the hepatopancreas across wild and cultured individuals. Bioinformatics led to the assembly of 15 nearly complete viral genomes. Interestingly there was more variation in virome composition across tissues than between environments, indicating tissue-specific viral communities. However, three viruses were more abundant in shrimp reared in aquaculture including Crangon crangon Bunya-like virus (CcBLV, *Bunyavirales*), Crangon crangon picorna-like virus1 (CcPLV1, *Picornavirales*), and Crangon crangon parvo-like virus (CcPaLV, *Parvoviridae*). It is suggested that CcBLV, a virus found primarily in muscle but also in relatively high abundance in two hepatopancreas samples within cultured individuals, may be a virus of concern due to reports of other bunya-like viruses associated with disease and mortalities in crustaceans (Cowley 2016). It should be noted that, despite the high abundance of this virus, all shrimp used in this study were apparently healthy. In any case, this study highlighted differences in viral abundance across rearing environments. This result was supported by Dupont et al. (2020) who reported that the environment was responsible for 35–48% of protistan, bacterial, and viral communities associated with Pacific oysters. Within the virome, 943 contigs were attributed to non-infectious environments versus 5 in the hatchery. None of these viruses could be identified taxonomically, but the diversity of the natural environment was greater than that of the hatchery. Within the bacteria, seven OTUs were enriched in the hatchery and included organisms previously reported in oysters including *Oleiphilus*, *Psychromonas*, *Porticoccus*, and members of the *Flavobacteriaceae* (Clerissi et al. 2020; Dubé et al. 2019; García Bernal et al. 2017; Yang et al. 2022). Oysters sampled from the natural environment harbored 23 indicator bacterial taxa, including 2 within the *Cryomorphaceae*, 8 within the

Flavobacteriaceae, and other OTUs identified as *Rhodobacteraceae* and *Methylophilaceae*, all of which have been identified specifically within the oyster microbiome (Arfken et al. 2017, 2021). This demonstrates that, although the communities are different, the bacteria that colonize captive and transplanted oysters are generally normal members of the bivalve microbiome. As stated previously, changes in taxa do not always translate to alterations of function, and metagenomic data should be analyzed at the gene level to determine if functional changes occur in captivity.

From the presented studies, is it apparent that the environment can alter the microbiome structure of viruses, bacteria, and protists. These changes should be more thoroughly explored to get a clear understanding of these impacts on the shellfish host. Specifically, in-depth functional analysis is necessary to deem whether potentially important functions are lost during environmental transitions.

10.5 Applied Metagenomics

Due to the value and expansion of shellfish aquaculture, interest in metagenomics in this industry has increased. Knowledge of the influence of development, rearing environment, and diet on the structure and function of the microbiome is becoming a commonly explored topic in most shellfish species. In addition, metagenomics has been applied to determine the potential for antibiotic resistance and horizontal gene transfer in these intensive culture environments.

Sequencing efforts indicate that the microbiome of shellfish changes during the development of the animal. Metagenomics and metataxonomics were performed on the gastrointestinal microbiome of Pacific white shrimp *Litopenaeus vannamei* after 30 and 60 days of rearing in aquaculture ponds (Gao et al. 2019). Shrimp lost diversity over time as indicated by 16S rRNA gene analysis, with *Proteobacteria* increasing in abundance and *Firmicutes* and *Tenericutes* decreasing. Early shrimp had greater variability in community structure indicating that, as the shrimp grew, they developed a microbiome that shared more core members. Indicator taxa for early rearing included *Photobacterium*, *Candidatus* Bacilloplasma, *Vibrionaceae*, *Mycoplasmatales*, *Clostridiales*, and *Enterovibrio*, versus *Donghicola*, *Pseudoalteromonas*, *Rhodobacteraceae*, and *Nautella* in shrimp remaining in ponds for longer periods. These younger communities are associated with bacterial taxa that contain potential pathogens versus potentially probiotic taxa in older individuals. This pattern is also suggested from metagenomics analysis, with greater metabolism (amino acid, energy, carbohydrate, cofactors, and vitamins), transport, and xenobiotics degradation in later shrimp versus more transcription/translation, and glycan biosynthesis and metabolism in early shrimp. The potential for the community to benefit the shrimp was shown through the capacity to generate digestive enzymes including amylase, lipase, and chitinase, with most of these enzymes produced by members of the *Vibrio* and *Pseudoalteromonas*. The community could also produce short-chain fatty acids (SCFA) which can provide energy to the host animal, aiding in digestion and growth. In comparing the Eastern oyster microbiome between age groups (Pimentel et al. 2021), *Mollicutes* were more abundant in the gastrointestinal tract of adults than in larvae. The nearly complete genome of one MAG was related to *Mycoplasma*, but its genome harbored unique metabolic pathways as compared to previously described strains, including pyruvate metabolism and the arginine deiminase pathway. The MAG could undergo glycolysis, but it did not have the full set of genes necessary for the TCA cycle, indicating that it likely obtains many nutrients from the oyster. The *Mollicutes* have been reported as a dominant and/or core member of the bivalve microbiome, and variable abundances among locations is often reported (Aceves et al. 2018; King et al. 2012; Pierce and Ward 2019; Valenzuela et al. 2021). These may be symbiotic relationships, but the dominance of these organisms may also represent dysbiosis, and further investigations are needed to elucidate these relationships.

Sequencing of environments has also gained traction in shellfish aquaculture as attempts to control pathogens have shifted from the use of antibiotics to using probiotics and other methods to maintain microbial balance and prevent the proliferation of potential pathogens. When sequencing the metagenomes of red swamp crayfish *Procambarus clarkii* reared in ponds and rice paddy fields, microbial communities within the gastrointestinal tract showed distinctive patterns (Chen et al. 2021). The bacterial phylum *Firmicutes* and the archaeal phylum *Euryarchaeota* were the most abundant groups in the gut communities, and both were significantly more abundant in crayfish grown in ponds. Just as van Eynde et al. (2020) found little differences between viromes of European brown shrimp in different culture environments, this study did not identify any differences in viral taxa, nor were there differences between fungal taxa. However, methane-producing *Archaea* were present in higher abundances in pond-raised crayfish. Within the genome's potential function, all genes that were significantly different between rearing environments were increased in pond-reared crayfish. These include genes related to antibiotic resistance, genetic information processing, and carbohydrate metabolism. Glycosyltransferases and carbohydrate esterases were also increased in pond crayfish. This study hypothesized that the glycosyltransferases were used in biofilm formation and that this characteristic may protect pathogens from other microorganisms within the gut microbiome. Combined with the increase in potentially pathogenic functions (i.e., antibacterial resistance, immune diseases), this may indicate a healthier microbiome composition and function in crayfish reared in rice paddy fields. The potential benefit of co-culturing crayfish and rice was further explored by Zhu et al. (2022). Samples were collected from sediment, water, and gastrointestinal contents of crayfish and crab grown in four different culture conditions: crab only, crayfish only, crab and crayfish co-culture, and crayfish-rice co-culture. Bacterial, archaeal, viral, and eukaryotic communities were compared for each sample type in each aquaculture system. Not surprisingly, the microbial communities were different based on sample type, with the environmental samples (water, sediment) having higher diversity than host-associated communities within the bacteria and viruses. Using network analysis, connectivity was consistently greater in crayfish-rice co-culture than the other systems. Within the crayfish gut microbiome, the co-culture of crayfish and rice had positive relationships between known probiotics (*Lactobacillus*, *Streptococcus*) and other microbes, whereas the positive connections in the other systems included *Mycoplasma*, a potential pathogenic genus. A community with high connectivity is viewed as healthier and more resilient to environmental perturbations. In fact, the microbial community of crayfish in the rice system was not correlated with any environmental factors, suggesting these communities were more stable. Together these studies indicate that the gastrointestinal communities of crayfish are influenced by rearing environment, and metagenomics suggest a more resilient microbiome in rice-crayfish co-cultures than when rearing crayfish alone. This can benefit the grower as they are able to sustainably produce two products, increasing the economic potential of the farm.

One technique for balancing microbial communities and maintaining diversity is to use biofloc technology. In these culture systems, aggregates of microbes and organic matter work symbiotically to improve water quality, remove waste compounds, and prevent proliferation of pathogens (El-Sayed 2021). Biofloc communities were analyzed in giant tiger shrimp *Penaeus monodon* aquaculture, with 60 individuals pooled into one sample and three replicate samples sequenced (Chen et al. 2022). Water and biofloc from the pond were also sequenced, with three replicates from each. Communities were analyzed with both targeted sequencing and metagenomics. The communities and their functions were different based on sample type. Organisms that were enriched in the gastrointestinal tract of shrimp as compared to both water and biofloc included unclassified members of the class *Clostridia*, orders *Cardiobacteriales* and *Clostridiales,* and family *Cardiobacteriaceae.* Members enriched in the biofloc included members of the phyla *Chloroflexi* and *Thaumarcheota,* the class *Nitrosophaeria*, the order *Nitrosopumilales,* and the family *Nitrosopumilaceae.* Members in the surrounding water included the class *Anaerolineae,* the orders *Caldilineales* and *Chitinophagales*, and the families *Caldilineaceae* and *Saprospiraceae.* As compared to the biofloc and water communities, the shrimp gastrointestinal tract had more genes associated with transport and catabolism, cell signaling, and metabolism (i.e., xenobiotics, lipid, glycan, energy, carbohydrates), but lower numbers of genes attributed to membrane transport, replication and repair, bacterial infectious disease, and antimicrobial drug resistance. This indicates that the microbial communities in shrimp within biofloc systems have greater metabolic potential to provide their host with food and energy and may also help resist disease. Thus, biofloc technology can help establish a healthy microbiome in these intensive culture systems.

Another potential influencer of the microbiome and therefore its benefit to the host is diet. Numerous studies have investigated the effects of diet on the microbiome through use of metataxonomics, but few have investigated the community at the metagenomics level. Lau et al. (2022) collected mud crabs *Scylla olivacea* from the wild and acclimatized them to aquaculture systems and various diets. The gastrointestinal microbiome was compared between feed treatments as well as between captive and wild individuals using both metataxonomics and metagenomics. The diets consisted of fish oil having a high abundance of long-chain polyunsaturated fatty acids (LC-PUFA) or linseed oil and corn oil having a high abundance of C18 PUFA, with an additional diet having linseed oil, corn oil, and an antibiotic (oxolinic acid). The inclusion of the antibiotic resulted in less digestive efficiency as well as lower survival. This diet also resulted in the lowest bacterial diversity whereas fish oil resulted in the highest. Crabs from all diets had significantly different microbiome structures. All captive animals showed an increase in *Vibrio* and a decrease in *Candidatus* Hepatoplasma. The diet containing antibiotics enriched *Carboxylicivirga* and *Sunxiuqinia*. Wild mud crabs had more genes associated with replication and repair, transcription and translation, and nucleotide metabolism. The captive diets led to an increased abundance of metabolic genes, with the fish oil diet having the least carbohydrate metabolism but greater cell motility, membrane transport, and xenobiotic degradation. The antibiotic diet had fewer genes associated with energy metabolism but was enriched in carbohydrate metabolism and transport and catabolism. All captive crabs had more genes associated with SCFA metabolism. *Vibrio* and *Shewanella* contributed the most to lipid metabolism in wild crabs, with *Ruegeria* also participating in captive individuals. Feeding the mixed oils resulted in greater abundances of *Vibrio* and *Shewanella*, which are hypothesized to make LC-PUFA from the C18-PUFA in the diet. The functions and diversity suggest a healthier microbiota and greater metabolic capacity in artificially fed crabs, perhaps due to specifically formulated diets. This study demonstrated that aquaculture conditions can alter the microbiome and its function to improve upon energy and food uptake for the host, allowing for greater growth rates and increased feed efficiency. Thus, a farmer can use less food to produce animals ready for market faster. Diets can be continuously manipulated to optimize these benefits and improve aquaculture efficiency.

A few studies have used metagenomics to investigate antibiotic resistance genes (ARGs) and their transfer in aquaculture environments. An ideal system would harbor fewer organisms with ARGs but also lower horizontal gene transfer events, indicating that the genes that are present are not being transferred between microbes. In comparing wild and captive crabs, Lau et al. (2022) reported similar distribution of ARGs regardless of the environment. This indicates that the aquaculture environment is not necessarily a greater carrier of these genes than natural ecosystems. Within a biofloc system, shrimp intestine contained more ARGs than the biofloc or the surrounding water. Even though the biofloc had the highest microbial diversity, it had the lowest number of ARGs compared to other sample types. Therefore, these complex community aggregates do not seem to increase the risk of antibiotic resistance. On the other hand, in crayfish-rice co-culture systems, more ARG types were found in water and sediment than in crab or crayfish gut (Zhu et al. 2022). When investigating horizontal gene transfer events associated with ARGs, the co-culture systems had significantly fewer transfer events than systems that did not include rice. This data, combined with that described above regarding a healthy microbiome structure, indicates that these co-culture systems are more sustainable than those without rice, and that these integrated systems provide less opportunity for the spread of antibiotic resistance through the microbial community.

10.6 Conclusions

Metagenomics has been used to investigate the potential function of shellfish symbionts and pathogens. Results support the influence of disease and environment on community structure that has been reported using targeted gene sequencing. However, metagenomics has indicated that a change in community structure does not always lead to a change in potential function. These analyses are increasingly important as climate change and anthropogenic stressors continue to impact wild shellfish species and as intensive shellfish aquaculture expands to meet the seafood demand of a growing human population. Although metagenomics provides a view of potential microbiome function and benefits for the host, it will be necessary to combine metagenomics with transcriptomics and proteomics to determine which genes are transcribed and translated. In addition, using these -omics technologies to investigate host gene expression will be vital for understanding the interactions that occur between microbiomes and their shellfish hosts, including how they influence immune function and digestive capability. Expanded use of metagenomics will improve our understanding of dysbiosis and the pathobiome, enabling the aquaculture industry to design systems and microbial manipulation strategies such as pre-and probiotics to maintain or restore a healthy microbiome structure and function. This technology, and those that arise as the field advances, have the potential to reduce losses and improve the productivity and sustainability of shellfish aquaculture worldwide.

References

Aceves AK, Johnson P, Bullard SA, Lafrentz S, Arias CR (2018) Description and characterization of the digestive gland microbiome in the freshwater mussel Villosa nebulosa (Bivalvia: Unionidae). J Molluscan Stud 84(3):240–246. https://doi.org/10.1093/mollus/eyy014

Alma L, Kram KE, Holtgrieve GW, Barbarino A, Fiamengo CJ, Padilla-Gamiño JL (2020) Ocean acidification and warming effects on the physiology, skeletal properties, and microbiome of the purple-hinge rock scallop. Comp Biochem Physiol A Mol Integr Physiol 240:110579. https://doi.org/10.1016/j.cbpa.2019.110579

Anderson DM, Fensin E, Gobler CJ, Hoeglund AE, Hubbard KA, Kulis DM, Landsberg JH, Lefebvre KA, Provoost P, Richlen ML, Smith JL, Solow AR, Trainer VL (2021) Marine harmful algal blooms (HABs) in the United States: history, current status and future trends. Harmful Algae 102:101975. https://doi.org/10.1016/j.hal.2021.101975

Ansorge R, Romano S, Sayavedra L, Porras MÁG, Kupczok A, Tegetmeyer HE, Dubilier N, Petersen J (2019) Functional diversity enables multiple symbiont strains to coexist in deep-sea mussels. Nat Microbiol 4(12):2487–2497. https://doi.org/10.1038/s41564-019-0572-9

Arfken A, Song B, Bowman JS, Piehler M (2017) Denitrification potential of the eastern oyster microbiome using a 16S rRNA gene based metabolic inference approach. PLoS One 12(9): e0185071. https://doi.org/10.1371/journal.pone.0185071

Arfken A, Song B, Allen SK, Carnegie RB (2021) Comparing larval microbiomes of the eastern oyster (Crassostrea virginica) raised in different hatcheries. Aquaculture 531:735955. https://doi.org/10.1016/j.aquaculture.2020.735955

Aßhauer KP, Wemheuer B, Daniel R, Meinicke P (2015) Tax4Fun: predicting functional profiles from metagenomic 16S rRNA data. Bioinformatics 31(17):2882–2884. https://doi.org/10.1093/bioinformatics/btv287

Bass D, Stentiford GD, Wang HC, Koskella B, Tyler CR (2019) The pathobiome in animal and plant diseases. Trends Ecol Evol 34(11):996–1008. https://doi.org/10.1016/j.tree.2019.07.012

Birkbeck TH, Feist SW, Verner-Jeffreys DW (2011) Francisella infections in fish and shellfish. In. J Fish Dis 34(3):173–187. https://doi.org/10.1111/j.1365-2761.2010.01226.x

Boardman CL, Maloy AP, Boettcher KJ (2008) Localization of the bacterial agent of juvenile oyster disease (Roseovarius crassostreae) within affected eastern oysters (Crassostrea virginica). J Invertebr Pathol 97(2):150–158. https://doi.org/10.1016/j.jip.2007.08.007

Cambon-Bonavita MA, Aubé J, Cueff-Gauchard V, Reveillaud J (2021) Niche partitioning in the Rimicaris exoculata holobiont: the case of the first symbiotic Zetaproteobacteria. Microbiome 9(1):87. https://doi.org/10.1186/s40168-021-01045-6

Cano I, van Aerle R, Ross S, Verner-Jeffreys DW, Paley RK, Rimmer GSE, Ryder D, Hooper P, Stone D, Feist SW (2018) Molecular characterization of an Endozoicomonas-like organism causing infection in the King Scallop (Pecten maximus L.). Appl Environ Microbiol 84(3): e00952-17. https://doi.org/10.1128/AEM.00952-17

Chen X, Fan L, Qiu L, Dong X, Wang Q, Hu G, Meng S, Li D, Chen J (2021) Metagenomics analysis reveals compositional and functional differences in the gut microbiota of red swamp crayfish, Procambarus clarkii, grown on two different culture environments. Front Microbiol 12: 735190. https://doi.org/10.3389/fmicb.2021.735190

Chen X, He Z, Zhao J, Liao M, Xue Y, Zhou J, Hoare R, Monaghan SJ, Wang N, Pang H, Sun C (2022) Metagenomic analysis of bacterial communities and antibiotic resistance genes in Penaeus monodon biofloc-based aquaculture environments. Front Mar Sci 8. https://doi.org/10.3389/fmars.2021.762345

Clerissi C, de Lorgeril J, Petton B, Lucasson A, Escoubas JM, Gueguen Y, Dégremont L, Mitta G, Toulza E (2020) Microbiota composition and evenness predict survival rate of oysters confronted to Pacific oyster mortality syndrome. Front Microbiol 11:311. https://doi.org/10.3389/fmicb.2020.00311

Coburn B, Sekirov I, Finlay BB (2007) Type III secretion systems and disease. Clin Microbiol Rev 20(4):535–549. https://doi.org/10.1128/CMR.00013-07

Cowley JA (2016) Bunyaviruses of crustaceans. In: Kibenge FSB, Godoy MG (eds) Aquaculture virology. Elsevier, pp 489–503. https://doi.org/10.1016/B978-0-12-801573-5.00035-8

de Souza Valente C, Wan AHL (2021) Vibrio and major commercially important vibriosis diseases in decapod crustaceans. J Invertebr Pathol 181:107527. https://doi.org/10.1016/j.jip.2020.107527

Destoumieux-Garzón D, Canesi L, Oyanedel D, Travers MA, Charrière GM, Pruzzo C, Vezzulli L (2020) Vibrio–bivalve interactions in health and disease. Environ Microbiol 22(10):4323–4341. https://doi.org/10.1111/1462-2920.15055

Dubé CE, Ky CL, Planes S (2019) Microbiome of the black-lipped pearl oyster pinctada margaritifera, a multi-tissue description with functional profiling. Front Microbiol 10 (JULY):1548. https://doi.org/10.3389/fmicb.2019.01548

Dubert J, Barja JL, Romalde JL (2017) New insights into pathogenic vibrios affecting bivalves in hatcheries: present and future prospects. Front Microbiol 8(MAY):1–16. https://doi.org/10.3389/fmicb.2017.00762

Dupont S, Lokmer A, Corre E, Auguet J-C, Petton B, Toulza E, Montagnani C, Tanguy G, Pecqueur D, Salmeron C, Guillou L, Desnues C, la Scola B, Bou Khalil J, de Lorgeril J, Mitta G, Gueguen Y, Escoubas J-M (2020) Oyster hemolymph is a complex and dynamic ecosystem hosting bacteria, protists and viruses. Animal Microbiome 2(1):12. https://doi.org/10.1186/s42523-020-00032-w

El-Sayed AFM (2021) Use of biofloc technology in shrimp aquaculture: a comprehensive review, with emphasis on the last decade. Rev Aquac 13(1):676–705. https://doi.org/10.1111/raq.12494

FAO (2022) The state of world fisheries and aquaculture 2022. Towards blue transformation. FAO. https://doi.org/10.4060/cc0461en

Gacutan RQ, Llobrera AT, Santago CB, Gutierrez PJ, Lio G (1977) A suctorean parasite of Penaeus monodon larvae. SEAFDEC Aquaculture Department Quarterly Research Report 1(1):6–11. http://hdl.handle.net/1834/9138

Gao S, Pan L, Huang F, Song M, Tian C, Zhang M (2019) Metagenomic insights into the structure and function of intestinal microbiota of the farmed Pacific white shrimp (Litopenaeus vannamei). Aquaculture 499:109–118. https://doi.org/10.1016/j.aquaculture.2018.09.026

García Bernal M, Trabal Fernández N, Saucedo Lastra PE, Medina Marrero R, Mazón-Suástegui JM (2017) Streptomyces effect on the bacterial microbiota associated to Crassostrea sikamea oyster. J Appl Microbiol 122(3):601–614. https://doi.org/10.1111/jam.13382

Ghadersohi A, Owens L (1999) Isolation, characterisation and DNA analysis of Mycoplasma spp. from moribund prawns Penaeus monodon cultured in Australia. Dis Aquat Org 35(1):53–61

Gómez-Gutiérrez J, Morales-Ávila JR (2016) Parasites and diseases. In: Siegel V (ed) Biology and ecology of Antarctic Krill. Springer, Cham, pp 351–386. https://doi.org/10.1007/978-3-319-29279-3_10

Guillou L, Viprey M, Chambouvet A, Welsh RM, Kirkham AR, Massana R, Scanlan DJ, Worden AZ (2008) Widespread occurrence and genetic diversity of marine parasitoids belonging to Syndiniales (Alveolata). Environ Microbiol 10(12):3349–3365. https://doi.org/10.1111/j.1462-2920.2008.01731.x

Hickey ME, Lee JL (2018) A comprehensive review of vibrio (Listonella) anguillarum: ecology, pathology and prevention. Rev Aquac 10(3):585–610. https://doi.org/10.1111/raq.12188

Hudson DA, Lester RJG (1994) Parasites and symbionts of wild mud crabs Scylla serrata (Forskal) of potential significance in aquaculture. Aquaculture 120:183

Infante-Villamil S, Huerlimann R, Jerry DR (2021) Microbiome diversity and dysbiosis in aquaculture. Rev Aquac 13(2):1077–1096. https://doi.org/10.1111/raq.12513

Iwai S, Weinmaier T, Schmidt BL, Albertson DG, Poloso NJ, Dabbagh K, DeSantis TZ (2016) Piphillin: improved prediction of metagenomic content by direct inference from human microbiomes. PLoS One 11(11):e0166104. https://doi.org/10.1371/journal.pone.0166104

Kakoolaki S, Afsharnasab M (2016) Prevalence and intensity of protozoan ectoparasite of the white leg shrimp (penaeus ind) in Helleh site, South of Iran. Iran J Aquat Anim Health 2(1):17–23

King GM, Judd C, Kuske CR, Smith C (2012) Analysis of stomach and gut microbiomes of the eastern oyster (Crassostrea virginica) from coastal Louisiana, USA. PLoS One 7(12):e51475. https://doi.org/10.1371/journal.pone.0051475

Langille MG, Zaneveld J, Caporaso JG, McDonald D, Knights D, Reyes JA, Clemente JC, Burkepile DE, Vega Thurber RL, Knight R, Beiko RG, Huttenhower C (2013) Predictive functional profiling of microbial communities using 16S rRNA marker gene sequences. Nat Biotechnol 31:814–821

Lau N-S, Ting SY, Sam K-K, M J, Wong SC, Wu X, Waiho K, Fazhan H, Shu-Chien AC (2022) Comparative analyses of Scylla olivacea gut microbiota composition and function suggest the capacity for polyunsaturated fatty acid biosynthesis. Microb Ecol. https://doi.org/10.1007/s00248-022-02046-0

Leitão JH (2020) Microbial virulence factors. Int J Mol Sci 21(15):1–6. https://doi.org/10.3390/ijms21155320

McCuaig B, Peña-Castillo L, Dufour SC (2020) Metagenomic analysis suggests broad metabolic potential in extracellular symbionts of the bivalve Thyasira cf. gouldi. Anim Microbiome 2(1):7. https://doi.org/10.1186/s42523-020-00025-9

Nakagawa S, Takai K (2008) Deep-sea vent chemoautotrophs: diversity, biochemistry and ecological significance. FEMS Microbiol Ecol 65(1):1–14. https://doi.org/10.1111/j.1574-6941.2008.00502.x

Numan M, Bashir S, Mumtaz R, Tayyab S, Rehman NU, Khan AL, Shinwari ZK, Al-Harrasi A (2018) Therapeutic applications of bacterial pigments: a review of current status and future opportunities. 3 Biotech 8(4):207. https://doi.org/10.1007/s13205-018-1227-x

Paillard C, le Roux F, Borrego JJ (2004) Bacterial disease in marine bivalves, a review of recent studies: trends and evolution. Aquat Living Resour 17(4):477–498. https://doi.org/10.1051/alr:2004054

Pelletier MC, Ebersole J, Mulvaney K, Rashleigh B, Gutierrez MN, Chintala M, Kuhn A, Molina M, Bagley M, Lane C (2020) Resilience of aquatic systems: review and management implications. Aquat Sci 82(2):1. https://doi.org/10.1007/s00027-020-00717-z

Pierce ML, Ward JE (2019) Gut microbiomes of the eastern oyster (Crassostrea virginica) and the blue mussel (Mytilus edulis): temporal variation and the influence of marine aggregate-associated microbial communities. MSphere 4(6):e00730-19. https://doi.org/10.1128/msphere.00730-19

Pimentel ZT, Dufault-Thompson K, Russo KT, Scro AK, Smolowitz RM, Gomez-Chiarri M, Zhang Y (2021) Microbiome analysis reveals diversity and function of Mollicutes associated with the eastern oyster, Crassostrea virginica. MSphere 6(3):e00227-21. https://doi.org/10.1128/msphere.00227-21

Ramachandran P, Reed E, Commichaux S, Strain E, Depaola A, Rikard S, Ottesen A (2018) Characterization of the microbiota of oyster larvae (Crassostrea virginica) and tank water from an aquaculture system with high and low larval survival rates. Genome Announc 6(25): e00597-18. https://doi.org/10.1128/genomeA.00597-18

Richard JC, Leis E, Dunn CD, Agbalog R, Waller D, Knowles S, Putnam J, Goldberg TL (2020) Mass mortality in freshwater mussels (Actinonaias pectorosa) in the Clinch River, USA, linked to a novel densovirus. Sci Rep 10(1):14498. https://doi.org/10.1038/s41598-020-71459-z

Romero X, Jimønez R (2002) Histopathological survey of diseases and pathogens present in redclaw crayfish, Cherax quadricarinatus (Von martens), cultured in Ecuador. J Fish Dis 25: 653–667

Sheik CS, Cleaves HJ, Johnson-Finn K, Giovannelli D, Kieft TL, Papineau D, Schrenk MO, Tumiati S (2020) Abiotic and biotic processes that drive carboxylation and decarboxylation reactions. Am Mineral 105(5):609–615. https://doi.org/10.2138/am-2020-7166CCBYNCND

Small HJ, Shields JD, Reece KS, Bateman K, Stentiford GD (2012) Morphological and molecular characterization of Hematodinium perezi (Dinophyceae: Syndiniales), a dinoflagellate parasite of the harbour crab, Liocarcinus depurator. J Eukaryot Microbiol 59(1):54–66. https://doi.org/10.1111/j.1550-7408.2011.00592.x

Valenzuela T, Rilling JI, Larama G, Acuña JJ, Campos M, Inostroza NG, Araya M, Altamirano K, Fujiyoshi S, Yarimizu K, Maruyama F, Jorquera MA (2021) 16s rrna–based analysis reveals differences in the bacterial community present in tissues of choromytilus chorus (Mytilidae, bivalvia) grown in an estuary and a bay in southern Chile. Diversity 13(5):209. https://doi.org/10.3390/d13050209

van Eynde B, Christiaens O, Delbare D, Shi C, Vanhulle E, Yinda CK, Matthijnssens J, Smagghe G (2020) Exploration of the virome of the European brown shrimp (Crangon crangon). J Gen Virol 101(6):651–666. https://doi.org/10.1099/jgv.0.001412

Vonaesch P, Anderson M, Sansonetti PJ (2018) Pathogens, microbiome and the host: emergence of the ecological Koch's postulates. FEMS Microbiol Rev 42(3):273–292. https://doi.org/10.1093/femsre/fuy003

Yang S, Li X, Zang Z, Li J, Wang A, Shi Y, Zhang X, Gu Z, Zheng X, Vasquez HE (2022) Effect of fresh and spray-dried microalgal diets on the growth, digestive enzymatic activity, and gut microbiota of juvenile winged pearl oyster Pteria penguin. Aquac Rep 25:101251. https://doi.org/10.1016/j.aqrep.2022.101251

Zeng S, Zhou R, Bao S, Li X, Deng Z, Hou D, Weng S, He J, Huang Z (2020) Identification of multigene biomarker for shrimp white feces syndrome by full-length transcriptome sequencing. Front Genet 11:71. https://doi.org/10.3389/fgene.2020.00071

Zhao Q, Gu H (2021) Gastrointestinal tract microbial community of Babylonia areolata and its diversity are closely correlated with the outbreak of disease. Aquac Res 53:1636–1648. http://www.drive5.com/usear

Zhu X, Ji L, Cheng M, Wei H, Wang Z, Ning K (2022) Sustainability of the rice-crayfish co-culture aquaculture model: microbiome profiles based on multi-kingdom analyses. Environ Microbiome 17(1):27. https://doi.org/10.1186/s40793-022-00422-4

Shellfish Microbiome and Probiotics: A Decade in Review

11

Andrea M. Tarnecki and Francisca Burgos

Abstract

Significant economic losses in shellfish aquaculture resulting from disease drive the development of sustainable prevention strategies. For aquatic animal production, antibiotic concerns have shifted strategies to microbial manipulation using probiotics. These live microbial additives are supplemented to the culture system through the water or feed to improve water quality, growth, digestion, immune status, disease resistance, microbiome structure and function. With *Bacillus* and lactic acid bacteria leading research and commercialization efforts, the probiotic feed market is valued at billions of dollars and provides production advantages to mollusks and crustaceans. As -omics and synthetic biology are incorporated into probiotic studies, understanding of shellfish–microbe interactions is rapidly expanding, leading to the incorporation of synbiotics, paraprobiotics, and live therapeutics that enhance probiotic benefits. Improvements in probiotic formulations continue to decrease mortalities and improve production efficiency, increasing the economic and environmental sustainability of the aquaculture industry. This chapter provides an overview of the past 10 years' research on probiotic selection, use, and benefits, as well as future directions for shellfish aquaculture.

A. M. Tarnecki (✉)
Auburn University Shellfish Lab, School of Fisheries, Aquaculture and Aquatic Sciences, Auburn University, Dauphin Island, AL, USA
e-mail: atarnecki@mote.org

F. Burgos
Escuela Superior Politécnica del Litoral, ESPOL, Facultad de Ingeniería Marítima y Ciencias del Mar, Guayaquil, Ecuador

A. Diwan et al. (eds.), *Microbiome of Finfish and Shellfish*,
https://doi.org/10.1007/978-981-99-0852-3_11

Keywords

Shellfish · Microbiome · Probiotics · Synbiotics · Therapeutics

11.1 Introduction

Shellfish aquaculture, including crustaceans and mollusks, produced 28.9 million tons worth US$111.3 billion in 2020, and has continued to grow worldwide each year (FAO 2022). Although difficult to estimate, diseases cost shellfish farmers billions of dollars on a yearly basis. For instance, in 2 years' time, a white spot syndrome outbreak in Asia resulted in US$6 billion in losses for shrimp farmers in the early 1990s (Lafferty et al. 2015). Disease prevention is imperative to the economic success and continued expansion of the shellfish aquaculture industry. Historically, pathogens have been controlled using antibiotics, often prophylactically, with the added benefit of improving growth rates. The indiscriminate use of antibiotics has led to an increase in antibiotic-resistant aquatic and human pathogens (Hossain et al. 2022) and has given consumers a negative perception of farmed seafood (López-Mas et al. 2021). Being of human health and environmental concern, numerous countries have instated regulations regarding antibiotic use in aquaculture (Hossain et al. 2022), and with the added pressure of consumer demand, the industry is pursuing other preventative measures for pathogen control.

One promising option for disease prevention is the use of probiotics, which are live beneficial microorganisms that provide aquatic system or host health benefits. Probiotics are typically either bacteria or yeast and may be added directly to the rearing water or incorporated into live or pelleted feeds, depending on the life stage of the animal. There is a substantial market for animal feeds incorporating probiotics, valued at US$4.4 billion in 2020 and growing (Markets and Markets 2021). With the reported benefits of probiotics including improved water quality, immune system stimulation, and enhanced digestion, leading to disease prevention, faster growth, and greater survival (Ringø 2020), research into beneficial microorganisms in aquaculture has increased exponentially in the past decade.

Since the 1970s, the use of probiotics as farm animal feed supplements has been well documented. In shellfish, probiotics research took off in the early 1990s (Ringø 2020). While a vast majority of the shellfish literature has been aimed at shrimp, investigations into probionts for other shellfish species are increasing. A Web of Science search[1] indicates there have been over 260 publications on probiotics and the microbiome in shellfish in the last 10 years (Fig. 11.1). Manuscripts dedicated to shrimp account for 56% of these communications, whereas all other shellfish

[1] Search performed on 02 Sept 2022. Search terms as follows for shellfish (shrimp term removed to determine publications associated with shrimp): Topic = probiotic? AND Topic = shellfish* OR crab* OR bivalve* OR oyster* OR crayfish* OR scallop* OR mussel* OR lobster* OR snail* OR mollus* OR abalone* OR clam* OR shrimp* AND Topic = microbio* OR microflora*; For fish: Topic = probiotic? AND Topic = fish* NOT shellfish* AND Topic = microbio* OR microflora*

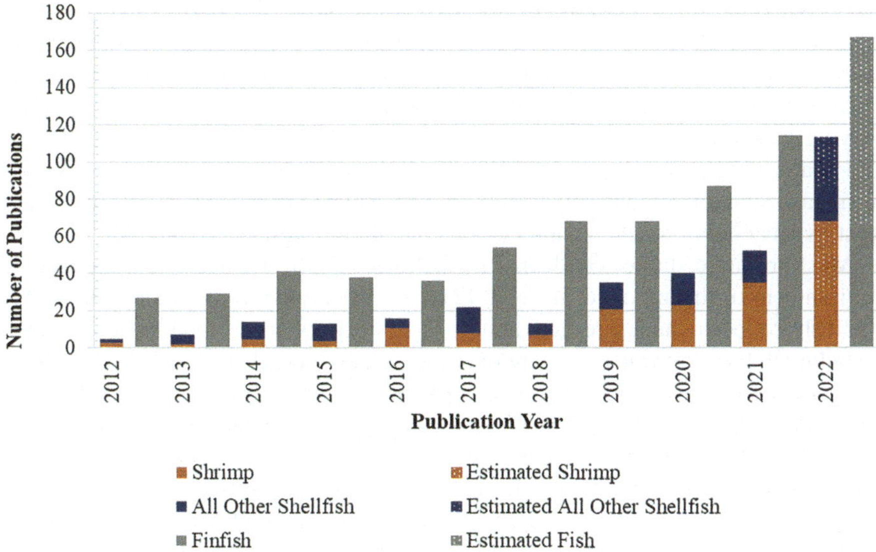

Fig. 11.1 Publications related to shellfish probiotics and the microbiome over the past 10 years

included in the search (mollusks and other crustaceans) represent the minority. In contrast, the fish literature includes over 560 publications, on track to reach 100 in 2022 alone. Therefore, probiotics research in fish is much further advanced than that on shellfish, with the non-shrimp shellfish lagging further behind. This chapter aims to provide an overview of shellfish probiotics research from the past decade (2012–2022), including discovery, implementation, mechanisms of action, and the future of microbial management in aquaculture.

11.2 Bioprospecting and Production

For many years, efforts have focused on collecting beneficial bacterial species from the environment that can be exploited for aquatic animal production. As a result, many strains of bacteria and yeast have been identified that enhance shellfish aquaculture. It is generally recognized that a microbe isolated from the host species of interest will exert the greatest influence on its physiology and environmental performance; however, many of them are not suited for commercialization due to impacts of industrial production on stability and viability.

11.2.1 Probiont Isolation

Aquatic organisms harbor a distinct microflora from their surrounding environment, even at larval stages (Dupont et al. 2020; Zhu et al. 2022; Zhang et al. 2022; Yu et al.

2019; Xu et al. 2022; Sawabe et al. 2016; Musella et al. 2020). Environmental conditions and host diet are known to influence the colonization of microbes in shellfish. However, host physiology (i.e., immune function, health status, taxonomy, genetics) also impacts the microbiome (Kroetsch et al. 2020) as evidenced by varying bacterial communities in species reared in the same environment (Pierce and Ward 2019; Vezzulli et al. 2018). The intimate interaction between environment, host, and microbiome suggests probiotics isolated from the host organism or at least its local environment should demonstrate the greatest effectiveness.

Most probiotic studies conducted in shellfish collect microbes from the rearing environment or the animal of interest. Early studies aiming to isolate probiotics for shellfish collected and screened hundreds of isolates from aquaculture systems including the rearing water, varying live stages, and microalgal feeds. Occasionally, bacteria are isolated from the natural environment including the water and sediments where the host organism resides. When sampling adult animals, the gastrointestinal tract is by far the most common place to collect bacterial isolates for probiotic screening, and this technique has been employed in crabs (Talpur et al. 2013; Wu et al. 2014; Yang et al. 2019), crayfish (Ambas et al. 2013; Didinen et al. 2016; Xu et al. 2021a), abalone (Amin et al. 2020; Grandiosa et al. 2018; Huddy and Coyne 2014; Xiaolong et al. 2019; Zhao et al. 2018), snails (Dushku et al. 2019; Dushku et al. 2020), mussels (Fajardo et al. 2014), oysters (Fajardo et al. 2014; Abasolo-Pacheco et al. 2016), scallops (Fajardo et al. 2014; Abasolo-Pacheco et al. 2016; de la Vega et al. 2020), and shrimp (Bolívar-Ramírez et al. 2021; Chen et al. 2021; Won et al. 2020). Samples may also be collected from other tissues such as hepatopancreas (Didinen et al. 2016; Fajardo et al. 2014), gill (Didinen et al. 2016) or hemolymph (Desriac et al. 2014; Offret et al. 2019; Sumithra et al. 2019). Nearly 850 isolates collected from hemolymph of oysters, scallops, mussels, and clams were screened for probiotic potential (Desriac et al. 2014), with 26 showing antimicrobial activity against aquatic pathogens. Putative probiotic strains have also been collected from mollusk and crab hemolymph (Offret et al. 2019; Sumithra et al. 2019), demonstrating that this non-sterile tissue may be an underexplored site for probiotic mining.

Although not as common, putative shellfish probiotics can be tested that were isolated from unrelated organisms, such as fishes (Du et al. 2019; Nguyen et al. 2014; Valipour et al. 2019). It should be noted that these studies were successful at identifying isolates with beneficial activities in crustaceans. Although the specific functions of an isolate can be expected to show a level of species-specificity, the shared bacteria among microbiomes of different aquatic species suggest probiotic benefits may reach beyond organisms from which they were originally isolated.

Regardless of the original source of probiotic isolates, the species-specificity of microbiomes and the unique microbial communities found within each tissue in aquatic organisms indicates a wealth of potential probiotic microbes that can be cultivated and screened from each species. Collecting from this rich source of beneficial bacteria is limited by biases associated with the composition of available microbiological media and the required growth factors of the bacteria present. Therefore, the repository of isolates for testing can be expanded by using a variety of media formulations and culture techniques such as in situ cultivation, nutrient dilution, and extended incubation times. However, identifying an effective probiont is only the beginning of the process, and upscaling production of probiotic strains for commercialization is a significant challenge for the industry.

11.2.2 Desired Probiotic Characteristics

Testing and selection of probiotic candidates (bacteria and yeast strains) covers a series of characteristics that determine if they will survive the aquaculture environment and benefit the host. Environmental tolerance makes the probiotic more desirable for use as shellfish can be grown at a variety of salinities and temperatures, and strains that can survive in these ranges can be used across geographic locations. Many probiotics impact the gastrointestinal tract via ingestion, and the ability to survive low pH and bile salts improves survival during passage through the stomach. In addition, strains may be tested for the ability to adhere to mucosa and form biofilms which increases their chance for establishing within the microbiome. Whether these screening mechanisms are used or not, probiotics are tested within aquaculture systems to ensure they provide an advantage to shellfish production such as improving survival, growth, feed efficiency, immune status, or stress resistance.

The most common screening method for probiotic selection is via antimicrobial assays of the potential probiotics against known shellfish and foodborne pathogens, particularly within the genus *Vibrio*. The pathogens that have been used to screen for putative probiotics in shellfish are listed as follows.

Gram-Negative: *Aeromonas caviae*, *A. hydrophila*, *A. salmonicida*, *A. veronii*, *Edwardsiella tarda*, *Escherichia coli*, *Photobacterium damselae*, *Pseudomonas fluorescens*, *P. putida*, *Roseovarius crassostreae*, *Salmonella enterica*, *Vibrio alginolyticus*, *V. anguillarum*, *V. anguillarum*-related (VAR), *V. cholerae*, *V. coralliilyticus*, *V. harveyi*, *V. owensii*, *V. parahaemolyticus*, *V. proteolyticus*, *V. splendidus*, *V. tapetis*, *V. tubiashii*, *V. vulnificus*, *Yersinia ruckeri*.

Gram-Positive: *Bacillus cereus*, *B. subtilis*, *Carnobacterium maltaromaticum*, *Clostridium perfringens*, *C. bifermentans*, *Enterococcus faecium*, *E. faecalis*, *Lactococcus garviae*, *Listeria innocua*, *L. monocytogenes*, *Staphylococcus aureus*, *β-Streptococcus*, *Vagococcus salmoninarum*.

This mode of probiotic selection assumes that bacteria with antagonistic activity in vitro will also inhibit the pathogen in vivo. Due to the presence of other microorganisms, host immune activity, and other compounds present in a more complex environment, it is not guaranteed that the interactions leading to pathogen inhibition in the lab will translate to benefits for the animal in an aquaculture system. Therefore, it is important to perform system-relevant trials to verify the effectiveness of probiotics selected based on their antimicrobial activity.

In the Peruvian scallop, an isolate of *Pseudoalteromonas* obtained from mollusk hemolymph demonstrated antagonistic activity against *V. harveyi* during the screening process. This strain, designated h*Cg*-6, conferred mature adult scallops protection against *V. harveyi* infection in experimental trials, with probiotic-supplemented animals having significantly higher survival 5 days post-challenge (Offret et al. 2019). Similarly, in juvenile Pacific calico scallop, exposure to the putative probiont *Pediococcus pentosaceus* G11 through supplementation of standard diets improved survival nearly 30% following challenge with *V. parahaemolyticus* (Yang et al. 2019). In the narrow-clawed crayfish, Wu et al. (Wu et al. 2014) confirmed the activity of two *Bacillus* probiotics isolated from the intestine of mud crabs (*B. pumilus* and *B. subtilis*), as both isolates inhibited *V. parahaemolyticus* in antagonism tests and improved survival of the crayfish following pathogen challenge. On the other hand, despite the supernatant of *L. plantarum* inhibiting *V. harveyi* and *V. anguillarum* in antagonism studies, there was no benefit of adding the bacterium to the feed of Pacific white shrimp (Thompson et al. 2022).

These studies indicate there is potential for isolation of effective probionts based on their antimicrobial activity, but this method may not always result in isolates that are beneficial in aquaculture systems. It should also be noted that these studies did not quantify the pathogen during these challenges, and therefore the probiotic may be working through another mechanism such as immune system stimulation, competitive exclusion, or production of other metabolites as opposed to direct antimicrobial inhibition.

11.2.3 Commercial Production

Production of probiotic feed is dominated by established companies that partake in the entire process, from prospecting new compounds to selling supplemented products (Speight et al. 2022). There are many challenges in selecting and manufacturing probiotics, including: (1) demonstration of safety, (2) retaining beneficial activity throughout the production process, (3) maintaining viability to increase shelf-life, and (4) identifying strains with properties that allow them to establish in harsh environments, such as those within the gastrointestinal tract. As these strains are produced and used in high concentrations, handled by humans, and applied to animals destined for consumption, manufacturers must demonstrate that they are safe. Probiotic safety is determined by the European Commission which provides Qualified Presumption of Safety (QPS) status or the US Food and Drug Administration which determines if an additive is Generally Recognized As Safe (GRAS). Currently, there are 17 species of yeast and 84 species of bacteria with QPS status, with 10 species each of the genera *Lactobacillus* and *Bacillus* included (EFSA Panel on Biological Hazards (BIOHAZ) 2022). The GRAS list, recognizing substances at the strain level and including specific bacterial products, currently has GRAS Notices for 26 *Lactobacillus* and 16 *Bacillus* strains. As numerous members within these two genera as presumed safe, these are often targets for commercial manufacturing and the most researched in shellfish (see Sect. 11.3: Commonly Used Probiotics).

Probiotics proven effects must be retained during large-scale commercialization. The industrial process therefore optimizes production strategies that result in high survival and ensure probiotic properties are not lost due to culture conditions (Forssten et al. 2011). Once these techniques are refined, various methods are employed to maintain viability during storage. Probiotic shelf-life is influenced by oxygen, temperature, and moisture (Forssten et al. 2011; Sarao and Arora 2017; Gao et al. 2021) and products must be packaged to reduce exposure to these elements. Typically, this involves freeze-drying to remove moisture. However, this process is potentially damaging for the microbial cells, particularly in groups that cannot produce resistant endospores. To improve viability, protectants such as trehalose, sucrose, skimmed milk, glycerol, and others are added to prevent desiccation damage (Berninger et al. 2018).

Following drying, probiotics are often covered in a protective layer in a process called encapsulation. Numerous compounds are used to encapsulate probiotics, including gellan and xanthan gum, chitosan, alginate, gelatin, and starch (Sarao and Arora 2017). Benefits of encapsulation include protection against elements during storage and resistance to conditions in the environment or animal, such as low pH or exposure to bile salts (Sarao and Arora 2017), that improve passage through the gastrointestinal tract and increase the chances of the probiotic establishing in the gut. The final capsules, tablets, or powders that are derived from drying and/or encapsulation maintain viability for up to 2 years (Forssten et al. 2011).

Powdered probiotics can be added directly to the rearing water or incorporated into feeds. Strains supplemented to the water column work via competitive exclusion, direct inhibition of pathogenic bacteria, and/or improvement of water quality. Commercial probiotics originate from both indigenous and exogenous microbiomes of aquatic animals but may also be selected due to evidence of effectiveness in livestock. The main strains of bacteria used as water probiotics include *Lactobacillus acidophilus, Bacillus subtilis, B. licheniformis, Nitrobacter* spp., *Aerobacter*, and the yeast *Saccharomyces cerevisiae.* In shrimp ponds strains of *Pseudomonas*, *Acinetobacter*, *Cellulomonas*, *Rhodopseudomonas*, and *Nitrosomonas* also act as bioremediators, enhancing decomposition of organic substances in the water column and on the pond bottom that could be toxic. Prior to use at aquaculture farms, it is common to activate powdered commercial probiotics to increase the number of active cells (Liu et al. 2018) and aid in establishment in intensive systems. Probiotics that perform in the gut can be incorporated directly into feeds using binders such as egg white and fish oil or encapsulated and dried using high or low temperature technologies (Liu et al. 2019) to improve viability.

11.3 Commonly Used Probiotics

11.3.1 Bacillus

The most reported bacteria used as probiotics in shellfish are within the genus *Bacillus*. *Bacillus* are Gram-positive rods capable of forming endospores during times of nutritional stress. These dormant structures are resistant to desiccation, ultraviolet light, extreme temperatures, and some chemicals. The production of endospores makes this group a desirable target for probiotic development as these structures help maintain viability throughout production and extend shelf-life. The group has demonstrated numerous benefits for aquatic species including production of antimicrobials, competitive exclusion of pathogens, immunomodulation, and improving digestion (Kuebutornye et al. 2019). Members of this genus are capable of degrading complex organic matter and can play a role in improving water quality and recycling nutrients. As *Bacillus* species are readily culturable from aquatic animals and their environments, as well as commercially available, they are widely used in aquaculture and much research has focused on these bacteria in shellfish.

The most widely used probiotics in mollusks and crustaceans are members of the *B. subtilis* group which include *B. subtilis, B. amyloliquefaciens, B. pumilus*, and *B. licheniformis*. Members of this group produce antimicrobials and other secondary metabolites that are widely used in industrial processes (Caulier et al. 2019; Harwood et al. 2018). As a result, this group is applied in agriculture to improve production of poultry (Zhang et al. 2021), swine (Li et al. 2021), fish and shellfish (Olmos et al. 2020), and plants (Dunlap et al. 2019). Specifically, within shellfish, the *B. subtilis* group has been tested in the oysters (Fdhila et al. 2017; Karim et al. 2013; Lai et al. 2020; Stevick et al. 2019; Sánchez-Ortiz et al. 2020), abalone (Xiaolong et al. 2019; Gao et al. 2018), crayfish (Ambas et al. 2013; Xu et al. 2021a), crab (Wu et al. 2014), lobster (Nguyen et al. 2014; Middlemiss et al. 2015), and shrimp (Won et al. 2020; Aribah et al. 2022; Riet et al. 2021).

Using a *B. pumilus* strain isolated from a marine sponge, Karim et al. (Karim et al. 2013) improved survival of Eastern oyster challenged with the pathogens *Roseovarius crassostreae* and *V. tubiashii*. Trials in abalone demonstrated both *B. amyloliquefaciens* and *B. licheniformis*, when combined with basal diets, improved survival, growth rate, and feed conversion ratios, as well as boosted immune function and protected against *V. parahaemolyticus* (Xiaolong et al. 2019; Gao et al. 2018). Similar benefits were reported in Pacific white shrimp using *Bacillus* from sponges, with *B. subtilis* outperforming *B. casei* and *B. altitudinis* (Aribah et al. 2022). *B. subtilis* also provided greater benefits to juvenile shrimp than *Lactococcus lactis*, although both strains increased growth, digestive capacity, immune response, and survival following *V. parahaemolyticus* challenge (Won et al. 2020). Interestingly, each probiotic had unique advantages, with *L. lactis* increasing intestinal villi height supporting greater surface area for nutrient absorption and *B. subtilis* increasing antioxidant activity. Strains of *B. pumilus* isolated from the intestinal tract of mud crabs also resulted in higher expression of immune-related genes (Wu et al. 2014). Although neither *B. subtilis* or *B. mycoides* impacted the growth or survival of marron crayfish before or after challenge with *V. mimicus*, there was evidence of increased recovery in total hemocyte counts and reduced bacteremia with probiotic supplementation (Ambas et al. 2013).

B. subtilis group probiotics have also been used to combat infections by viruses, including white spot syndrome virus (WSSV) in the red swamp crayfish (Xu et al. 2021a; Lai et al. 2020). In addition to reducing mortality to the virus, *B. amyloliquefaciens* strains lowered the number of viral DNA detected in challenged individuals (Lai et al. 2020) and boosted immunity and digestive enzyme activity (Xu et al. 2021a; Lai et al. 2020). Some studies have genetically modified members of this group to enhance protection against WSSV (Riet et al. 2021; Dekham et al. 2022) and these are detailed further under Sect. 11.5: Beyond Probiotics in this chapter.

As shellfish are often maintained in systems that require live algal feed during early life stages, interactions between probiotics and microalgae should be investigated. Sánchez-Ortiz (Sánchez-Ortiz et al. 2020) tested *B. licheniformis* and *B. subtilis* with algal species used in oyster spat culture: *Isochrysis galbana* and *Chaetoceros calcitrans*. These coculture experiments indicated that probiotics did not impact the growth of *I. galbana* but did reduce the growth of *C. calcitrans* when compared to a non-supplemented control. This suggests a negative interaction between some probiotics and algae; however, the authors stated that despite the reduction in growth for *C. calcitrans*, larvae fed this probiotic-supplemented species grew larger and had greater survival than controls. Thus, larvae more effectively digest or absorb nutrients from the algae when given probiotics. Probiotic-microalgae interactions are not often examined but warrant further research to ensure supplements do not negatively impact feeding during these early life stages.

11.3.2 Lactic Acid Bacteria

The second most common group of bacteria used in shellfish probiotic trials is the lactic acid bacteria (LAB). These bacteria are Gram-positive, but do not form endospores and may be either rods or cocci. LAB in shellfish increase feed efficiency, provide digestive enzymes and nutrients to their host, increase growth rate, improve stress and disease resistance, stimulate the immune system, and improve water quality (Ringø et al. 2020). However, they have complex growth requirements (Ficoseco et al. 2018) which can make them more expensive to produce than other probionts. Genera that have been employed in shellfish include *Lactobacillus* (Talpur et al. 2013; Didinen et al. 2016; Dushku et al. 2019; Dushku et al. 2020; de la Vega et al. 2020; Nguyen et al. 2014; Valipour et al. 2019; Foysal et al. 2021; Foysal et al. 2020; Li et al. 2022), *Enterococcus* (Grandiosa et al. 2020), and *Pediococcus* (Yang et al. 2019).

When supplemented to Catarina scallop, *Lactobacillus plantarum* increased growth rate, decreased antioxidant activity, and reduced gene expression associated with immune function (de la Vega et al. 2020), which may indicate a reduced level of stress in the animal. *L. plantarum* isolated from snail adhered to mucosa while promoting phagocytic and bactericidal activity in hemolymph (Dushku et al. 2019). This probiotic upregulated gene expression for receptors that activate the immune response. In a follow-up study, this strain reestablished a healthy microbiome following dysbiosis from *Listeria monocytogenes* challenge, while reducing clinical signs of disease and mortality from the pathogen (Dushku et al. 2020). *L. plantarum* supplemented to the blue swimming crab through the water not only led to lower mortalities, but also reduced *Vibrio* counts and increased digestive enzyme activity (Talpur et al. 2013). A combination of *L. acidophilus* and *L. plantarum* fed to marron crayfish increased protein and energy levels in the tail muscle, improved hepatocyte and gut morphology, and increased hemocyte counts, but had no effect on growth (Foysal et al. 2020). *L. plantarum* exerted immune and digestive enzyme effects when fed to the narrow-clawed crayfish (Valipour et al. 2019).

In Pacific white shrimp, *L. plantarum* improved survival, growth, and feed efficiency (Bolívar-Ramírez et al. 2021; et al. 2016). *L. pentosus* increased survival following *V. parahaemolyticus* challenge and was demonstrated by qPCR to colonize the gastrointestinal tract where it reduced the relative abundance of *Vibrio* (Du et al. 2019). This study also showed that, when the surface proteins of the probiotic were removed, it was unable to confer the same benefits as it may have lost the ability to colonize the intestine. *L. lactis* was effective at preventing mortalities in Pacific white shrimp following challenge with *V. alginolyticus* and *V. parahaemolyticus* while increasing weight gain and immune function (Won et al. 2020).

Probiotic trials using *Enterococcus* and *Pediococcus* demonstrated similar benefits to those of *Lactobacillus*. Both *L. reuteri* and *P. acidilactici* improved water quality and resulted in greater growth in Pacific white shrimp (Wu et al. 2022; Khademzade et al. 2020). Yang et al. (Yang et al. 2019) tested *E. faecalis* and *P. pentosaceus* in mud crab. Only the *Pediococcus* strain improved survival and increased lysozyme activity following a *V. parahaemolyticus* challenge. However, both probiotics led to an increase in antioxidant capacity. *P. pentosaceus* at 10^9 CFU/g of feed was the most effective at improving survival to WSSV in Pacific white shrimp, accompanied by benefits including higher weight gain, digestive enzyme activity, and expression of immune-related genes (Le et al. 2022). *E. faecium* was found to colonize the gastrointestinal tract and improve survival when administered via live feed enrichment to Pacific white shrimp; however, there was no impact on water quality, growth rate, or feed conversion ratios (Du et al. 2021).

11.3.2.1 Other Probiotic Organisms

Streptomyces (García Bernal et al. 2017; García-Bernal et al. 2019), *Clostridium* (Foysal et al. 2019), and *Marinilactibacillus* (Panigrahi et al. 2021) are other genera of Gram-positive bacteria tested as probiotics. *Streptomyces* strains collected from marine sediments altered overall growth and antioxidant activity in oyster species (García-Bernal et al. 2019). *C. butyricum* fed to marron crayfish enhanced immune activity and growth (Foysal et al. 2019). Adding either *Marinilactibacillus* or Gram-negative *Novosphingobium* to Indian prawn tanks enhanced water quality, improved overall survival, boosted digestive enzyme activity, and increased expression of innate immune factors, with these benefits improving in combination with biofloc systems as compared to clear water (Panigrahi et al. 2021).

Within the Gram-negative bacteria, the most tested genus is *Phaeobacter*. These bacteria are members of the Roseobacter clade and provide numerous benefits in aquaculture through the production of bactericidal compounds such as tropodithietic acid (TDA) and inhibition of aquatic pathogens (Sonnenschein et al. 2021). Porsby et al. (Porsby and Gram 2016) demonstrated that *P. inhibens* and the TDA it produced prevented the growth of *V. vulnificus*. Although this study was designed to eliminate a human pathogen, the ability of this species to impede *V. vulnificus* suggests potential to prevent *Vibrio* diseases of aquatic animals. In abalone, *P. daeponensis* protected against *V. harveyi* and increased immune response (Zhao et al. 2018). A strain of *Phaeobacter* improved survival in Eastern oyster during exposure to *R. crassostreae* and *V. tubiashii* (Karim et al. 2013).

Few other recent studies have used Gram-negative microorganisms as probionts. *Vibrio midae* colonized the gut and increased growth of South African abalone (Huddy and Coyne 2014) and *V. diabolicus* improved survival against *V. parahaemolyticus* in Pacific white shrimp (Restrepo et al. 2021). Importantly, as many *Vibrio* species cause disease in shellfish and horizontally transfer genes within the microbiome, there is potential for commensal strains to acquire virulence genes and become pathogenic (Restrepo et al. 2021). The study by Restrepo et al. (Restrepo et al. 2021) demonstrated that *V. diabolicus* strain ILI could acquire the virulence plasmid vp-AHPND from *V. parahaemolyticus*, but the transformed strain did not cause mortality in shrimp, indicating a lack of other factors necessary to cause disease. *Pseudoalteromonas* improved survival following challenge with *V. harveyi* when supplemented to European abalone (Offret et al. 2019). *Shewanella* showed promise as a probiotic in Pacific white shrimp when combined with feeds at 10^6 or 10^7 CFU/g, resulting in faster growth, reduced feed conversion ratios, and greater resistance to ammonia stress (Wei et al. 2021). Lai et al. (Lai et al. 2022) supplemented Oriental prawn feed with *Psychrobacter* at 10^9 CFU/100 g of feed and observed a decrease in mortality following challenge with *A. hydrophila* from $>60\%$ in controls to $<25\%$ in treated animals.

A few studies have explored the use of yeast probionts in an aquaculture setting. *Candida cretensis* increased growth rate and improved immune status in Catarina scallop (de la Vega et al. 2020). *Saccharomyces cerevisiae* increased growth rate, enhanced immune function, and improved survival in crayfish challenged with *Citrobacter fruendii* (Xu et al. 2021b).

Although numerous taxa prove effective as probionts in shellfish, characteristics such as lack of QPS/GRAS status and instability during industrial scale-up may limit their commercialization potential. It is likely due to the proven safety and production scalability of *Bacillus* and *Lactobacillus* that limited research has focused on other probiotic species.

11.3.3 The Consortia Effect

The probiotic consortia effect has been well documented. As bacteria provide strain-specific advantages to the host, mixing multiple species into one probiotic concoction allows the host to benefit from the characteristics of each strain in the product. Combining probiotic strains may improve the functionality of the supplement by providing antimicrobial activity against multiple pathogens or providing benefits beyond disease resistance, including improving water quality, digestibility, immune function, or protection against contaminants. Indeed, a meta-analysis indicated that the use of a consortium was more effective at increasing survival and growth rate in shrimp than the use of a single strain (Toledo et al. 2019).

Bacillus are often combined with other microorganisms to generate a multi-strain probiotic formulation. A mixed probiotic composed of *B. subtilis*, *B. cereus*, and *B. coagulans* decreased toxicity of indigo dye and improved resistance towards *V. harveyi* in the Pacific oyster (Fdhila et al. 2017), providing protection against both a xenobiotic and a pathogen. *B. tequilensis* with *B. endophyticus* increased growth and immune function in Cortez oysters (García-Bernal et al. 2019). When *B. pumilus*, *B. cereus*, and *L. plantarum* were added through the diet of ornate spiny lobster, the authors reported improved growth, feed conversion ratio, and survival following a challenge with *V. owensii* (Nguyen et al. 2014). A mixed probiotic including *Enterococcus*, *Exiguobacterium*, and *Vibrio* improved growth (Grandiosa et al. 2018) and resistance to *V. splendidus* challenge while reducing oxidative stress and increasing hemocytes in abalone (Grandiosa et al. 2020).

A consortia of *B. subtilis* and *S. cerevisiae* supplemented to Pacific white shrimp resulted in greater growth, better feed conversion, and improved survival following challenge with *V. parahaemolyticus* (Yao et al. 2021). In shrimp, *B. subtilis*, *B. licheniformis*, and *B. pumilus* at a concentration of 0.05–1% of feed resulted in similar benefits as well as enhanced immunity, digestive enzyme activity, and increased intestinal villi height (Hien et al. 2022). However, when these species were combined with *L. plantarum*, there was no apparent benefit to shrimp production (Bolívar-Ramírez et al. 2021). Similarly, adding yeast species to a mix of five *Bacillus* probiotics did not improve their performance (Nimrat et al. 2021). When testing a commercial probiotic containing numerous species of *Lactobacillus* (*L. plantarum*, *L. delbrueckii* subsp. *bulgaricus*, *L. acidophilus*, *L. rhamnosus*), *E. faecium*, and numerous other live organisms (*Bifidobacterium bifidum*, *Streptococcus salivarus* subsp. *thermophilus*, *Aspergillus oryza*, *Candida pintolopesii*) combined with host-isolated bacterium *Hafnia alvei*, the authors reported no influence of probiotics on growth or survival in stage II narrow-clawed crayfish (Didinen et al. 2016); however, other studies have reported no influence on these two parameters while seeing improvement in recovery from pathogen challenge (Ambas et al. 2013), and these results do not rule out the ability of this mix to aid in disease recovery.

Providing a consortium of microbes often results in greater benefits than using a mono-culture supplement. However, the use of multiple probiotic species or strains

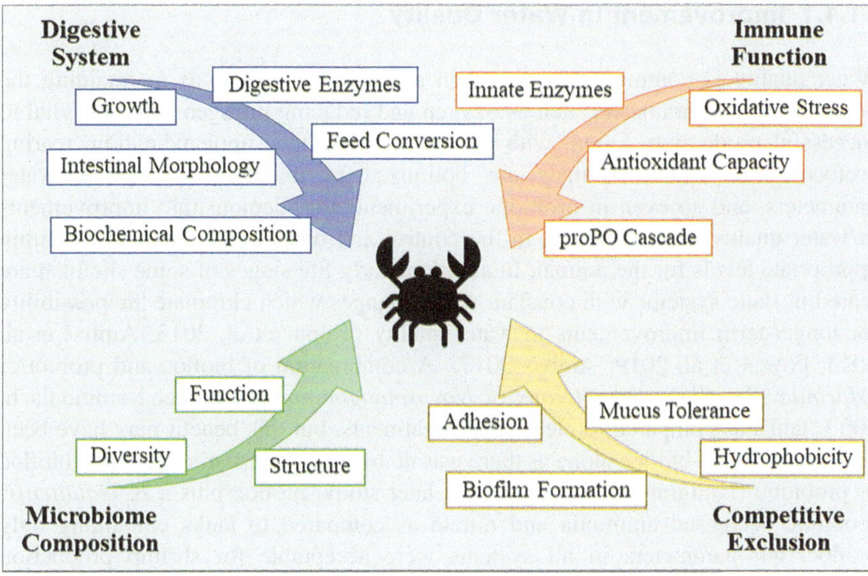

Fig. 11.2 Probiotic mechanisms of action and how they are measured in shellfish

leads to greater probiotic production costs; therefore, it is not in the farm's best interest to use multiple strains if they do not provide additional benefits.

11.4 Mechanisms of Action

Disease in aquaculture leads to massive economic losses each year and as such, a desired benefit of probiotics is to protect against aquatic disease. As stated, it is typical for probiotic screenings to start with antimicrobial assays to demonstrate antimicrobial capacity of putative probiotics, followed by a test to confirm these activities in a culture environment. In addition, the effectiveness of probiotics may be determined by the ability of the animal to survive stressors with probiotic supplementation, including exposure to varying salinity and temperature, xenobiotics, and transport. Improved survival is without a doubt a benefit in many of these studies, and much effort has gone into identifying the mechanism(s) by which probiotics exert their effects. The studies to date indicate a few potential areas where probiotics are beneficial (Fig. 11.2) including improvement of water quality, competitive exclusion, digestion and nutrient assimilation, immunomodulation, and microbiome manipulation.

11.4.1 Improvement in Water Quality

Water quality is commonly measured in aquaculture studies, as maintaining the proper levels of parameters such as oxygen and reducing nitrogen wastes is vital to successful production. Often, with or without probiotic supplementation, rearing protocols for aquatic animals are optimized to maintain appropriate water parameters, and so even in probiotic experiments that demonstrate improvements in water quality, measurements in the control and other treatments are all within appropriate levels for the animal. In addition, early life stages of some shellfish are reared in static systems with constant water changes which eliminate the possibility for longer-term improvements in water quality (Talpur et al. 2013; Ambas et al. 2013; Foysal et al. 2019; Ambas 2017). A combination of biofloc and probiotics (*Marinilactibacillus piezotolerans* or *Novosphingobium* sp.) reduced ammonia in 500 L tanks as compared to clear water treatments, but this benefit may have been attributable to the biofloc alone as there was no treatment consisting of solely biofloc or probiotic (Panigrahi et al. 2021). In a later study, biofloc plus a *B. tequiliensis* probiotic decreased ammonia and nitrate as compared to tanks containing only biofloc, but parameters in all systems were acceptable for shrimp production (Panigrahi et al. 2022). *P. acidilactici* was able to decrease ammonia whereas *L. reuteri* reduced nitrite in an 81 L system (Wu et al. 2022), but a combination of *L. plantarum* and a consortium of *Bacillus* spp. increased nitrite in a 9000 L biofloc system (Bolívar-Ramírez et al. 2021). Microbes are key contributors to nutrient cycling in aquatic environments and so, with the right strain and concentration, it is expected that probiotics can improve water quality in intensive aquaculture.

11.4.2 Competitive Exclusion

There are numerous advantages to a probiotic that can adhere to the surfaces of the host organism of interest. For these strains, the probiotic directly interacts with body tissues impacting digestion, nutrient absorption, immune function, and adherent microbiome. At a more basic level, when probiotics colonize surfaces, they competitively exclude opportunistic pathogens by competing for space and nutrients at that site. If the probiotic produces antimicrobials, the exclusion capacity may extend to areas adjacent to where that bacterium is adhered. Therefore, a probiotic that colonizes a host, be it on the outer surfaces or in the digestive tract, is desirable. Investigations into the adherence of probiotics demonstrate that not all strains colonize shellfish.

Hydrophobic properties on bacterial cell surfaces allow microbes to interact with hydrocarbons associated with mucosa and epithelial cells, and therefore to move from the gastrointestinal water-based lumen and attach to the mucosal surface (Krasowska and Sigler 2014). Mucosal binding leads to a greater concentration of the microbe in these layers, preventing pathogen attachment and providing a surface for dispersion of beneficial compounds produced by the probiotic (Mcshane et al. 2021). During investigations of probiotic bacteria for the Kumamoto oyster,

Abasolo-Pacheco et al. (Abasolo-Pacheco et al. 2016) confirmed two isolates highly hydrophobic, identified as *L. plantarum* and *B. flexus*. These bacteria increased growth and protein and carbohydrate content when supplemented to juveniles via the rearing water. Although testing for hydrophobicity during the growth phase of bacteria can indicate the potential for adhesion, these results should be confirmed in the shellfish rearing system.

Adherence alone is not sufficient for a probiotic to establish in the host mucosa. Mucosal surfaces harbor innate immune compounds such as enzymes and antimicrobials that can inhibit growth or kill the microbe. In addition, if the targeted surface is the gastrointestinal tract, the microbe must be able to survive passage through the low pH in the stomach. Some authors have exposed putative probiotics to mucosa from external and gastrointestinal tissues and to varying pH. In common garden snails, 20 LAB isolates from the digestive tract were resistant to mucosa from the pedal and gastrointestinal tract as well as low pHs (Dushku et al. 2019). However, LAB isolates from the esophagus and crop were moderately tolerant to these conditions and did not grow well in the presence of mucus from the pedal. Many of these isolates were also hydrophobic and capable of forming biofilms until they reached the stationary phase, at which point they lost these qualities. The authors indicate that the continued ability to form biofilms at the stationary phase was found in isolates that are tolerant to the mucosa and pH tested, suggesting these LAB strains showed the best promise for probiotic trials. A subset of these were identified and all were classified as *L. plantarum*. Two of these LAB isolates were tested for their ability to adhere to the intestinal mucosa, one from each isolation source (gastrointestinal tract versus esophagus/crop), with only the intestinal isolate capable of adhering to the tissue. Three *Bacillus* isolates collected from mud crabs and proven to inhibit *V. parahaemolyticus* were able to survive exposure to bile and low pH with *B. subtilis* and *B. pumilus* having significantly higher hydrophobicity than *B. cereus* (Wu et al. 2014). These studies indicate the interaction between bacterial strains and mucosal conditions are not only strain-specific, but also site-specific, lending evidence that ideal probiotics are isolated from the tissue where interaction is desired.

Adherence is not always a prerequisite for success of a probiotic and transient microbes influence shellfish physiology. Although a small percentage of the probiotic *Pseudoalteromonas* h*Cg*-6 temporarily established in the hemolymph of abalone and its concentration decreased over time, this interaction was sufficient to protect the animal from the pathogen *V. harveyi* in challenge experiments (Offret et al. 2019). In these cases, further experiments investigating their mode of action will determine if continued supplementation is required to maintain the benefits demonstrated.

11.4.3 Digestion and Nutrient Assimilation

One of the greatest reported benefits of probiotic use in shellfish is increased growth. This advantage results from probiotics that secrete or stimulate production of enzymes that aid in digestion of proteins (proteases), lipids (lipases), and carbohydrates (amylases, cellulases), increasing access for the host to metabolic products that can be absorbed and incorporated into biomass. Interestingly, the interactions between probiotics and shellfish lead to alterations in gut morphology that provide greater surface area for nutrient absorption. These interactions alter the biochemical composition of the animal and confirm nutrient uptake from the presence of probiotic strains, resulting in better growth rates and improved feed conversion ratios.

The contribution of probiotics to digestion has been examined via direct screening of isolates for digestive enzyme production or through measurement of these enzymes in the animals receiving the treatment. In either case, the enzymes most examined include trypsin, amylase, and lipase for the ability to degrade proteins, carbohydrates, and lipids, respectively. *Bacillus* strains isolated from the digestive tract of different bivalve species produced these enzymes, as well as cellulase, and led to increased growth in Kumamoto oysters (Sánchez-Ortiz et al. 2020). Two *Lactobacillus* strains did not produce these enzymes but resulted in higher protein and carbohydrate content in the oyster. These strains did not impact growth unless they were given in combination with the *Bacillus* strains. *Lactobacillus* isolated from rainbow trout increased the activity of protease, lipase, amylase, and alkaline phosphatase (another protease), but these also had no impact on growth in the narrow-clawed crayfish (Valipour et al. 2019), whereas higher enzyme activity in Pacific white shrimp is consistently associated with increased growth rate and enhanced feed efficiency (Won et al. 2020; Le et al. 2022; Lai et al. 2022; Hien et al. 2022; Ren et al. 2022; Liu et al. 2022; Abidin et al. 2022; Liao et al. 2022). In testing *Bacillus* and *Phaeobacter*, only *Bacillus* and a combination of the two strains resulted in an increase in lipase as well as greater final weight and shell length (Zhao et al. 2018). These studies indicate that probiotics, especially *Bacillus*, show promise for translating digestive enzyme activity into biomass.

Biochemical composition of the animal can also be altered when probiotics are given, and these changes indicate improved digestive function. Using metabolomics, Grandiosa et al. (Grandiosa et al. 2018) measured greater expression of amino acids, fatty acids, and organic acids in abalone fed probiotics, with these individuals also having greater weight gain. Histological investigations have demonstrated the ability of probiotics to alter intestinal morphology by increasing microvilli concentration and length (Chen et al. 2021; Won et al. 2020; Foysal et al. 2021; Foysal et al. 2020; Ren et al. 2022; Abidin et al. 2022), resulting in greater surface area for nutrient absorption and faster growth and lower feed conversion ratios in crustaceans. Improvements in digestion and nutrient assimilation may also result in faster metamorphosis and molting (Foysal et al. 2019). Indeed, the interaction between shellfish and their microbiomes can impact development, digestion, and growth in cultured species.

11.4.4 Immunomodulatory Capability

Hemolymph is the fluid in shellfish that serves a combined role of the blood and lymph found in higher organisms. It transports oxygen, compounds, and immune cells throughout the body. Shellfish rely solely on an innate immune system to prevent infection and do not have antibodies. Their immunity consists of both cell-mediated and humoral components. The immune cells are termed hemocytes and in crustaceans, there are three types (Bouallegui 2021). Hyalinocytes (HCs) generally lack granules and participate in phagocytosis. Semi-granular cells (SGCs) have granules and some are capable of phagocytosis but these cells play a larger role in encapsulation, which involves surrounding pathogens that are too large to be ingested via phagocytosis. Phagocytosis and encapsulation are both part of the cell-mediated immune system. HGCs and granular cells (GCs) are responsible for humoral immunity, including release of antimicrobial peptides, lectins for opsonization of pathogens, and the prophenoloxidase (proPO) cascade (Bouallegui 2021). The proPO cascade is triggered by MAMPs (microbe-associated molecular patterns) such as lipopolysaccharide (LPS) and peptidoglycan, which activate serine proteases, resulting in the formation of the phenoloxidase (PO) enzyme. This enzyme forms melanin which surrounds pathogens and results in the generation of reactive oxygen species (ROS) which destroy the invading cells (Tassanakajon et al. 2018) and trigger the synthesis of antimicrobial peptides. Within the mollusks, hemolymph cells are grouped morphologically into hyalinocytes (agranular hemocytes) or granulocytes. The specific roles associated with each type are still in question as both cells types are diverse in structure and function (Al-Khalaifah and Al-Nasser 2019). However, the innate immune system of mollusks has the same types of protective mechanisms as that of crustaceans, including phagocytosis, encapsulation, opsonization, and the proPO cascade.

An increase in hemocytes indicates activation of the immune system and enhanced pathogen fighting capability. Increased levels of hemocytes following probiotic administration have been reported in crayfish (Ambas et al. 2013; Valipour et al. 2019; Lai et al. 2020; Foysal et al. 2021; Foysal et al. 2020; Foysal et al. 2019; Ambas 2017), crab (Yang et al. 2019), abalone (Grandiosa et al. 2018; Dushku et al. 2019; Gao et al. 2018; Jiang et al. 2013), and shrimp (Khademzade et al. 2020; Hien et al. 2022; Ferreira et al. 2015) but have not been investigated in bivalve studies.

Reactive oxygen species (ROS) are compounds that include hydroxyl radicals, superoxide anions, and hydrogen peroxide. These compounds form from oxygen metabolism, environmental interactions, and respiratory burst activity from hemocytes. ROS are reactive and are used by the immune system to destroy pathogens; however, there is a balance, and an excess of ROS can lead to deleterious effects on host cells. Therefore, shellfish produce antioxidant enzymes such as superoxide dismutase (SOD), glutathione peroxidase (GPx), and catalase (CAT) to remove excess ROS. When the antioxidant enzymes do not remove enough ROS, the organism is said to be under oxidative stress. Most studies report increases in respiratory burst activity as an indicator of enhanced immune response (Wu et al. 2014; Xiaolong et al. 2019; Gao et al. 2018; Jiang et al. 2013) and greater activity or gene expression associated with antioxidant enzymes to demonstrate performance of the antioxidant system (Wu et al. 2014; Yang et al. 2019; Xu et al. 2021a; Grandiosa et al. 2018; Chen et al. 2021; Won et al. 2020; Foysal et al. 2020; Foysal et al. 2019; Lai et al. 2022; Ambas 2017; Panigrahi et al. 2022; Jiang et al. 2013). However, SOD activity showed variation in studies, particular among narrow-clawed crayfish which can likely be attributed to differences in experimental protocols. For example, Lai et al. (Lai et al. 2020) reported a decrease in SOD activity in probiotic-treated animals versus controls. A second study investigated the effect of this probiotic but reported an increase in SOD activity prior to WSSV challenge (Xu et al. 2021a). Xu et al. (Xu et al. 2021a) fed the crayfish for 28 days prior to the challenge whereas Lai et al. (Lai et al. 2020) only measured SOD after 24 and 48 h. Thus, increased SOD activity may occur >2 days after probiotic administration. Despite lower SOD activity, the probiotic in Lai et al. (Lai et al. 2020) was still effective at preventing mortality from WSSV. In the Catarina scallop, a mixed *Bacillus* probiotic resulted in decreased activity of SOD, CAT, and GPx (de la Vega et al. 2020). This contrasted with *Lactobacillus* which decreased only CAT and GPx and a yeast (*Candida maris*) which reduced only SOD. Despite having less antioxidant effects, the *Lactobacillus* and *Candida* strains increased scallop growth. SOD activity was increased by one strain of *Streptomyces* in Kumamoto oysters (RL8), but all three tested strains increased weight gain (García-Bernal et al. 2019). In Cortez oysters, one *Streptomyces* strain (V4) decreased SOD activity without altering weight gain, while the other two increased it, but only N7 also provided greater weight gain. A *Bacillus* mix also increased SOD and weight gain in this host species, but not in the other. In shrimp, *B. subtilis* and *P. pentosaceus* increased SOD activity, whereas *L. lactis* did not (Won et al. 2020). As evidenced by these studies, probiotics have different modes of action and the ability to alter innate enzyme activity varies at the strain level.

Microorganisms have microbe-specific molecules such as LPS, peptidoglycan, and β-1,3-glucans that act as MAMPs and are recognized as foreign by receptors on hemocytes (pattern recognition receptors, PRRs) including Toll-like receptors (TLR) and lectins. This recognition results in the activation of the proPO cascade. It also leads to the production of chemicals called cytokines, such as interleukins (IL), which activate hemocytes and regulate inflammation. Some studies report that probiotics upregulate PRRs associated with microbial recognition (Dushku et al. 2019; Chen et al. 2021), activate proPO genes (Wu et al. 2014; Yang et al. 2019; Won et al. 2020; Foysal et al. 2020; Foysal et al. 2019; Hien et al. 2022), or both (Xu et al. 2021a; Du et al. 2019; Foysal et al. 2021; Panigrahi et al. 2021; Lai et al. 2022; Panigrahi et al. 2022), representing increased capacity for recognizing foreign invaders alongside an active immune response. Activation of proPO should also increase the concentrations of ILs, as confirmed in crayfish studies (Foysal et al. 2020; Foysal et al. 2019), as well as antimicrobial compounds which are discussed below. So far, proPO activation has been studied primarily in crustaceans, leaving a knowledge gap for the interaction between probiotics and these factors in mollusks.

Numerous enzymes are involved in bactericidal activities in the innate immune system and probiotics enhance the production of these compounds. These include lysozyme (LYZ) which degrades peptidoglycan in the bacterial cell wall, cathepsin, and acid phosphatase (ACP) which are lysosomal proteases that degrade microbes following phagocytosis, alkaline phosphatase (AKP) which plays a role in lysozyme production along with many other beneficial functions within the host (maintains mucosal barrier integrity, generates tolerance towards the microbiome, and acts as an anti-inflammatory), and myeloperoxidase (MPO) which aids in release of ROS to destroy microbes. Probiotic supplementation generally increase LYZ activity (Xu et al. 2021a; Won et al. 2020; Valipour et al. 2019; Khademzade et al. 2020; Le et al. 2022; Foysal et al. 2019; Lai et al. 2022; Xu et al. 2021b; Panigrahi et al. 2022), lysosomal proteases (Xu et al. 2021a; Foysal et al. 2021), antimicrobials (Chen et al. 2021; Foysal et al. 2021; Abidin et al. 2022), and the multifunctional enzyme AKP (Chen et al. 2021; Le et al. 2022; Lai et al. 2022; Yao et al. 2021; Ren et al. 2022). In Catarina scallop, all tested probiotics significantly decreased LYZ activity (de la Vega et al. 2020). The authors in this case state that higher LYZ activity is a bioindicator of stress. It has been suggested that high LYZ levels are associated with stress whereas decreased levels may indicate an immunocompromised state (Coates and Söderhäll 2021). Similarly, studies often report a decrease in malondialdehyde (MDA), a metabolite that is an indicator of cellular damage (Yao et al. 2021; Amoah et al. 2019; Amoah et al. 2020; Xie et al. 2019). As baseline enzyme activity levels have not been established for many shellfish species, this may be an important step to understanding the true consequences of altered innate immunity parameters.

11.4.5 Microbiome Manipulation

The microbiome plays a significant role in the health of shellfish. Just as probiotic strains can confer disease resistance through competitive exclusion and production of antimicrobials, influence digestion, and nutrient absorption, and alter innate immune function, the other members of the microbial community are active in these functions. If the microflora is altered from its normal healthy composition, a condition known as dysbiosis, the community is dominated by opportunistic or primary pathogens and infection and disease follow (Holt et al. 2021). As probiotics establish in the aquaculture system or on the host tissues, they alter the microbiome structure via competition and antimicrobial production. Therefore, insight into the influence of probiotics on the microbiome is important to understand their overall influence on shellfish health.

One sign of a dysbiotic microbiome is a reduction in microbial diversity (Infante-Villamil et al. 2021), releasing more niches for pathogen colonization. Numerous studies have reported an increase in microbial diversity upon supplementation with probiotics (Chen et al. 2021; Foysal et al. 2021; Foysal et al. 2020; Foysal et al. 2019; Xu et al. 2021b; Xie et al. 2019). In juvenile Cortez oysters, probiotic supplementation resulted in a more diverse microbiome than controls which loss significant diversity over 30 days (García Bernal et al. 2017). This maintenance of diversity increases resiliency against pathogenic invasion. In some studies, diversity indices are lower in probiotic-supplemented individuals despite having a greater number of species (Du et al. 2019; Amoah et al. 2020). These communities have more dominant species within their microbiome than non-treated animals. It is currently unclear what leads to these differing patterns of diversity. However, as microbes share functions within the community, a change in diversity does not necessarily shift functional capacity of the microbiome.

Although recognized as a marker of dysbiosis in humans (Rizzatti et al. 2017), the phylum *Pseudomonadota* (formerly *Proteobacteria*) are commonly found in association with shellfish and often make up a significant portion of their microbial community. However, all the Gram-negative bacteria used for screening probiotics fall into this phylum (see Sect. 11.2.2 Desired Probiotic Characteristics) whereas most of the probiotics are members of the phylum *Bacillota* (formerly *Firmicutes*). Therefore, it is expected that probiotics would reduce the abundance of *Pseudomonadota* in favor of *Bacillota*. In practice, this does sometimes occur (Xu et al. 2021a; Du et al. 2021; Restrepo et al. 2021; Xu et al. 2021b; Abidin et al. 2022), including the reduction of potentially pathogenic genera found within this phylum such as *Aeromonas* and *Vibrio* (Foysal et al. 2021; Du et al. 2021; Foysal et al. 2019; Abidin et al. 2022). In some cases, an increase in *Bacillota* is also reported (Xu et al. 2021a; Foysal et al. 2021; Xu et al. 2021b). Contrarily, numerous studies describe an increase in *Pseudomonadota* when using probiotics that are beneficial in shrimp (Du et al. 2019; Panigrahi et al. 2021; Wei et al. 2021; Amoah et al. 2019; Amoah et al. 2020), and so phylum-level changes are not sufficient indicators for probiotic effectiveness.

When using a *Clostridium* probiotic which falls into the *Bacillota*, Foysal et al. (Foysal et al. 2019) observed an increase in the *Fusobacteriota* (i.e., *Fusobacteria*), particularly the genus *Hypnocyclicus*. This genus is found in other shellfish species (Lau et al. 2022) and is correlated with antioxidant activity in fish (Siddik et al. 2020). In this study on marron crayfish, probiotic supplementation altered predicted functions of the microbiome including carbohydrate, amino acid, and fatty acid metabolism, production of secondary metabolites, and degradation of xenobiotics (Lau et al. 2022), suggesting *Hypnocyclicus* may be a beneficial member of the shellfish microbiome. Other investigations reported higher abundances of other genera including *Lactobacillus* and *Bacillus*, both when they are the selected probiotic (Chen et al. 2021; Foysal et al. 2021; Foysal et al. 2020) and when they are not (Panigrahi et al. 2021; Xu et al. 2021b). Xu et al. (Xu et al. 2021b) also reported an increase in *Cetobacterium* which has probiotic effects in freshwater finfish aquaculture (Xie et al. 2022a; Xie et al. 2022b; Zhou et al. 2022), coinciding with a decrease in *Bacteroides*. This negative correlation was also reported by Hao et al. (Hao et al. 2017) testing plant- and animal- based diets and suggests these organisms are associated with dietary protein and fiber content which were likely altered with the inclusion of the yeast probiotic used in narrow-clawed crayfish. Two studies in shrimp reported an increase in the genus *Tamlana* with probiotic treatment (Restrepo et al. 2021; Amoah et al. 2020). Members of this genus are associated with nitrogen cycling and can produce antifungal and protease enzymes (Bibi et al. 2018; Sun et al. 2019) and may be beneficial members of the shrimp gastrointestinal microbiome.

As each probiotic will interact with the environment, host, and microbiome differently, and each host species harbors a unique microbial community, it is expected that the influence of probiotics will vary among studies. However, some patterns do emerge among the experiments that have characterized probiotic effects on the shellfish microbiome. When supplemented with *Lactobacillus* probiotics, relative abundances of taxa within the *Actinomycetota* increase in crayfish (Foysal et al. 2021; Foysal et al. 2020), Chinese mitten crab (Li et al. 2022), and Kumamoto oyster (García Bernal et al. 2017). This phylum also increased in shrimp fed *Bacillus* strains (Amoah et al. 2020; Cheng et al. 2020). This group is well known for their ability to produce secondary metabolites such as antimicrobials and are a target for probiotic development in the aquaculture sector (Das et al. 2008). Most of these studies also reported an increase in the abundance of *Planctomycetota* whose members are capable of anaerobic ammonium oxidation (annamox) and are responsible for substantial removal of nitrogen from aquatic systems (Kuypers et al. 2005). These shifts in microbial community composition have implications for maintaining water quality and shellfish health in aquaculture systems.

Despite the known importance of the microbiome to the health of shellfish through its interaction with the digestive and immune systems, little work has been done to date regarding the functions of the microbiome during probiotic treatment. Predicted metagenomic functions show potential for enhanced metabolism including that of amino and fatty acids and carbohydrates, as well as protection against xenobiotics (Li et al. 2022; Lau et al. 2022). These roles are supported by studies that report greater digestive enzyme production, as well as the ability of probiotics to degrade and protect against synthetic dyes (Fdhila et al. 2017). This is an area of research that is expected to expand in the coming years.

Overall, microbiome analyses in probiotic studies demonstrate changes in microbial diversity, accumulation of taxa described as probiotics, and reduction of potentially pathogenic groups. These studies demonstrate the ability of probiotics to alter the shellfish microbiome and prevent dysbiosis. However, this field of research is relatively new outside of shrimp and would benefit from more advanced genomics techniques including metagenomics, metatranscriptomics, and metaproteomics to generate a full picture of the microorganisms present and their roles in shellfish health.

11.5 Beyond Probiotics

As technologies are developed that expand knowledge on shellfish–microbe interactions, techniques for developing effective additives have advanced beyond typical probiotics. These include the use of synbiotics, paraprobiotics/postbiotics, and living therapeutics.

11.5.1 Synbiotics

Synbiotics, defined as supplements that combine prebiotics (food additives that enhance growth of beneficial gut microorganisms) and probiotics, are the most studied alternative to using probiotics alone. In the narrow-clawed crayfish, both combinations of LAB and oligosaccharide prebiotics (*Enterococcus* with xylooligosaccharide, *Pediococcus* with mannanoligosaccharide) decreased mortality following *Nocardia brasielience* and *V. harveyi* challenges (Safari et al. 2017). The *Enterococcus* mix also increased immune function (hemocyte count, proPO, LYZ, AKP) and protected against *A. hydrophila*. A consortium probiotic containing *L. acidophilus*, *B. subtilis*, and *S. cerevisiae* was tested against a synbiotic including prebiotic fructooligosaccharides. The synbiotic improved growth, immune and digestive enzyme activity, and heat stress resistance as compared to treatment with the probiotic mix alone (Wan et al. 2022). The synbiotic-fed Chinese mitten crab exhibited higher growth rates and greater resistance to transport stress than those fed probiotics alone. In Pacific white shrimp, a synbiotic composed of *B. amyloliquefaciens*, *B. pumilus*, *B. subtilis*, and chitosan performed better than feed coated with chitosan alone, providing greater growth benefits, influencing intestinal morphology, and improving survival following *V. parahaemolyticus* challenge (Kewcharoen and Srisapoome 2022). Similarly, Abidin et al. (Abidin et al. 2022) demonstrated better performance of a feed-supplemented synbiotic (*L. acidophilus* at 10^7 CFU/g plus moringa leaf water extract at 2.5 g/kg) than the probiotic or moringa leaf extract alone, improving Pacific white shrimp survival following a challenge with *V. alginolyticus* while enhancing growth rates, intestinal surface area, digestive enzyme activity, and immune response. Similarly, shrimp fed a combination of cacao pod husk at 5 g/kg and *L. plantarum* at 10^7 CFU/g results in greater growth and immune function and prevented mortality due to low temperature stress and *V. alginolyticus* challenge (Kuo et al. 2021).

Synbiotics do not always improve shellfish aquaculture. In testing sodium humate at 3 mg/L and a probiotic containing *Lysinibacillus* and *B. subtilis* each at 10^5 CFU/mL, added directly to the rearing water of Pacific white shrimp, Liao et al. (Liao et al. 2022) reported a decrease in benefits of the synbiotic as compared to the prebiotic alone, but they did indicate that the probiotic had the ability to improve water quality. Boonanuntanasarn et al. (Boonanuntanasarn et al. 2016) compared the performance of β-glucan alone and in combination with microencapsulated *B. subtilis* or *P. acidilactici* but found no synbiotic effect. Similarly, there was no combined effect of using *L. lactis* with astaxanthin (Liu et al. 2022), with these two studies recommending using solely the prebiotic. Although the prebiotic-probiotic combination must be optimized by selecting bacterial strains and compounds that have proven combined effectiveness and blending each at the right dose, synbiotics show promise for amplifying benefits of a single product in shellfish aquaculture.

11.5.2 Paraprobiotics

Paraprobiotics, also called postbiotics, are supplements that consist of inactivated cells or their metabolites. Most shellfish studies that have investigated these products are performed in shrimp. Some of these benefits are retained even when a heat-killed formulation of the probiotic is administered. For example, Ren et al. (Ren et al. 2022) found that killed *L. plantarum* supplemented with two compound probiotics (each containing *Lachnospiraceae*, *Porphyromonadaceae*, *Lactobacillus*, and other unspecified microbes) was more effective at altering immune function, intestinal morphology, and feed efficiency in Pacific white shrimp than the compound probiotic alone, specifically with the *L. plantarum* administered at 1% with the compound probiotics at 15% and 2% of the feed. On the other hand, Cheng et al. (Cheng et al. 2020) did not find any differences in growth or survival when Pacific white shrimp were fed soybean meal fermented by *B. subtilis* or an antimicrobial peptide originating from the probiotic strain. Studies have reported shifts in the gut microbiome of shellfish fed diets containing postbiotics (Ren et al. 2022; Cheng et al. 2020; Zheng et al. 2020; Sha et al. 2016), suggesting the potential for impacts on host physiology.

11.5.3 Living Therapeutics

The production of living therapeutics involves using synthetic biology to generate genetically engineered probiotics that are often tailored to prevent or cure disease. Research in this area has been dedicated to controlling viral disease in shrimp through activation of the RNA interference (RNAi) mechanism. This mechanism is present in eukaryotic organisms to control expression of messenger RNA (mRNA). Briefly, long double-stranded RNA (dsRNA) molecules that are homologous to a target gene are cleaved into small RNAs that, with the help of an enzyme, attach to and degrade the homologous target mRNA. The goal of living therapeutics is to administer a genetically modified dsRNA with homology to viral genes to a host, which then activates the RNAi mechanism preventing the viral particles from replicating. Direct injection of these dsRNAs has improved survival of Pacific white shrimp to infectious myonecrosis virus (IMNV) and WSSV (Feijo et al. 2015; Anil et al. 2015); however, injection is time consuming and it was recognized that oral delivery through a vector, such as a bacterium, would be a more effective method (Anil et al. 2015). This mechanism would require the use of a microbe with QPS or GRAS status that can be produced in large scale to ensure safety and efficacy (Riet et al. 2021). Riet et al. (Riet et al. 2021) altered the genetics of a strain of *B. subtilis* to express a long dsRNA that targets the WSSV protein VP28. This strain was combined with a pelleted feed at 10^{10} CFU/g and fed to shrimp for 15 days prior to WSSV challenge. The living therapeutic decreased mortality from 100% to 66% and significantly reduced the viral load as compared to controls. The authors suggest that the high mortality rate even with treatment may be due to the modified probiotic not colonizing the gastrointestinal tract and state that a bacterium isolated directly from the shrimp may perform better. A study by Dekham et al. (Dekham et al. 2022) observed a difference in mortality rates when shrimp were administered *L. lactis* or *L. plantarum* that had been altered to produce anti-WSSV virus dsRNAs. Despite both modified bacterial strains significantly reducing viral loads in comparison to controls, the modified *L. lactis* strain was able to reduce mortality from 90% to 33% at high doses, whereas the modified *L. plantarum* strain performed similarly to wild-type *L. plantarum*, with mortality rates of about 70%. The study indicated the *L. lactis* strain was able to increase expression of genes associated with RNAi and innate immune function and therefore provided greater protection against WSSV. Therefore, the effective use of living therapeutics will require not only identifying the best carrier for these genetic manipulations, but also the ideal dose. However, the legal issues, consumer perception, and biocontainment concerns will need to be addressed before genetically modified organisms can be applied to aquaculture production systems.

11.6 Conclusions

The use of probiotics in shellfish aquaculture particularly that of shrimp has become routine. With proven benefits ranging from faster growth to disease and stress resistance, these practices stand to decrease production time and disease loss, resulting in improved economic sustainability. As research continues to unravel the interactions between probiotics and host health, improved formulations can be manufactured that exploit living organisms as well as compounds and metabolites that enhance their effectiveness. As the field advances, incorporation of molecular analyses including transcriptomics, proteomics, and metagenomics will provide in-depth information regarding the physiological benefits and mechanisms, such as genetic engineering, for improving microbial manipulations in shellfish aquaculture.

References

Abasolo-Pacheco F et al (2016) Isolation and use of beneficial microbiota from the digestive tract of lions-paw scallop Nodipecten subnodosus and winged pearl oyster Pteria sterna in oyster aquaculture. Aquac Res 47(10):3042–3051

Abidin Z et al (2022) Effect of dietary supplementation with Moringa oleifera leaf extract and Lactobacillus acidophilus on growth performance, intestinal microbiota, immune response, and disease resistance in whiteleg shrimp (Penaeus vannamei). Fish Shellfish Immunol 127:876–890

Al-Khalaifah H, Al-Nasser A (2019) Immune response of Molluscs. In: Molluscs. IntechOpen

Ambas I (2017) Health Status of Marron, Cherax cainii (Austin, 2002) Fedcustomized probiotic Bacillus mycoides. J Aquac Mar Biol 6(4)

Ambas I, Suriawan A, Fotedar R (2013) Immunological responses of customised probiotics-fed marron, Cheraxtenuimanus, (Smith 1912) when challenged with Vibrio mimicus. Fish Shellfish Immunol 35(2):262–270

Amin M et al (2020) Isolation and screening of lactic acid bacteria associated with the gastrointestinal tracts of abalone at various life stages for probiotic candidates. Aquac Rep 17:100378

Amoah K et al (2019) Dietary supplementation of probiotic Bacillus coagulans ATCC 7050, improves the growth performance, intestinal morphology, microflora, immune response, and disease confrontation of Pacific white shrimp, *Litopenaeus vannamei*. Fish Shellfish Immunol 87:796–808

Amoah K et al (2020) Administration of probiotic Bacillus licheniformis induces growth, immune and antioxidant enzyme activities, gut microbiota assembly and resistance to Vibrio parahaemolyticus in Litopenaeus vannamei. Aquac Nutr 26(5):1604–1622

Anil K et al (2015) Double stranded RNA simultaneously targeting four white spot syndrome virus (WSSV) genes provides protection against WSSV in Litopenaeus vannamei. Int J Mar Sci Ocean Technol 2(2):5–10

Aribah D, Widanarni, Wahyudi AT (2022) The effectiveness of marine bacterial microcapsules in controlling vibriosis disease caused by the infection of Vibrio parahaemolyticus in white shrimp Litopenaeus vannamei. Aquaculture 549:737795

Berninger T et al (2018) Maintenance and assessment of cell viability in formulation of non-sporulating bacterial inoculants. Microb Biotechnol 11(2):277–301

Bibi F et al (2018) Halophytes-associated endophytic and rhizospheric bacteria: diversity, antagonism and metabolite production. Biocontrol Sci Tech 28(2):192–213

Bolívar-Ramírez NC et al (2021) Performance of Pacific white shrimp fed probiotics Lactobacillus plantarum and Bacillus spp. in a biofloc system. Bol Inst Pesca 47

Boonanuntanasarn S et al (2016) Effects of dietary supplementation with β-glucan and synbiotics on growth, haemolymph chemistry, and intestinal microbiota and morphology in the Pacific white shrimp. Aquac Nutr 22(4):837–845

Bouallegui Y (2021) A comprehensive review on crustaceans' immune system with a focus on freshwater crayfish in relation to crayfish plague disease. Front Immunol 12:667787

Caulier S et al (2019) Overview of the antimicrobial compounds produced by members of the Bacillus subtilis group. Front Microbiol 10(FEB):302

Chen L et al (2021) Effects of Bacillus velezensis supplementation on the growth performance, immune responses, and intestine microbiota of Litopenaeus vannamei. Front Mar Sci 8

Cheng AC et al (2020) Intestinal microbiota of white shrimp, Litopenaeus vannamei, fed diets containing Bacillus subtilis E20-fermented soybean meal (FSBM) or an antimicrobial peptide derived from B. subtilis E20-FSBM. Aquac Res 51(1):41–50

Coates CJ, Söderhäll K (2021) The stress–immunity axis in shellfish. J Invertebr Pathol 186:107492

Das S, Ward LR, Burke C (2008) Prospects of using marine actinobacteria as probiotics in aquaculture. Appl Microbiol Biotechnol 81(3):419–429

de la Vega MRV et al (2020) Reducing stress by improving performance of hatchery-reared Catarina scallop (Argopecten ventricosus) spat with different genera of beneficial microorganisms: a biochemical and molecular analysis. Aquac Rep 17:100298

Dekham K et al (2022) Probiotics expressing double-stranded RNA targeting VP28 efficiently protect shrimps from WSSV infection. Aquac Rep 23

Desriac F et al (2014) Exploring the hologenome concept in marine bivalvia: Haemolymph microbiota as a pertinent source of probiotics for aquaculture. FEMS Microbiol Lett 350(1): 107–116

Didinen B et al (2016) Effect of lactic acid bacteria and the potential probiotic Hafnia alvei on growth and survival rates of narrow clawed crayfish (Astacus leptodactylus Esch, 1823) stage II juveniles. Iran J Fish Sci 15:1307–1317

Du Y et al (2019) Understanding the roles of surface proteins in regulation of Lactobacillus pentosus HC-2 to immune response and bacterial diversity in midgut of Litopenaeus vannamei. Fish Shellfish Immunol 86:1194–1206

Du S et al (2021) Enterococcus faecium are associated with the modification of gut microbiota and shrimp post-larvae survival. Anim Microbiome 3(1):88

Dunlap CA, Bowman MJ, Rooney AP (2019) Iturinic lipopeptide diversity in the bacillus subtilis species group – important antifungals for plant disease biocontrol applications. Front Microbiol 10(AUG):1794

Dupont S et al (2020) Oyster hemolymph is a complex and dynamic ecosystem hosting bacteria, protists and viruses. Anim Microbiome 2(1)
Dushku E et al (2019) Probiotic properties and immunomodulatory activity of gastrointestinal tract commensal bacterial strains isolated from the edible farmed snail Cornu aspersum maxima. Fish Shellfish Immunol 92:792–801
Dushku E et al (2020) Listeria monocytogenes induced dysbiosis in snails and rebiosis achieved by administration of the gut commensal Lactobacillus plantarum Sgs14 strain. Fish Shellfish Immunol 104:337–346
EFSA Panel on Biological Hazards (BIOHAZ) (2022) Update of the list of QPS-recommended biological agents intentionally added to food or feed as notified to EFSA 15: suitability of taxonomic units notified to EFSA until September 2021. EFSA J 20(1):e07045
Fajardo P et al (2014) Bacteria isolated from shellfish digestive gland with antipathogenic activity as candidates to increase the efficiency of shellfish depuration process. Food Control 46:272–281
FAO (2022) The state of world fisheries and aquaculture 2022. Towards Blue Transformation
Fdhila K et al (2017) Culture conditions improvement of Crassostrea gigas using a potential probiotic Bacillus sp strain. Microb Pathog 110:654–658
Feijo RG et al (2015) RNAi-based inhibition of infectious myonecrosis virus replication in Pacific white shrimp Litopenaeus vannamei. Dis Aquat Org 114:89–98
Ferreira GS et al (2015) Microbial biofloc as source of probiotic bacteria for the culture of Litopenaeus vannamei. Aquaculture 448:273–279
Ficoseco CA et al (2018) Safety and growth optimization of lactic acid bacteria isolated from feedlot cattle for probiotic formula design. Front Microbiol 9:2220
Forssten SD, Sindelar CW, Ouwehand AC (2011) Probiotics from an industrial perspective. Anaerobe 17(6):410–413
Foysal MJ et al (2019) Marked variations in gut microbiota and some innate immune responses of fresh water crayfish, marron (Cherax cainii, Austin 2002) fed dietary supplementation of Clostridium butyricum. PeerJ 2019(8):e7553
Foysal MJ et al (2020) Lactobacillus acidophilus and L. plantarum improve health status, modulate gut microbiota and innate immune response of marron (Cherax cainii). Sci Rep 10(1)
Foysal MJ et al (2021) Lactobacillus plantarum in black soldier fly (Hermetica illucens) meal modulates gut health and immunity of freshwater crayfish (Cherax cainii). Fish Shellfish Immunol 108:42–52
Gao X et al (2018) Effects of a probiotic (Bacillus licheniformis) on the growth, immunity, and disease resistance of Haliotis discus hannai Ino. Fish Shellfish Immunol 76:143–152
Gao J et al (2021) Probiotics in the dairy industry - advances and opportunities. Compr Rev Food Sci Food Saf 20(4):3937–3982
García Bernal M et al (2017) Streptomyces effect on the bacterial microbiota associated to Crassostrea sikamea oyster. J Appl Microbiol 122(3):601–614
García-Bernal M et al (2019) Growth and antioxidant response of juvenile oysters Crassostrea sikamea and Crassostrea corteziensis treated with Streptomyces strains. Arquivo Brasileiro de Medicina Veterinaria e Zootecnia 71(6):1993–1998
Grandiosa R et al (2018) Multi-strain probiotics enhance immune responsiveness and alters metabolic profiles in the New Zealand black-footed abalone (Haliotis iris). Fish Shellfish Immunol 82:330–338
Grandiosa R et al (2020) Immune response in probiotic-fed New Zealand black-footed abalone (Haliotis iris) under Vibrio splendidus challenge. Fish Shellfish Immunol 104:633–639
Hao YT et al (2017) Succession and fermentation products of grass carp (Ctenopharyngodon idellus) hindgut microbiota in response to an extreme dietary shift. Front Microbiol 8 (AUG):1585
Harwood CR et al (2018) Secondary metabolite production and the safety of industrially important members of the Bacillus subtilis group. FEMS Microbiol Rev 42(6):721–738

Hien TTT et al (2022) Effects of dietary supplementation with pro-A on growth performance, feed utilization, immune responses, and intestinal microbiota of whiteleg shrimp (Litopenaeus vannamei). Aquac Rep 24:101125

Holt CC et al (2021) Understanding the role of the shrimp gut microbiome in health and disease. J Invertebr Pathol 186:107387

Hossain A et al (2022) Antibiotics, antibiotic-resistant bacteria, and resistance genes in aquaculture: risks, current concern, and future thinking. Environ Sci Pollut Res 29(8):11054–11075

Huddy RJ, Coyne VE (2014) Detection and localisation of the abalone probiotic Vibrio midae SY9 and its extracellular protease, VmproA, within the digestive tract of the South African abalone, Haliotis midae. PLoS One 9(1):e86623

Infante-Villamil S, Huerlimann R, Jerry DR (2021) Microbiome diversity and dysbiosis in aquaculture. Rev Aquac 13(2):1077–1096

Jiang HF et al (2013) Effects of dietary supplementation of probiotic Shewanella colwelliana WA64, Shewanella olleyana WA65 on the innate immunity and disease resistance of abalone, Haliotis discus hannai Ino. Fish Shellfish Immunol 35(1):86–91

Karim M et al (2013) Probiotic strains for shellfish aquaculture: protection of eastern oyster, crassostrea virginica, larvae and juveniles againsl bacterial challenge. J Shellfish Res 32(2): 401–408

Kewcharoen W, Srisapoome P (2022) Potential synbiotic effects of a Bacillus mixture and chitosan on growth, immune responses and VP(AHPND) resistance in Pacific white shrimp (Litopenaeus vannamei, Boone, 1931). Fish Shellfish Immunol 127:715–729

Khademzade O et al (2020) The effects of water additive Bacillus cereus and Pediococcus acidilactici on water quality, growth performances, economic benefits, immunohematology and bacterial flora of whiteleg shrimp (Penaeus vannamei Boone, 1931) reared in earthen ponds. Aquac Res 51:1759–1770

Krasowska A, Sigler K (2014) How microorganisms use hydrophobicity and what does this mean for human needs? Front Cell Infect Microbiol 4(AUG):112

Kroetsch SA et al (2020) The effects of taxonomy, diet, and ecology on the microbiota of riverine macroinvertebrates. Ecol Evol 10(24):14000–14019

Kuebutornye FKA, Abarike ED, Lu Y (2019) A review on the application of bacillus as probiotics in aquaculture. Fish Shellfish Immunol 87:820–828

Kuo H-W, Chang C-C, Cheng W (2021) Synbiotic combination of prebiotic, cacao pod husk pectin and probiotic, Lactobacillus plantarum, improve the immunocompetence and growth of Litopenaeus vannamei. Fish Shellfish Immunol 118:333–342

Kuypers MMM et al (2005) Massive nitrogen loss from the Benguela upwelling system through anaerobic ammonium oxidation

Lafferty KD et al (2015) Infectious diseases affect marine fisheries and aquaculture economics. Annu Rev Mar Sci 7:471–496

Lai Y, Luo M, Zhu F (2020) Dietary bacillus amyloliquefaciens enhance survival of white spot syndrome virus infected crayfish. Fish Shellfish Immunol 102:161–168

Lai X et al (2022) Effects of the probiotic Psychrobacter sp. B6 on the growth, digestive enzymes, antioxidant capacity, immunity, and resistance of Exopalaemon carinicauda to Aeromonas hydrophila. Probiot Antimicrob Proteins

Lau N-S et al (2022) Comparative analyses of Scylla olivacea gut microbiota composition and function suggest the capacity for polyunsaturated fatty acid biosynthesis. Microb Ecol

Le B et al (2022) Dietary supplementation with Pediococcus pentosaceus enhances the innate immune response in and promotes growth of Litopenaeus vannamei shrimp. J Fish Dis 45: 1343–1354

Li HH, Jiang XR, Qiao JY (2021) Effect of dietary Bacillus subtilis on growth performance and serum biochemical and immune indexes in weaned piglets. J Appl Anim Res 49(1):83–88

Li T et al (2022) Response of Chinese mitten crab (Eriocheir sinensis) intestinal microbiota to commercial probiotic application. Aquac Res

Liao W et al (2022) Effects of sodium humate and probiotics on growth performance enzyme activity and microbial environment of Litopenaeus vannamei in high-density zero-water exchange systems. Front Mar Sci 9

Liu Z et al (2018) Effect of a commercial microbial agent on the bacterial communities in shrimp culture system. Front Microbiol 9:2430

Liu H et al (2019) Protective approaches and mechanisms of microencapsulation to the survival of probiotic bacteria during processing, storage and gastrointestinal digestion: a review. Crit Rev Food Sci Nutr 59(17):2863–2878

Liu L et al (2022) Dietary supplementation of astaxanthin is superior to its combination with Lactococcus lactis in improving the growth performance, antioxidant capacity, immunity and disease resistance of white shrimp (Litopenaeus vannamei). Aquac Rep 24

López-Mas L et al (2021) Farmed or wild fish? Segmenting European consumers based on their beliefs. Aquaculture 532:735992

Markets and Markets (2021) Probiotics in animal feed market by livestock (poultry, ruminants, swine, aquaculture, pets), Source (bacteria, yeast), Form (dry, liquid), and Region (North America, Europe, Asia Pacific, South America, & Rest of World) - Global Forecast to 2026, in AGI 3616

Mcshane A et al (2021) Mucus. Curr Biol 31:R931–R947

Middlemiss KL et al (2015) Combined effects of UV irradiation, ozonation, and the probiotic Bacillus spp. on growth, survival, and general fitness in European lobster (Homarus gammarus). Aquaculture 444:99–107

Musella M et al (2020) Tissue-scale microbiota of the Mediterranean mussel (Mytilus galloprovincialis) and its relationship with the environment. Sci Total Environ 717:137209

Nguyen VD et al (2014) Screening of marine bacteria with bacteriocin-like activities and probiotic potential for ornate spiny lobster (Panulirus ornatus) juveniles. Fish Shellfish Immunol 40(1): 49–60

Nimrat S et al (2021) Dietary administration of Bacillus and yeast probiotics improves the growth, survival, and microbial community of juvenile whiteleg shrimp, Litopenaeus vannamei. J Appl Aquac 33(1):15–31

Offret C et al (2019) Protective efficacy of a Pseudoalteromonas strain in European abalone, Haliotis tuberculata, infected with Vibrio harveyi ORM4. Probiotics Antimicrob Proteins 11(1):239–247

Olmos J et al (2020) Bacillus subtilis, an ideal probiotic bacterium to shrimp and fish aquaculture that increase feed digestibility, prevent microbial diseases, and avoid water pollution. Arch Microbiol 202(3):427–435

Panigrahi A et al (2021) Bioaugmentation of biofloc system with enzymatic bacterial strains for high health and production performance of Penaeus indicus. Sci Rep 11(1)

Panigrahi A et al (2022) A biofloc-based aquaculture system bio-augmented with probiotic bacteria bacillus tequilensis AP BFT3 improves culture environment, production performances, and proteomic changes in Penaeus vannamei. Probiotics Antimicrob Proteins 14(2):277–287

Pierce ML, Ward JE (2019) Gut microbiomes of the eastern oyster (Crassostrea virginica) and the blue mussel (Mytilus edulis): temporal variation and the influence of marine aggregate-associated microbial communities. mSphere 4(6):e00730-19

Porsby CH, Gram L (2016) Phaeobacter inhibens as biocontrol agent against Vibrio vulnificus in oyster models. Food Microbiol 57:63–70

Ren X et al (2022) Effects of dietary heat-killed Lactobacillus plantarum and compound probiotics on the growth performance, antioxidative capacity, intestinal morphology and microbiota of Pacific white shrimp Litopenaeus vannamei. Aquac Res 53(9):3516–3530

Restrepo L et al (2021) Microbial community characterization of shrimp survivors to AHPND challenge test treated with an effective shrimp probiotic (Vibrio diabolicus). Microbiome 9(1): 88

Riet J et al (2021) Bacillus subtilis expressing double-strand RNAs (dsRNAs) induces RNA interference mechanism (RNAi) and increases survival of WSSV-challenged Litopenaeus vannamei. Aquaculture 541:736834

Ringø E (2020) Probiotics in shellfish aquaculture. Aquac Fish 5(1):1–27

Ringø E et al (2020) Lactic acid bacteria in shellfish: possibilities and challenges. Rev Fish Sci Aquac 28(2):139–169

Rizzatti G et al (2017) Proteobacteria: a common factor in human diseases. Biomed Res Int 2017: 9351507

Safari O, Paolucci M, Motlagh HA (2017) Effects of synbiotics on immunity and disease resistance of narrow-clawed crayfish, Astacus leptodactylus leptodactylus (Eschscholtz, 1823). Fish Shellfish Immunol 64:392–400

Sánchez-Ortiz AC et al (2020) Probiotic bacterium and microalga interaction on rearing Kumamoto oyster Crassostrea sikamea spat. Curr Microbiol 77(10):2758–2765

Sarao LK, Arora M (2017) Probiotics, prebiotics, and microencapsulation: a review. Crit Rev Food Sci Nutr 57(2):344–371

Sawabe T et al (2016) Individual Apostichopus japonicus fecal microbiome reveals a link with polyhydroxybutyrate producers in host growth gaps. Sci Rep 6

Sha Y et al (2016) Bacterial population in intestines of litopenaeus vannamei fed different probiotics or probiotic supernatant. J Microbiol Biotechnol 26(10):1736–1745

Siddik MAB et al (2020) Influence of fish protein hydrolysate produced from industrial residues on antioxidant activity, cytokine expression and gut microbial communities in juvenile barramundi Lates calcarifer. Fish Shellfish Immunol 97:465–473

Sonnenschein EC et al (2021) The Roseobacter-group Bacterium Phaeobacter as a safe probiotic solution for aquaculture. Appl Environ Microbiol 87(5):e0258120

Speight RE et al (2022) Platforms to accelerate biomanufacturing of enzyme and probiotic animal feed supplements: discovery considerations and manufacturing implications. Anim Prod Sci

Stevick RJ et al (2019) Bacterial community dynamics in an oyster hatchery in response to probiotic treatment. Front Microbiol 10(MAY):1060

Sumithra TG et al (2019) A glimpse towards cultivable hemolymph microbiota of marine crabs: untapped resource for aquatic probiotics/antibacterial agents. Aquaculture 501:119–127

Sun Y et al (2019) Assessing key microbial communities during nitrogen transformation in composting of cow manure using Illumina high-throughput sequencing. Waste Manag 92:59–67

Talpur AD et al (2013) Indigenous lactobacillus plantarum as probiotic for larviculture of blue swimming crab, Portunus pelagicus (Linnaeus, 1758): effects on survival, digestive enzyme activities and water quality. Aquaculture 416-417(416-417):173–178

Tassanakajon A et al (2018) Shrimp humoral responses against pathogens: antimicrobial peptides and melanization. Dev Comp Immunol 80:81–93

Thompson J et al (2022) Antagonistic activity of lactic acid bacteria against pathogenic Vibrios and their potential use as probiotics in shrimp (Penaeus vannamei) culture Front Mar Sci 9

Toledo A et al (2019) Impact of probiotics on growth performance and shrimp survival: a meta-analysis. Aquaculture 500:196–205

Valipour A et al (2019) Dietary lactobacillus plantarum affected on some immune parameters, air-exposure stress response, intestinal microbiota, digestive enzyme activity and performance of narrow clawed crayfish (Astacus leptodactylus, Eschscholtz). Aquaculture 504:121–130

Vezzulli L et al (2018) Comparative 16SrDNA gene-based microbiota profiles of the Pacific oyster (Crassostrea gigas) and the Mediterranean mussel (Mytilus galloprovincialis) from a shellfish farm (Ligurian Sea, Italy). Microb Ecol 75(2):495–504

Wan JJ et al (2022) Changes in the growth performance, antioxidant enzymes and stress resistance caused by dietary administration of synbiotic (fructooligosaccharide and probiotics) in juvenile Chinese mitten crab, *Eriocheir sinensis*. Aquac Int 30(1):467–481

Wei C et al (2021) Effects of dietary Shewanella sp. MR-7 on the growth performance, immunity, and intestinal microbiota of Pacific white shrimp. Aquac Rep 19:100595

Won S et al (2020) Evaluation of potential probiotics bacillus subtilis WB60, Pediococcus pentosaceus, and Lactococcus lactis on growth performance, immune response, gut histology and immune-related genes in whiteleg shrimp, Litopenaeus vannamei. Microorganisms 8(2): 281

Wu HJ et al (2014) Enhancement of the immune response and protection against Vibrio parahaemolyticus by indigenous probiotic Bacillus strains in mud crab (Scylla paramamosain). Fish Shellfish Immunol 41(2):156–162

Wu YS et al (2022) Effects of dietary Lactobacillus reuteri and Pediococcus acidilactici on the cultured water qualities, the growth and non-specific immune responses of Penaeus vannamei. Fish Shellfish Immunol 127:176–186

Xiaolong G et al (2019) Effects of the probiotic Bacillus amyloliquefaciens on the growth, immunity, and disease resistance of Haliotis discus hannai. Fish Shellfish Immunol 94:617–627

Xie JJ et al (2019) Effects of dietary mixed probiotics on growth, non-specific immunity, intestinal morphology and microbiota of juvenile pacific white shrimp, *Litopenaeus vannamei*. Fish Shellfish Immunol 90:456–465

Xie M et al (2022a) Growth performance, hepatic enzymes, and gut health status of common carp (Cyprinus carpio) in response to dietary Cetobacterium somerae fermentation product. Aquac Rep 23:101046

Xie M et al (2022b) Stabilized fermentation product of Cetobacterium somerae improves gut and liver health and antiviral immunity of zebrafish. Fish Shellfish Immunol 120:56–66

Xu L et al (2021a) Screening of intestinal probiotics and the effects of feeding probiotics on the digestive enzyme activity, immune, intestinal flora and WSSV resistance of Procambarus clarkii. Aquaculture 540

Xu Y et al (2021b) Effects of dietary saccharomyces cerevisiae yfi-sc2 on the growth performance, intestinal morphology, immune parameters, intestinal microbiota, and disease resistance of crayfish (Procambarus clarkia). Animals (Basel) 11(7):1963

Xu Y et al (2022) Study on microflora structure of Macrobrachium rosenbergii larvae and environmental water at different developmental stages. J Freshw Ecol 37(1):360–369

Yang Q et al (2019) Lactic acid bacteria, Enterococcus faecalis Y17 and Pediococcus pentosaceus G11, improved growth performance, and immunity of mud crab (Scylla paramamosain). Fish Shellfish Immunol 93:135–143

Yao W et al (2021) Effects of dietary synbiotics supplementation methods on growth, intestinal health, non-specific immunity and disease resistance of Pacific white shrimp, *Litopenaeus vannamei*. Fish Shellfish Immunol 112:46–55

Yu Z et al (2019) The differences of bacterial communities in the tissues between healthy and diseased yesso scallop (Patinopecten yessoensis). AMB Express 9(1)

Zhang S et al (2021) Dietary supplementation with Bacillus subtilis promotes growth performance of broilers by altering the dominant microbial community. Poult Sci 100(3)

Zhang T et al (2022) Monitoring bacterial community dynamics in abalone (Haliotis discus hannai) and the correlations associated with aquatic diseases. Water 14(11):1769

Zhao J et al (2018) Effects of dietary supplementation of probiotics on growth, immune responses, and gut microbiome of the abalone Haliotis diversicolor. Aquaculture 493:289–295

Zheng X et al (2020) The effect of Lactobacillus plantarum administration on the intestinal microbiota of whiteleg shrimp Penaeus vannamei. Aquaculture 526:735331

Zhou W et al (2022) Effect of dietary supplementation of Cetobacterium somerae XMX-1 fermentation product on gut and liver health and resistance against bacterial infection of the genetically improved farmed tilapia (GIFT, Oreochromis niloticus). Fish Shellfish Immunol 124:332–342

Zhu X et al (2022) Sustainability of the rice-crayfish co-culture aquaculture model: microbiome profiles based on multi-kingdom analyses. Environ Microbiome 17(1)

Vieira FdN et al (2016) Use of probiotic-supplemented diet on a Pacific white shrimp farm. Rev Bras Zootec 45(5):203–207

Impact of Climate Change on the Gut Microbiome of Fish and Shellfish

12

Arvind Diwan, Sanjay N. Harke, and Archana Panche

Abstract

The impact of global warming on animal and plant biodiversity is well-known. Consequently, the impact of climate change on the gut microbiota of fish and shellfish is bound to be there and getting momentum, and in the recent past, considerable research work has been carried out in such niche areas of science. It is well known that the impact of climate change on the gut microbiota could lead to possible dysbiosis with strong consequences on hosts and ecosystems. These changes may be caused by a number of environmental factors particularly rising temperatures, changes in water salinities, pH, turbidity, depletion of oxygen levels, pollutants, etc. that may lead to an increase in disease prevalence in the habitat ecosystems. Hence, understanding the factors affecting host–microbiota interactions with cultivable organisms would improve predictions on biodiversity responses to climate change. Several environmental factors which act as stressors can disrupt the strong relationship between hosts and their associated microbiomes leading to dysbiosis and enhancing the chances of host diseases. Hosts and their associated microbiomes form a coherent biological entity termed "holobionts"—that must be studied together to properly understand biological systems and how they will be impacted by climate change. This is particularly crucial for habitat-forming organisms because impacts on these holobionts can affect entire ecosystems. There are several reports indicating that the capacity of organisms to adapt to new habitat conditions basically depends on their plasticity attributes, of which their gut commensal microbiota might be an essential impact factor. Particularly in aquatic organisms, which inhabit the aquatic environment,

A. Diwan (✉) · S. N. Harke
Institute of Biosciences and Technology, MGM University, Aurangabad, Maharashtra, India

A. Panche
Centre for Biosustainability, Technical University of Denmark, Copenhagen, Denmark

A. Diwan et al. (eds.), *Microbiome of Finfish and Shellfish*,
https://doi.org/10.1007/978-981-99-0852-3_12

the complex and dynamic microbiota have significant effects on health and development. However, an understanding of the relative contribution of internal sorting and colonization processes of the microbiota in the gut system is still not clear and research investigation is required in this particular area on a priority basis. Besides environmental factors, pollutants in the form of contaminants and toxins present in the environment may also influence the fish gut microbiota, and also other kinds of chemicals such as pesticides, heavy metals, and antibiotics can also induce gut microbiota dysbiosis associated with changes in the intestinal mucus layer and inflammation in the digestive tract in fish, thereby reducing the ability to absorb nutrients.

Keywords

Climate impact · Microbiome · Extrinsic · Intrinsic · Trophic · Seasons

12.1 Introduction

The importance of the gut microbiome for the maintenance of health and improving the quality of fish and shellfish produce in captive conditions is now widely accepted (Egerton et al. 2018; Butt and Volkoff 2019) and has been extensively studied for potential applications in aquaculture farming and the growth of the fish industry (Llewellyn et al. 2014; Yukgehnaish et al. 2020; Diwan et al. 2021). However, there is a lack of information concerning environmental factors and their effects on the microbial composition of the fish gut system and its interactions with the host. The fish gut microbiome is associated with a variety of beneficial functions for the host, such as mediating and stimulating optimal gastrointestinal development (Bates et al. 2007), producing and supplying vitamins to the host (Roeselers et al. 2011; Tsuchiya et al. 2008), and improving the host's nutrient uptake by providing additional enzymatic activities (Llewellyn et al. 2014). In particular, herbivorous fish heavily rely on their gut microbiome for degrading cellulose and other complex polysaccharides to enhance the nutritional value of organic carbon sources (Di Maiuta et al. 2013; Van Kessel et al. 2011). Aside from optimized nutrient uptake, the gut microbiome is also involved in immunomodulation and maintaining mucosal tolerance (Bates et al. 2007), enhancing the host's resilience against infectious diseases (Llewellyn et al. 2014) and the production of anticarcinogenic and anti-inflammatory compounds (Hamer et al. 2009). The gut microbiome is often considered an "**extra organ**," as it plays a key role in intestinal development and physiology, as well as overall development, growth, and health (Feng et al. 2018). Recent studies suggest that the gut microbiota is involved in energy homeostasis by regulating feeding, digestive and metabolic processes, as well as the immune response (Burokas et al. 2015; Goncalves and Gallardo-Escárate 2017; Johnson and Foster 2018). Consequently, deciphering gut microbiome dynamics in cultured fish species will play an indispensable role in promoting animal health and aquaculture productivity (Yukgehnaish et al. 2020; Perry et al. 2020; Diwan et al. 2021).

The changes in climatic conditions that are occurring all over the world are now considered to be the greatest threat to our biodiversity and ecological niches, but their impacts on the bacterial communities associated with plants and animals remain largely unknown. It has been reported that an increase in the temperature by 2–3 °C than normal, causes a 34% loss of populations' microbiota diversity, with possible negative consequences for host survival (Bestion et al. 2017). The impact of climate change is seen in global declines of marine habitats including the fishery through physiological effects and through changes to ecological interactions, with projected effects of ocean warming and acidification which are likely to exacerbate such impacts in coming decades. Interactions between habitat formers and their microbiomes are fundamental for host functioning and resilience, but how such relationships will change in future conditions is largely unknown (Qiu et al. 2019). Generally, the gut microbiota is shaped by host traits and environment and, in turn, plays multiple essential functions for hosts, including digestion, immunity, and other physiological activities in the life cycle (Folke et al. 1998; Rosenberg and Falkovitz 2004). Consequently, changes in the gut microbiota could lead to potential dysbiosis with strong consequences for hosts and ecosystems (Harvell et al. 2002). These changes may be caused by a number of environmental factors particularly rising temperatures that lead to an increase in disease prevalence (McFall-Ngai et al. 2013). Understanding the factors affecting host–microbiota interactions would improve predictions on biodiversity responses to climate warming (Rosenberg et al. 2007).

The composition of the microbiome community that colonizes the body surface and gut system of animals is influenced by a number of environmental factors and can have a crucial impact on various physiological activities of the host (Tremaroli and Bäckhed 2012; McFall-Ngai et al. 2013). An important role of the microbial community present in the host is to protect animals from pathogens by inhibiting their pathogenic effect (Kamada et al. 2013), although sometimes commensal bacteria can become pathogenic (Garnier et al. 2007; Olson et al. 2014). It is important to know how environmental factors affect host–microbiota interactions since these factors can be used to control and mitigate the impact of infectious diseases and improve animal health and survival. Aquatic animals are in a medium that favors exposure to microbes because the diet, water, and sediments all harbor high levels of microorganisms that can colonize the body surfaces and the gut system. The gut microbiota of fish and shellfish is always dependent on the microbiome present in the ambient aquatic environment and the large volume of water that flows through the body and tissues (Gatesoupe 1999).

Folke et al. (1998) while working on the impact of climate change on marine biodiversity emphasized that understanding the mechanisms of climate-driven shifts is very important and critical to properly managing and conserving marine ecosystems. Harley et al. (2012) reported that ocean warming and acidification can have direct physiological effects on habitat-forming species, negatively affecting their performance and survival. Several workers also confirmed that the changes in marine climatic conditions bring the indirect effects of these stressors via changes in species' interactions (Vergés et al. 2014; Bellard et al. 2012). However, there is not much information regarding the studies which focus on interactions among microorganisms, such as competition and predation, and so also on interactions between macro- and microorganisms (Rosenberg and Falkovitz 2004; Harvell et al. 2002). While it is clear that microorganisms are critically important for the normal development and functioning of host animals, environmental stressors can disrupt the strong relationship between hosts and their associated microbiomes leading to dysbiosis and host disease (Rosenberg et al. 2007; Egan et al. 2013). Hosts and their associated microbiomes form a coherent biological entity—"holobionts"—that must be studied together to properly understand biological systems and how they will be impacted by climate change. This is particularly crucial for habitat-forming organisms because impacts on these holobionts can affect entire ecosystems (Wahl et al. 2015).

Climate change has created a significant impact on shrimp aquaculture farming and it has been mentioned that it is difficult to select one single factor of climate change which is responsible for creating such an impact (De Silva and Doris 2009). Depletion of dissolved oxygen, decrease in salinity and frequent occurrence of algal blooms can seriously affect farm production (Nagothu et al. 2012). Similarly, high temperatures could also increase the intensity and frequency of disease outbreaks of shrimp and adversely affects water quality (Harvell et al. 2002; Vilchis et al. 2005). In such changed climatic circumstances when there is dysbiosis in the captive ecosystem, it will have a vital impact on the microbial composition of the gut system in both fish and shrimp. When there is a deterioration in the quality of the gut microbial community it will have a direct effect on the health of the shrimp. In certain geographical locations, due to the changes in climatic conditions animals inhabiting such habitats have been reported to have a poor quality of microbes present in their gut system. There are several reports indicating that the capacity of organisms to adapt to new habitat conditions basically depends on their plasticity attributes, of which their gut commensal microbiota might be an essential impact factor. Particularly in aquatic organisms, which inhabit the aquatic environment, the complex and dynamic microbiota have significant effects on health and growth development. However, an understanding of the relative contribution of internal sorting of the microbiomes and colonization processes of the same is still unclear.

In order to understand the impact of rapid environmental changes on microbial communities of the gut system in shrimp, several workers have carried out investigations on the gut microbiota of native and habitat-exchanged shrimp using 16S rRNA amplicon sequencing technology (Chen et al. 2017). It has been reported that the gut microbial composition of shrimp living in the lake to river waters significantly differed from those of shrimp living in the river to lake waters. From these findings, they have suggested that shrimp gut microbiota has high plasticity when hosts are exposed to environmental changes even for a short duration. Furthermore, the changes in the external environment might influence the gut microbiome not just by providing environment-associated microbes directly, but also by interfering with the composition of indigenous gut bacteria indirectly (Chen et al. 2017). In this chapter, efforts have been made to collect comprehensive information on various aspects of the impact of climate change on fish and shellfish gut microbiomes.

12.2 Fish Gut Microbiome

Several workers have analyzed the gut microbiome composition of varieties of fish based on molecular tools like gene sequence analysis and NGS technology and it has been observed that the gut microbiome community contains the following bacterial groups: Proteobacteria, Fusobacteria, Firmicutes, Bacteroidetes, Actinobacteria, Clostridia, Bacilli, and Verrucomicrobia. Among these, Proteobacteria, Firmicutes, and Bacteroidetes represent up to 90% of the fish intestinal microbiome composition in different species (Ringo et al. 2006; Desai et al. 2012; Li et al. 2013; Givens 2012). The concept of the presence of the core of microbiomes in the fish gut system has been suggested by some workers based on their role, particularly with respect to the physiological and digestive functions, immunity, and the preferred salinity levels in the surrounding water (Roeselers et al. 2011). However, further research investigations on the core gut microbiome are the need of hour to facilitate the development of safe and effective methods for manipulating the composition of the gut microbiome to promote the health and growth of fish (Turnbaugh et al. 2015; Roeselers et al. 2011; Wu et al. 2012). The Next Generation Sequencing (NGS) approaches offer the ability to obtain a large number of sequence reads in a single run, yielding more accurate and reliable sequence databases of microbiome species. The advanced genetic tools in molecular biology thus allow researchers to apply a massively parallel sequencing approach to address the core gut microbiome concept and to examine the relationship between gut microbiome composition and functional aspects of fish species with different or similar features, such as gastrointestinal physiology and feeding behaviors.

Studies on the gut microbiome of fish have indicated that there exists intra and inter-species microbial diversity depending on the changes in environmental factors. Factors that influence this diversity include life stage (Hansen and Olafsen 1999), trophic level (Clements et al. 2007), diet (Cordero et al. 2015), season (Hovda et al. 2012), habitat (Bano et al. 2007), captive condition (Dhanasiri et al. 2011), sex (Dhanasiri et al. 2011), and phylogeny (Miyake et al. 2015). The meta-analysis studies on the factors that affect the gut microbiome composition of fish stated that trophic level, habitat, and possibly host phylogeny are the most likely influencers (Sullam et al. 2012). In the following sections, attempts have been made to discuss in detail how different environmental factors affect the fish gut microbiome.

12.2.1 The Influence of Environmental and Host-Related Intrinsic Factors

The relationship between the environmental factors and the structural composition of the microbial community of the fish gut system is very important and crucial for improving various physiological functions in the fish including performance in captive conditions (Wu et al. 2012). Nayak (2010) while working microbial composition of freshwater fish mentioned that fish acquires microbes in their gut tract system through surrounding aquatic ecosystems. Identical observation has been reported by Wu et al. (2012) while working on bacterial composition present in the intestinal lumen of grass carp. Furthermore, they confirmed that the fish always depends on their microbial composition in the surrounding environment like sediments and the quality of water. Sullam et al. (2012) carried out studies on meta-analysis based on 25 bacterial 16S rRNA gene sequence libraries obtained from the intestinal microbiomes of different fish species (e.g., salmon, trout, and zebrafish) to address the relationship between the gut microbiota of fish to biotic and abiotic variables such as trophic levels and habitats. They reported that there is an increased intensity of bacterial order belonging to Aeromonadales in freshwater fishes and Vibrionales in saltwater fishes. In another investigation of the intestinal microbiome of zebrafish and inhabiting surrounding water under fed and starved conditions, Semova et al. (2014) observed that some bacterial taxa in the fish intestine were found at similar frequencies in the water, while other taxa were enriched specifically in the intestine. These findings suggest that the fish gut microbiome composition more closely resembles the surrounding environment. By using metagenomics tools along with the 16S rRNA analysis, Xing et al. (2013) characterized the taxonomic and metabolic profiles of the captive adult turbot (*Scophthalmus Maximus*) and investigated the microbiome relationship with the surrounding aquatic environment. The results of these findings indicated that both the gut system and mucus samples from the fish were largely made up of Proteobacteria and Firmicutes. It is also reported that the gut system might possess bacteria that are initially associated with seawater and some workers have also mentioned besides local habitat influence, other host-related factors can also shape fish gut microbiome composition (Larsen et al. 2014; Ye et al. 2014; Li et al. 2014). Roeselers et al. (2011) while working on fish microbiome composition characterized and compared the bacterial group and structure profile among laboratory-reared zebrafish in different geographic locations with that of zebrafish collected from their natural habitat. It was observed that regardless of differences in the life histories and local environments of the two groups of zebrafish, a core set of bacterial genera, including the classes Gamma-Proteobacteria and Fusobacteria were present in both domesticated and wild-caught zebrafish. Li et al. (2012) while working on the microbiome composition of freshwater fish observed that the fish reared in a similar environment possess very distinct dominant intestinal microbiota compositions. However, some workers have reported differences in the microbial composition of fish gut systems reared under a common environment (Franchini et al. 2014). Hence, supporting pieces of evidence available on both aspects of microbial composition indicate that their composition may differ even though the fish habitats are similar and composition may not differ much in a similar environment. This concept has been again substantiated by the works of Li et al. (2014) on carp and Navarrete et al. (2012) and Franchini et al. (2014) on rainbow trout. Roeselers et al. (2011) mentioned that the host may also contribute to the bacterial community assemblage by selecting microbial populations that include specialized bacteria to aid in the digestion and absorption of nutrients from a variety of food sources. This suggests that surrounding the environment and other ecological factors create an impact on host physiology and also on gut anatomy when assessing relationships between microbiota composition and host.

During the early developmental stages of fish particularly at different embryonic developmental phases, there are no microbes, and even during hatching time also the presence of any microbe has not been reported. However, gut microbes appear after post-hatch and they originate from surrounding environments (Li et al. 2017). After hatching, fish are exposed to change environmental factors which can greatly influence the gut microbiota throughout their lifespan. For example, in fish inhabiting freshwater and marine ecosystems, their gut microbiota differs significantly and bacterial sp. like *Aeromonas* and *Pseudomonas* predominate in the gut of freshwater fish, whereas in marine fish, Vibrio is the most common genus (Vatsos 2016). Schmidt et al. (2015) while working on the gut microbiome community of fish black molly, *Poecilia sphenops* observed that an increase in salinity induces changes in dominant bacterial taxa in the microbiomes, and in rainbow trout, an increase in temperature results in an increase in microbial growth (Schmidt et al. 2015; Huyben et al. 2018). Similarly, in the case of fish Mahseer, *Tor tambroides* Mohd Nosi et al. (2018) found that the gut microbial diversity at different stages of growth of fish like at larval, juvenile, and adult stages have higher gut microbiota diversity than fingerling and yearling stages. This change in microbial diversity might be due to changes and modifications in the diet and feeding strategies at different stages of development and growth. Fish of different ages and sexual stages might have different nutrient requirements and might adjust their diets and feeding rates to obtain adequate energy intake. As the diet and gut microbiota composition change with age, is likely that the nature of the contribution of the gut microbiota to host homeostasis also changes. However, further research is needed to explain this potential relationship between fish age, fish gut microbiota, and host energy homeostasis (Butt and Volkoff 2019).

Seasonal changes, accompanied by changes in environmental factors particularly high temperature, higher salinity, light intensity, and low pH might induce alterations in food consumption due to variations in nutrient loads in the water column and as a consequence, the composition of the microbiota may get affected. It has been reported that in the case of tilapia fish the gut microbiome load decreases in the winter season compared to other seasons and in hybrid tilapia, it is observed that bacterial species such as Pseudomonas, Micrococcus, and Flavobacterium are present only in the winter season (Al-Harbi and Naim Uddin 2004). When fish are cultured in the captive condition they show a different type of microbial composition in their gut system. For example, in Atlantic salmon, fish held in two different holding conditions have different microorganism compositions within their gut microbiota (Dehler et al. 2017). Pollutants and other toxic contaminants present in the environment may also influence the fish microbiota and other kinds of environmental chemicals such as pesticides heavy metals and antibiotics can induce gut microbiota dysbiosis associated with changes in the intestinal mucus layer and inflammation in fish, thereby reducing the ability to absorb nutrients (Meng et al. 2018; Kan et al. 2015; Giri et al. 2018).

12.2.1.1 Host-Specific Intrinsic Factors

A number of workers have reported that host-specific intrinsic factors also influence the microbial composition of the digestive tract in the case of fish. Intrinsic factors such as phylogeny, genetics and sex, age/life stages, and diet/feeding habits also determine microbial composition among fish. Genetics/Sex Genetic background of the fish generally determines the type of gut microbiota present in the body of fish and intra- and inter-specific variations in microbiota have also been demonstrated in the number of fish (Egerton et al. 2018). In rainbow trout, Navarrete et al. (2012) mentioned that the presence of some specific bacterial groups in the digestive tract is due to different habitats and different diets of fish. Bolnick et al. (2014) reported that the sex of the fish may influence the gut microbiota through sex-specific host–microbe interactions, diet preferences, or immune responses. Furthermore, it is also observed that the gut microbial composition was somewhat different between the sexes of the three spine stickleback, *Gasterosteus aculeatus,* and Eurasian perch *Perca fluviatilis,* but in zebrafish, there was not much difference in microbial composition between the sexes (Liu et al. 2016a, b). There is still some controversy that whether host genetics has got any influence on gut microbiome composition in fish but there are reports substantiating this concept and it has been considered today that host genetics is responsible for shaping the fish gut microbiota. However, many time environment also plays important role in shaping the gut microbiome composition suggesting that environmental factors can overcome differences in host genetics. For example, channel catfish, *Ictalurus punctatus,* and blue catfish, *Ictalurus furcatus* raised under similar environmental and husbandry conditions have been reported to have similar gut microbiota compositions (Bledsoe et al. 2018). Other factors like age and sexual maturity also determine the type of microbiome composition present in the fish and significant differences have been reported in juvenile and matured adult fish by a number of workers. The intensity of microbiota present in the gut system of juvenile zebrafish was found to be higher than elderly adult fish (Yan et al. 2016). Similarly, in southern catfish, gut microbial diversity increases as the host become aged (Zhang et al. 2018a, b). In Atlantic salmon (Salmo salar), intestinal microbiota compositions vary between embryonic stages, with embryonic communities having lower richness and diversity compared to those of hatchlings. Mohd Nosi et al. (2018) observed that in mahseer fish, *Tor tambroides* the gut system showed higher microbiota diversity at larval, juvenile, and adult stages of fish than in fingerling and yearling stages. This change in microbiome diversity is speculated because of a change in feed and diet, it is well-established fact that fish at different stages of growth and at sexual maturity their nutritional requirement is different. As the diet and gut microbiota composition change with age is likely that the nature of the contribution of the gut microbiota to host homeostasis also changes. However, further research is needed to explain this potential relationship between fish age, fish gut microbiota, and host energy homeostasis (Ye et al. 2017). The type of feed and feeding habits of fish also have a greater influence on microbial composition and diversity. It has been observed that those fish that are carnivorous generally show low microbial diversity compared to herbivorous and omnivorous fish (Wang et al. 2017). Egerton et al. (2018) while working on fish gut microbiota observed that herbivorous fish have more abundant microbiota in their gut tract and the dominant bacteria found were *Clostridium, Citrobacter, Leptotrichia, Cetobacterium,* and *Halomonas* in omnivorous fish and Clostridium, Cetobacterium, and Halomonas in carnivorous species (Kashinskaya et al. 2018; Li et al. 2014; Liu et al. 2016a, b). This trend has been found in both marine and freshwater fish, suggesting that the trophic level is likely one of the most influential factors affecting the gut microbiota composition.

The gut microbiota can also vary within species of the same trophic level. In the case of silver carp, *Hypophthalmichthys molitrix*; bighead carp, *Hypophthalmichthys nobilis*; grass carp, *Ctenopharyngodon Idella*; and common carp, *Cyprinus carpio* when these fish are cultured in the same environmental conditions their microbial composition exhibits inter-specific differences and this is because of relative abundance of the cellulose-degrading bacteria belonging to phyla Firmicutes in their digestive tracts (Liu et al. 2016a, b). Many times change in the diet composition also influence microbial profile in the gut tract of several fish. Extreme dietary changes, such as fasting also shape the fish's gut microbiota. During times of fasting, morphological changes in the gastrointestinal tract occur due to the reduced nutrient uptake, which may account for changes observed in the gut microbiota (Bruce et al. 2018). Furthermore, the depletion of nutrients induces changes in gut microbiota composition to favor bacterial species/communities that use more diverse energy sources and are capable of survival under limited nutrient conditions (Vatsos 2016). Xia et al. (2014) while working on the grouper microbiome reported that Proteobacteria were predominantly present in fasting conditions whereas in fed conditions Firmicutes were predominant. A similar type of observation has been made in Asian sea bass in which fasting condition has induced the dominant presence of Bacteroidetes and depletion of Betaproteobacteria. Mente et al. (2018) while working on the microbiome of rainbow trout reported that a short time feeding after 3 h is not sufficient to make any significant differences in bacterial composition.

12.2.1.2 Impact of Trophic Level

In an aquatic ecosystem, the food chain or trophic levels of natural food web configuration are always get influenced by several environmental factors. The food web configuration and available food density at different water levels influence the microbial composition of the gut system of the animals inhabiting different trophic levels of the ecosystem (Miyake et al. 2015; Li et al. 2017). In terms of investigating the relationship of gut microbial composition with trophic level, several workers have reported earlier such investigations on flatfish (Liston 1956) and salmon (Yoshimizu and Kimura 1976). Later, some researchers showed for the first time that marine herbivorous fish like surgeonfish and sea chub possessed distinct symbiotic gut microbiota that aided fermentative digestion (Fishelson et al. 1985; Rimmer and Wiebe 1987). Since then, anaerobic bacterial species, belonging to the phylum Firmicutes and class Clostridia, have been repeatedly observed in the digestive tracts of herbivorous fish (Mouchet et al. 2012). However, more corroborating research studies found that the microbiota of herbivores was distinct from that of fish with other diets and strains of Firmicutes dominated the gastrointestinal microbial communities of these fish. Similar observations have also been made by other workers while working on the gut microbiome of different species of herbivorous fish such as brown-spotted spine foot, *Siganus stellatus*, butterfish, *Odax pullus*, daisy parrotfish, *Chlorurus sordidus*, dusky parrotfish, *Scarus niger*, marble fish, *Aplodactylus arctidens* and zebra perch, *Hermosilla azurea* (Mountfort et al. 2002; Fidopiastis et al. 2006; Clements et al. 2007; Miyake et al. 2015). It is reported that fish that are found in other trophic levels have less characteristic dominance of microbial species in their digestive tract when compared to herbivores. Uchii et al. (2006) while working on the gut microbial composition of herbivore freshwater fish mentioned that these fish contained different unique intestinal bacterial communities. In non-herbivorous fish which are inhabiting in different trophic marine environments, it has been observed that in such fish Proteobacteria were found to be more dominant than Firmicutes in their gut microbial composition (Miyake et al. 2015). Several studies have been carried out on gut microbiome of temperate pelagic fish such as mackerel, *Scomber scombrus*, and herring as well as other tropical planktivores such as pipefish, *Syngnathus scovelli*, sardines, *Sardinella longiceps*, damselfish, and cardinal fish noticed the dominance of Gram-negative bacteria, viz., *Vibrio, Pseudomonas, Psychrobacter, Achromobacter, Shewanella, Alteromonas, Endozoicomonas, Vibrionaceae,* and *Pasteurellaceae* in their microbial composition (Hansen et al. 1992; Ransom 2008; Svanevik and Lunestad 2011; Parris et al. 2016). In carnivorous fish, most of the workers have observed the dominance of Vibrio species and photobacterium in gut microbial composition. Besides these bacterial species, another bacterial group dominance was *Aeromonas, Cetobacterium, Bacillus, Mycoplasma, Acinetobacter, Vibrio* spp., *Enterobacter* spp., *Pseudomonas* spp., and *Aeromonas* spp., which were isolated from the marine fish gut tract (Ray et al. 2012). It is mentioned that all these bacterial species are the core components of the fish gut microbiome which produces protease enzymes. This is especially important as new species of bacteria continue to enter the aquaculture sector and diet manipulation becomes common practice as a means to improve the health and performance of the culturable organisms in captive conditions. Hence, the relationship between environmental factors influencing trophic levels and further the impact of trophic levels on gut microbial composition in fish inhabiting different trophic zones becomes vital for research and developmental studies. Other aspects of the microbial composition of different species of fish inhabiting in similar water ecosystems will be very interesting and considerable work has been reported dealing with such studies. It has been revealed that a larger influence of host selective pressures and trophic levels occurs on the gut microbiota than the influence of environmental factors as different cohabiting species. For instance, cohabiting species of silver carp, grass carp, bighead carp, and blunt snout bream revealed distinct gut microbiome composition providing for the influence of host-specific factors on the gut microbial communities (Li et al. 2012). Identical results have been reported in paddlefish, *Polyodon spathala,* and bighead carp, *Aristichthys nobilis* with similar feeding behavior and reared in similar ponds (Li et al. 2014). Thus host-associated factors outplay the influence of environmental factors. However, more investigations are required to establish the relationship between the influences of trophic levels on the gut microbial composition of finfish inhabiting different environmental ecosystems.

12.2.1.3 Impact of Season

Seasonal changes in the environment particularly in freshwater ecosystems have a greater impact on gut microbiome composition in fish. Considerable literature that is available has highlighted seasonal variations and temperature changes as important parameters for fish gut microbial composition (Nayak 2010; Sullam et al. 2012; Ringø et al. 2016). During summer and autumn months a number of workers have reported that in freshwater fish there is a bacterial predominance in their digestive gut tract system (Macmillan and Santucci 1990; Al-Harbi and Naim Uddin 2004; Hagi et al. 2004; Naviner et al. 2006). The impact of seasonal variations on the gut microbial composition of fish has been observed by the number of workers and so also changes in plankton availability have a greater impact on the gut bacterial composition in fish. Hovda et al. (2012) did their first study directly to investigate the impact of seasonal variations in gut microbiota in marine fish. They analyzed the gut microbiota of adult Atlantic salmon using 16S rRNA DNA sequencing techniques during different seasons of the year. It was reported that the water temperature varied between 5.5 and 18.8 °C during the sampling period. Although few bacterial species were only recorded at some of the sample time points, however, overall the variation reported was not statistically significant. Contrarily, a more recent study on salmon did find a relationship between seasonal water temperatures and shifts in gut microbial composition (Neuman et al. 2016). In this study, increasing temperatures (up to 21 °C) were associated with a disappearance of lactic acid bacteria (LAB) and *Acinetobacter* spp. and an increase in *Vibrio* spp. The loss of protective LAB and an increase in potentially virulent *Vibrio* spp. could have a negative impact on host health and has the potential to become an important issue with sea temperatures rising and stocks of wild salmon declining. Further research is required to determine the effects of seasonal variation and temperature changes on the gut microbiota of marine fish. Figure 12.1 illustrates the overall impact of the different environmental and other biological factors on the composition of the gut microbiome of fish.

12.2.2 Microbial Composition of Fish in Wild and Captive Conditions

Several studies have been carried out to investigate whether there is any difference in the gut microbial composition in the fish that is caught from wild environment and those caught from captive farming systems. In the recent past due to improvements in aquaculture technologies fish produced in captive conditions are of better quality with good growth. Due to a controlled environment, balanced feed diet, and disease control technologies, there is substantial improvement in the quality production of fish at the farm gate. Breeding and rearing of fish in captive conditions generally involve the manipulation of multiple factors, including the culture environment. However, many times in order to earn more business activities and profit, intense

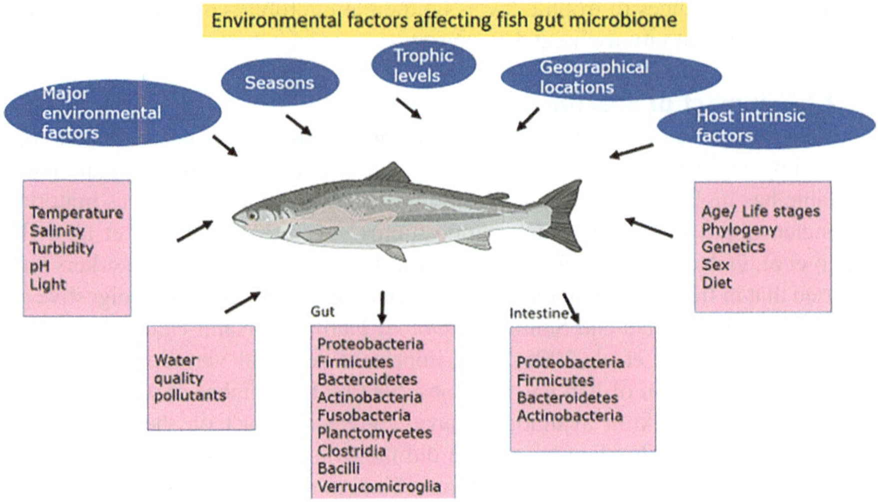

Fig. 12.1 Impact of various intrinsic and extrinsic factors on fish gut microbiome composition

stocking densities and increased stress levels can lead to the deterioration of water quality and the spread of infectious diseases which is a major problem for the aquaculture industry (Verschuere et al. 2000). For the prevention of the spread of infectious pathogens, antibiotics have been used very frequently and sometimes very liberally and even prophylactically (Cabello 2006). Because of the deterioration of water quality and the prevalence of bacterial and viral diseases number of researchers have observed a reduction in microbial gut diversity in many aquaculture species (Navarrete et al. 2008; Defoirdt et al. 2011). Today, as regulations on the use of antibiotics in aquaculture system are becoming more stringent in many countries, research to find out the alternative methods of disease control is being prioritized. However, the aquaculture industry continues to expand and such regulations are overlooked in many countries on a global scale. Artificial diets and increased food intake levels, often with number of other stress factors, can cause alterations in the microbiome composition in the fish gut intestinal tract (Clements et al. 2014). Holben et al. (2002) while studying the gut microbial composition of Atlantic salmon mentioned that farmed Salmon showed greater microbial diversity compared to those collected from natural habitats. Interestingly, a novel Mycoplasma Phylotype was found to dominate in wild Atlantic salmon and pen-reared fish in Scotland, whereas the farmed fish in Norway were dominated by Acinetobacter. Llewellyn et al. (2016) while studying the changes in the gut microbiota of salmon throughout the life cycle observed that all stages were dominated by Proteobacteria and were enriched for Tenericutes. These studies suggest the presence of a core group of microbiota that can persist often in spite of changing several factors. Roeselers et al. (2011) also reported that there were significant similarities in the gut microbiome found in zebra fish collected from their natural habitat and those reared for generations in lab facilities.

It has been also reported that carnivorous fish when it is fed with a plant-based diet develops a different type of microbial composition in their digestive tract, and the whole nutritional physiological process like feed efficiency, growth rates, and whole-body composition of fish and nitrogen retention is found to be negatively affected (Fournier et al. 2004). Identical results were reported in a study performed on red sea bream, *Pagrus major*, whereby the experimental diet with low fishmeal and high plant-protein levels caused significant reductions in feed conversion and protein efficiency ratio, digestibility of protein, and disease resistance against bacterial infection (Khosravi et al. 2015). A number of workers have observed that when fish, Salmon is fed with a plant-based diet, in due course of time it develops intestinal disorders (Penn et al. 2011; Sahlmann et al. 2013). This kind of disorder might be due to changed gut microbial composition in response to a changed diet. Similar results have been frequently reported by other workers also in conjunction with alterations in the gut microbiota composition (Bakke-McKellep et al. 2007; Green et al. 2013). Gut microbial changes related to plant-protein diets have also been recorded in other carnivorous fish species. Now it is a well-established fact that when certain plant ingredients are fed to carnivorous fish, the gut microbiota composition also gets affected and intestinal disorders occur and hence concerted efforts are being made to reduce the negative impacts of these ingredients through supplementation and further modification of the compound diets (Krogdahl et al. 2010). Changes in gut microbiota composition attributed to a captive state have also been reported in freshwater fish (Bucio et al. 2006) as well as other marine animals (Nelson et al. 2013). However, knowledge of the gut microbiome composition in wild marine fish requires more attention to provide a baseline for comparative purposes to better understand the effects of captive rearing.

For the maintenance of healthy microbiome composition in the digestive tract of fish, water quality is the most important parameter irrespective of whether the fish lives in natural waters or is cultivated in captive conditions. Due to anthropogenic and other activities, inland water bodies, as well as seas and oceans all over the world, are getting heavily polluted and this kind of environmental impact on the microbial composition of fish becomes a great concern for the fisheries sector. Hence, a comparison of the water quality and host gut microbiome is expected to improve the understanding of environmental influence. Besides anthropogenic factors, other fluctuating physico-chemical parameters such as salinity, temperature, pH, oxygen, and CO_2 levels, etc. also have a great influence on the gut microbial composition of fish and other animals. The salinity of water also largely determines the microbial composition in the fish gut as revealed from an analysis of a large number of species with different ecological needs for survival (Sullam et al. 2012). Studies on wild and captive invasive carp species also establish the environment as a key factor in shaping the gut microbiome (Eichmiller et al. 2016). On the contrary, studies also suggest otherwise that the abundant microbial taxa in surrounding waters are not found in the gut of habitant fish and vice versa suggesting a much stronger influence of host-associated factors than the environment (Schmidt et al. 2015). Species that are known to survive in different stressed habitats are considered appropriate to assess if a change in the quality of surrounding waters other than microbial composition also affects the host gut microbiota. For instance, the endogenous microbiota in Amazonian Tambaqui fish, *Colossoma macropomum* which is tolerant to significant variation in pH levels has been shown to be significantly altered at experimental low pH levels; however, it also displays a strong resilience once pH levels are restored (Sylvain et al. 2016).

Hassenrück et al. (2021) carried out studies on the effect of thermal stress on the gut microbiome of juvenile milkfish, *Chanos chanos*. In their study, it was observed that the gut microbiome composition of fish adapted to 330 °C was totally distinctly different from those of fish collected from normal tank water temperature (260 °C). The gut microbiome community of thermally stressed fish was dominated by Cetobacterium, Enterovibrio, and Vibrio. However, it was further mentioned that after a few days microbiota composition of both temperature fish was almost similar. From these, it was concluded thermal stress-induced changes in the milkfish gut microbiome may contribute to the host's habituation to elevated temperature over time. Several workers have reported similar findings in other species of fish in response to temperature changes. For instance, rearing temperature and the relative abundance of Firmicutes were negatively correlated in rainbow trout (Huyben et al. 2018). In Atlantic salmon, increasing water temperatures resulted in a shift from generally beneficial lactic acid bacteria to potentially pathogenic *Vibrio* spp. with putative negative implications for host health (Neuman et al. 2016). Similar findings were published for yellowtail kingfish juveniles, mummichog, and pinfish (Soriano et al. 2018; Givens 2012). While the gut microbiome composition has been previously investigated in wild-caught, outdoor pond, and open-cage farmed milkfish (Rasheeda et al. 2017; Prayitno and Sarwan 2015; Thangamani and Rajendran 2016), the effect of elevated water temperatures on the gut microbiome and the associated consequences for the fitness and welfare of the host remain unknown.

An increase in the environmental temperature seriously affects various physiological activities of fish, particularly feeding, growth, immunity, and disease resistance capacity. It has been reported that these changes are closely related to the destruction of intestinal barrier function, the change of intestinal microbiota, and metabolic dysfunction. Zhou et al. (2022) carried out studies on the effect of heat stress on gut microbiota composition and metabolism analysis in rainbow trout to link possible correlations between microbiome and metabolic functions. It is observed that fish exposed to high temperatures (22.5 °C, 23.5 °C, and 24.5 °C for 24 h, respectively) showed a decrease in relative abundance and diversity of intestinal microbiota significantly, while some bacterial groups such as Mycoplasma, Firmicutes, and Tenericutes increased substantially when it was compared to normal fish (16 °C). Furthermore, while correlating metabolic activities with microbiome they observed that there were significant changes in amino acid levels, vitamins, and short-chain fatty acids in the serum of rainbow trout under acute heat stress which were strongly correlated with the decrease of relative abundance of various intestinal microbiota, especially Morganella, Enterobacter, Lactobacillus, Lawsonia, and Cloacibacterium. In addition, it was also observed that acute heat stress seriously affected the intestinal structure and barrier function, and also caused pathological damage to epithelial cells of the intestine. From these studies, it was concluded that the gut microbiome of rainbow trout under acute heat stress could mediate metabolite transfer through the gut barrier by affecting its integrity. They could also notice significant changes in gut morphology, permeability, antioxidant capacity, and pro-inflammatory cytokine levels under heat shock. A growing number of studies have shown that increased ambient temperature reduces the relative abundance and diversity of the gut microbiome of animals, which may have more negative effects on hosts. Therefore, it is necessary to explore the changes of intestinal microbiota under heat stress to understand the regulatory mechanism of heat stress and protect the intestinal health of rainbow trout from the negative effects of rising water temperatures.

12.2.2.1 Variations Within a Species at Different Geographical Locations

In order to find out any changes in the microbial composition in the digestive tract of fish belonging to the same species found in different geographical locations, Roeselers et al. (2011) while working on zebra fish reported that the core set of microbial composition in gastrointestinal tract will not change due to host-specific selective pressures even if there is a change in the environmental factors. Although some differences occur in microbial composition, however, the low diversity that has been observed in the gut microbiome of different fish species of ecologically and geographically different origins with respect to the bacterial phyla, further warrants the argument of host selective pressures on the gut microbiome (Liu et al. 2016a, b; Ringo et al. 2006; Givens et al. 2014; Sullam et al. 2015; Roeselers et al. 2011).

12.2.3 Colonization and the Factors Affecting the Fish Gut Microbiome

Two types of colonization of fish gut microbiome have been reported in the literature. One type is the stochastic or neutral organization of microbes and the second one is a deterministic or non-neutral assemblage of microbiota which may be dependent on host selective pressures or active dispersal by the host and host–microbe and microbe–microbe interactions. It has been reported that during early developmental stages of zebrafish, *Danio rerio*, herbivorous fish, *Ctenopharyngodon idellus*, carnivorous Chinese perch, *Siniperca chuatsi*, and catfish *Silurus meridionalis* the gut microbial colonization is governed by seeding from surrounding environments and later progressively changes to the non-neutral processes as the host matures to become adult (Yan et al. 2016: Burns et al. 2016). It is mentioned that the digestive tract of the host fish provides shelter for certain types of microbes as an adaptation for this kind of selective microbes. Ley et al. (2006) critically reviewed the demands imposed on microbial flora by the digestive tract depending on different functions to be carried out by the microorganisms in the host. Some microbiomes are attached to the mucosal epithelium of gut wall for the specific functions, and some microbiomes must possess cell surface molecules for adhesion; the selection of some other microbiomes are due to their efficient utilization of all the nutrients calls for the production of enzymes and there are other microbes armed with genetic tools for adapting to the ecological plasticity offered by the gastrointestinal tract and for immunity against bacteriophages. Nayak (2010) mentioned that there are two types of microbiomes present in the digestive tract of fish, viz., "autochthonous" and "allochthonous." Autochthonous microbes are those which are able to meet ecological demands, grow, and survive in the digestive tract, and are able to appropriate most niche spaces to become permanent dwellers in the host. While others, known as "allochthonous" are visitors of the gut tract that derive from the surrounding environment. Thus, microbes associated with the host in diverse ways which may be simply fortuitous at first and may become obligatory later. The adherence of the bacteria to the epithelial cells is considered an essential factor in determining their effector functions within the host. Rieu et al. (2016) mentioned that there is a relationship between the biofilm-forming ability of bacteria and the successful colonization and functions of gut flora. Furthermore, it is reported that the ability to form biofilms helps bacteria to survive in hostile environments available in the digestive tract and is seen in cases of both pathogenic and probiotic bacteria (Kalia et al. 2017; Sanchez et al. 2016).

Furthermore, there are shreds of evidence that the microbial composition also differs in different regions along the gut intestinal tract, the foregut communities being significantly different from hindgut communities (Ye et al. 2014). However, the degree to which these two population types differ is not completely understood and needs further clarification. The interplay of a variety of factors determines the constitution of the fish gut microbiome. The quality of surrounding water and microbial communities directly influence the gut microbiome of fish (Sullam et al. 2012). Strong shreds of evidence are also available on other factors such as host genetics, developmental stage, immune status, and other host-specific pressures on the gut microbiome selection (Burns et al. 2016; Ye et al. 2014; Rawls et al. 2006; Hennersdorf et al. 2016). In addition, the diet also shapes the gut microbiome which displays differential composition with differences in dietary intake (Miyake et al. 2015; Scott et al. 2013; Bolnick et al. 2014). Hence, the factors influencing the gut microbiome may be broadly grouped into three classes, which are (1) ecology and environmental conditions, (2) host-specific, and (3) trophic level and/or feeding behavior.

12.2.3.1 Environment and Host–Microbe Interactions

The microbiota associated with the fish gut tract has a great influence of several environmental factors both intrinsic and extrinsic. However, beneficial and harmful effects on fish health depend on the type of microbiota present in the gut. When the beneficial micro-organisms dominate in the gut system a state of "normobiosis" prevails whereas any perturbations in its normal composition may result in "dysbiosis" in which state the harmful microbes predominate giving rise to a diseased condition. There are a number of intrinsic factors also responsible for shaping microbial composition in the fish gut system and depending on the dominance of either beneficial or harmful microbiota, host biology has also equally gets affected. It has been reported that three ways interaction involving host–microbe–

environment operates to maintain homeostasis in the fish gut. The specific microorganisms which are always present as a core group in the fish gut tract generally have the ability to modulate the expression of several cellular genes that endows the host with important functions such as metabolism, nutrient absorption, immune response generation, and gut tract maturation (Hooper and Gordon 2001). Similar findings have been obtained using zebrafish models (Rawls et al. 2004; Kanther and Rawls 2010; Chessman and Guillemin 2007) and in other fish species where microorganisms demonstrated to regulate metabolism (Li et al. 2013). In the case of genetically modified fast-growing common carp, *Cyprinus carpio* it was observed that the gut flora of fish particularly the presence of Firmicutes plays an important functional role in accelerating the growth process (Li et al. 2013). Similarly in several herbivorous fish, the microbial species are known to help in the digestion of cellulose (Wu et al. 2012; Nayak 2010; Saha et al. 2006), influence innate immune responses in fish (Gomez and Balcazar 2008), and improve the host-immune system for better protection against invading pathogens. These findings provide strong perspectives on the interactions between resident microorganisms and their host.

12.2.4 Impact of Climate Change on Shellfish Gut Microbiome

Like fish and other animals, climate change impact also occurs in the microbial composition of shellfish, particularly in the shrimp gut system. In aquaculture farming all over the world, shrimp is considered as most important candidate species for culture in captive conditions. In a culture system, whether it is intensive culture, semi-intensive or extensive, shrimp are exposed to varied climatic conditions in all culture operation systems and both biotic and abiotic factors influence the microbial composition of not only the gut tract but also other body parts. Due to the poor quality of water and the presence of pollutants or contaminants in the water, there are several incidences of the outbreak of diseases in aquaculture farming, and shrimp mortality both in wild and culture water bodies has been reported. In such dead shrimp when the microbial composition of the gut tract and other body parts was analyzed, several workers reported totally different microbiota compared to healthy shrimp (Diwan et al. 2021). This indicates that the microbial composition of the gut tract of shrimp is very important for the maintenance of health, growth, and survival irrespective of their habitat locations. While analyzing the microbial composition of varieties of shrimp during different phases of growth and with reference to water quality conditions, several researchers have mentioned changes in the microbiota composition of shrimp species, such as *Penaeus monodon, Fenneropenaeus chinensis, Penaeus penicillatus, Penaeus merguiensis,* and *Litopenaeus vannamei* (Liu et al. 2011; Rungrassamee et al. 2014; Wang et al. 2014; Huang et al. 2014; Oxley et al. 2002). Rungrassamee et al. (2014) while working on the microbiota composition of shrimp *P. monodon* collected from wild and domesticated waters reported that there was not much difference in the microbiota composition of shrimp collected from two different environmental conditions. Though several studies carried out in shrimp indicated similar observations but it is necessary to understand the microbial composition of healthy and diseased shrimp. This kind of knowledge also helps us to know how environmental impact occurs on microbial composition in shrimp.

Chen et al. (2017) carried out studies on the environmental impact on the microbial composition of the gut tract of shrimp *Macrobrachium nipponense* to understand how microbial communities differ in response to rapid environmental change and for this purpose they surveyed and studied the environment and gut microbiota of native and habitat-exchanged shrimp. It has been observed that microbial diversity increased in shrimp when they were transferred from lake to river waters and microbial diversity was decreased when shrimp were transferred from rivers to lake waters. From these studies, it was concluded that shrimp gut microbiota has high plasticity when its host faces environmental changes, even over short timescales. Furthermore, the changes in the external environment might influence the gut microbiome not just by providing environment-associated microbes directly, but also by interfering with the composition of indigenous gut bacteria indirectly. It is also observed that change in the environmental factors affects the physiological status of the animals and in response to this, there is a change in gut microbial composition and also the gene expression pattern of microbes which leads to phenomic plasticity (Alberdi et al. 2016; Bordenstein and Theis 2015). A recent study on the gut microbiota of freshwater shrimp showed that the host genetics was a main contributor to the divergence of shrimp gut microbiomes between two closely related species, while the host habitat type seemed to play a critical role when comparing the microbial variation of hosts belonging to different lineages (Tzeng et al. 2015). When habitat exchange of shrimp occurs, i.e., from lake to river and vice versa, many symbiotic microbes were also changed during the acclimation process. It is observed that in Lake Shrimp there is a low abundance of *Flavobacteriales* in the gut microbiome composition but turned out to be the dominant group in Lake Shrimp reared in the river environment for 1 month. Coincidently, the gut commensal *Flavobacteriaceae* was found to reside in all shrimp dwelling in different rivers (Tzeng et al. 2015), implying a critical role of this bacterial family in the gut of shrimp living in river habitats. Figure 12.2 shows the overall impact of the different environmental and other biological factors on the composition of the gut microbiome of shrimp.

12.2.5 Microbial Composition in Wild and Domesticated Shrimp

The microbial composition of the gut tract and other body parts of some of the commercial shrimp collected from wild/natural waters has been analyzed by a number of workers (Diwan et al. 2021; Granados et al. 2017; Chen et al. 2017). Rungrassamee et al. (2014) made an in-depth analysis of microbiota in shrimp *P. monodon* collected from the wild and compared it with domesticated shrimp. Earlier also Chaiyapechaia et al. (2011) and Rungrassamee et al. (2013) also reported the intestinal microbial composition of *P. monodon* collected from wild waters. Comparing the differences in intestinal bacteria of wild-caught and domesticated and

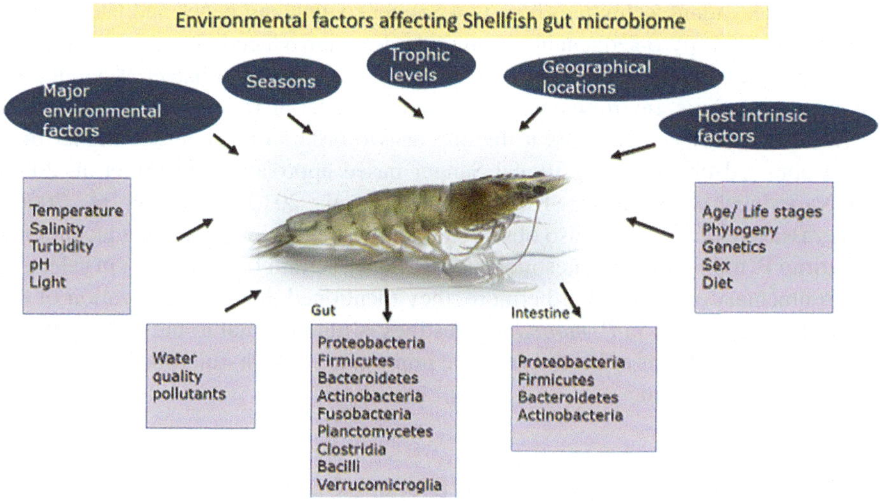

Fig. 12.2 Impact of various intrinsic and extrinsic factors on Penaeid shrimp gut microbiome composition

cultured shrimp can reveal changes in intestinal bacteria as shaped by aquaculture farming protocols. These changes have implications for shrimp health and consequently, the application of this knowledge may improve survival rates of not only *P. monodon* but other shrimp under domestication. The bacterial community belonging six genera, viz., *Proteobacteria* (*Vibrio, Photobacterium, Pseudomonas, Sphingomonas, Novospingobiumcommon,* and *Undibacterium*), two genera from Firmicutes (*Fusibacter and Lactobacillus*) and one genus from *Bacteroidetes* (*Cloacibacterium*) have been reported in the intestine of *P. monodon*. However, there are several reports which concluded that there is not much difference in the core group composition of intestinal microbiota in shrimp collected from the wild and shrimp from captive conditions. Among shared bacteria, the highly prevalent ones are *Vibrio, Photobacterium, Sphingomonas,* and *Novosphingobium* in the intestine collected from shrimp *P. monodon* post-larvae and juveniles. It has been observed that *Vibrio* and *Photobacterium* are commonly associated with shrimp living in marine waters (Soto-Rodriguez et al. 2015). The prevalence of the above-reported bacteria in *P monodon* has also been observed in other shrimp-like *Fenneropenaeus chinensis, P. merguiensis,* and in lobster, *Nephrops norvegicus* (Oxley et al. 2002; Liu et al. 2011; Meziti et al. 2012). The similarities of the core group of microbial communities in wild and cultured shrimp suggest that the prevalence of the presence of specific bacterial communities is due to selective pressure from the environment within the host gut. It has been reported that the mucosal immune system in the host digestive tract plays a very important role in maintaining commensal bacteria (Jarchum and Pamer 2011; Schluter and Foster 2012). It has been mentioned that most intestinal bacterial population analyses in penaeid shrimp have been characterized by both culture-dependent and culture-independent techniques based on Sanger sequencing genes and later a high-throughput 454 pyro-sequencing technique was introduced for the analysis of intestinal microbial diversity which enabled a greater depth to bacterial population analysis in various environments, for instance, bacterial richness in the deep sea environment has been reported to be one or two orders of magnitude higher than those obtained under the traditional Sanger based approaches (Oxley et al. 2002; Chaiyapechaia et al. 2011; Rungrassamee et al. 2013; Sogin et al. 2006). Rungrassamee et al. (2014) also observed that by using pyro-sequencing technology in shrimp P. monodon, the intestine possesses only certain bacterial taxa in selective environmental conditions. Furthermore, they mentioned that characterization of the bacterial composition in *P. monodon* intestines is fundamental to the understanding of the beneficial effects of bacteria and the importance of host–microbe interaction in a non-model organism.

12.2.6 Microbial Composition in Diseased Shrimp

There are several reports about the microbiome composition of the digestive tract in shrimp collected from natural waters as well as from captive conditions. It is mentioned that the gut and hepatopancreatic microbiota and their potential functions in wild-type shrimp are unknown and a lot of research in this particular area is warranted. Therefore, analysis of microbial community from such wild-type shrimp can be used as reference material for comparison with healthy and diseased cultured shrimp in hatcheries (Rungrassamee et al. 2014). However, information regarding microbial composition in the gut of diseased shrimp is very limited and in recent years attempts have been made by several researchers (Holt et al. 2021; Huang et al. 2020a, b; Zoqratt et al. 2018; Dai et al. 2019, 2020) covering some of these aspects. It is reported that in most of the studies in the recent past, the gut microbiome sequence data in penaeid shrimp has been analyzed using a high-throughput sequencing method in order to investigate the varied diversity of gut-associated microbiota in shrimp grown under a range of conditions across the world. The microbiota belonging to the phylum Proteobacteria has been reported to be the most dominant group of this community present in the gut tract of shrimp which is highly diverse in terms of physiology, morphology, and genetics (Hakim et al. 2015; Holt et al. 2020; Huang et al. 2014; Meziti et al. 2010; Rungrassamee et al. 2013, 2014; Zhang et al. 2014). Another group of bacteria observed was Gamma-proteobacteria, the largest class in the phylum, which is often described as the most common bacteria in the gut tract of giant tiger shrimp, *Penaeus monodon* (Chaiyapechara et al. 2012; Rungrassamee et al. 2013, 2014, 2016) and Pacific white shrimp, *Litopenaeus vannamei.* In diseased shrimp suffering from tail necrosis, red disease, and loose shell syndrome many Vibrio species which produce chitinolytic enzymes may create pathogenic symptoms due to the absence of sufficient chitin substrate in their diet. In such a situation the presence of the dominant group of vibrio bacteria will have a negative effect on the health of shrimp as chitinolytic enzymes produced by vibrio digest available chitin in the animals causing loose shell syndrome disease and necrosis (Sugita and Ito 2006; Liu and Lee 2002; Jayasree et al. 2006). As such, several *Vibrio* spp. have caused huge losses to the aquaculture industry, with vibriosis often causing mass mortalities among shrimp (Lavilla-Pitogo et al. 1998; Manilal et al. 2010). Despite this, Vibrio spp. is often described as the dominant genus within the shrimp gut microbiota and many inhabit harmoniously with the host. In penaeid shrimp, the elongated gut system is divided into three main parts, i.e., the foregut, midgut, and hindgut. All these parts have different histological structures and functions. A number of studies have focused microbiome composition of the midgut rather than other parts of the digestive tract (Ceccaldi 1989). However, a study from Vietnam reported that Alphaproteobacteria along with Planctomycetales dominated the stomach of healthy shrimp *L. vannamei* (Chen et al. 2017). Microbial profiles are likely influenced by the elongated length of the gut itself as different morphologies and functions along the gut will induce differential pressures on the selection of microbiome species. These internal pressures are perhaps the reasons for the high similarity of microbiome composition in wild-caught and domesticated shrimp *P. monodon* (Rungrassamee et al. 2014). In the shrimp, *L. vannamei* it was observed that the gut tracts collected from different farms also showed a similar composition of microbiome community despite differences in the microbiome community structure of their respective rearing waters (Zoqratt et al. 2018). In an earlier study, wild-type *L. vannamei* from Mexico was shown to harbor a more diverse bacterial community compared to healthy cultured animals and, unlike *P. monodon*, contained substantial proportions of Cyanobacteria (Granados et al. 2017, 2018). Huang et al. (2014) observed the dominance of Gammaproteobacteria in the guts of *L. vannamei* at different life stages in a holding facility, with the exception of 2-month-old juveniles which mainly harbored Bacteroidetes. Aquaculture practices, such as indoor- vs. pond-based culture can also influence the microbial composition of the gut tract of shrimp (Landsman et al. 2019) as the integration of multi-trophic protocols in aquaculture farming, improve the shrimp productivity (Omont et al. 2020). Overall, the growing wealth of evidence suggests that both environmental and internal, host-associated factors can contribute to the determination of microbial communities.

While discussing the environmental impact on the microbial composition of the gut of shrimp, the incidences of mass mortality in aquaculture farming due to bacterial and viral diseases have been reported by a number of workers (Seibert and Pinto 2012; Stentiford et al. 2012). The information pertaining to the presence of pathogens and their impact on the microbial composition in the digestive gut tract of shrimp and subsequently microbial influence on metabolic processes within the host is very limited. Very little is understood with regard to the changes in the gut microbiome and invasion by the pathogen and the mechanisms involved. Several workers have reported changes in gut microbiome structure which could also facilitate the progression of enteric pathogens that rely on translocation through the gut epithelia to initiate infection in the target tissue (Dai et al. 2019; Bass et al. 2019; Huang et al. 2020b). Characterization of the interplay between the microbiota and the host-immune system is becoming increasingly interesting and well-defined (Sekirov et al. 2010). Although there are key differences between the vertebrate and invertebrate immune systems, the gut microbiota likely has an important role to play in maintaining the health of the shrimp.

When there is dominance of pathogenic bacteria in the gut tract and other body parts of shrimp due to non-favorable environmental factors number of physiological changes occur in the body. Zheng et al. (2021) while working on shrimp *P. vannamei* reported that the respiratory glycoprotein hemocyanin is a crucial host factor that modulates microbial composition and diversity in the hepatopancreatic tissues. Furthermore, they observed that those shrimp where hepatopancreatic tissue was atrophied due to the dominance of pathogenic vibrio infection showed a low concentration of hemocyanin and the gastrointestinal tract also did not show a high density of microbiomes. Similarly, it was reported that shrimp having depleted amounts of hemocyanin had mitochondrial depolarization, increased reactive oxygen species (ROS) levels, and dysregulation of several energy metabolism-related genes. However, in the shrimp when hemocyanin level is suppressed together with ROS scavenger (*N*-acetylcysteine) it was observed that the shrimp showed improved microbial diversity and decreased *Vibrio* dominance in the hepatopancreatic tissues. Several other workers also mentioned that when there is damage to the hepatopancreatic tissue due to pathogenic bacterial infection, also affects the microbial composition of the digestive tract. It is well known that the hepatopancreas is an essential organ for digestion and absorption of nutrients, and it plays a significant role in innate immunity in invertebrates. Therefore, any changes in the hepatopancreatic tissue are bound to have an effect on the microbial composition of the digestive tract. There are very limited studies on the microbial composition of the hepatopancreas in shrimp (Cheung et al. 2015). Moreover, the hepatopancreas can be drastically affected by pathogens such as *V. parahaemolyticus*, which causes the devastating early mortality syndrome, also called acute hepatopancreatic necrosis disease (AHPND), in shrimp (Lee et al. 2015; Tran et al. 2013; Joshi et al. 2014). It is mentioned that due to the constant interaction between the shrimp digestive system and its surrounding environment, the microbiomes of the intestine and hepatopancreas always play important role in disease resistance mechanisms (Granados et al. 2017).

12.2.7 Changes in the Gut Microbiome and the Incidence of Disease

A number of workers have analyzed the microbiome composition of the digestive tract of shrimp collected from pond seawater and compared it with those collected from the wild. It has been noticed that the bacterial community structure and dynamics in pond seawater were totally different when compared with the microbial composition of those shrimp's digestive tract collected from pond seawater during different periods of sampling which suggests the varied development process of microbiomes in different environments (Chen et al. 2017). Identical observations were also made earlier by Cardona et al. (2016) and Johnson et al. (2008) while working on the bacterial community of pond seawater and the digestive tract of the shrimp *L. vannamei*. This is because the pond seawater was in a higher oxidizing condition due to continuous oxygenation from mechanical aeration and microbial photosynthesis, whereas the shrimp stomachs harbored populations of several anaerobic bacteria such as Desulfobacterales and Caldilineales with far higher abundance than seawater. Furthermore, it is also observed that a number of environmental and physiological factors also affect the microbiome of shrimp's digestive tract, and also during different stages of the growth cycle, the shrimp shows the different types of microbiome communities in their stomach and intestinal tract (Chen et al. 2017). The gradual shift in the microbiome diversity of shrimp stomachs could be also attributed to the transit of bacterial population due to a change in the feeding conditions of the infected shrimp (Xiong et al. 2017). Chen et al. (2017) reported that when shrimp was diseased with acute hepatopancreatic necrosis disease (AHPND), the ability to select bacterial residents in the stomach markedly decreased, thus the stochastic process that is associated with environmental factors became more important in shaping the stomach microbiome composition. When stomach content was analyzed for microbial composition in shrimp infected with AHPND, it was observed that bacterial species like Weissella and Vibrio were the dominant microbes in the stomach of shrimp. The *Weissella* and some *Vibrio* species are facultatively anaerobic populations and several strains of these bacteria have been reported to be probiotic and reduce the infectivity and persistence of digestive pathogens (Castex et al. 2014; Nam et al. 2002; Cai et al. 1998). In healthy and diseased shrimp when microbiota was compared, it was noticed that Weissella was very much dominant in shrimp infected with AHPND which indicated that this particular bacterium is a potential indicator for the health status of the stomach microbiomes in white shrimp. Chen et al. (2017) also mentioned that when eutrophication takes place in pond seawater cyanobacteria populations become prominent in the shrimp pond and the dominance of cyanobacteria was reported to be associated with several environmental factors such as light, salinity, temperature, and nutrient levels, and it could regulate the water quality in the pond seawater. The presence of cyanobacteria which have been reported to be dominant in coastal seawater or brackish water generally exerts toxic effects on marine invertebrates (Martins et al. 2007). From these observations, it was concluded that the occurrence of AHPND in shrimp could be a consequence of changes in stomach bacterial communities with disruptions in species-to-species co-occurrence patterns, which could be useful for developing microbial early warning systems for shrimp farming for more efficient management and for minimizing the outbreak of AHPND (Chen et al. 2017).

12.2.8 Gut Microbiota of Oysters and Mussels

The information about the gut microbiome composition of oysters and mussels is limited. Like in other organisms, several environmental factors are also responsible for shaping the gut microbial composition in oysters and mussels. Host-related intrinsic factors are also playing a significant role in the organization of the microbiome community in these shellfish. In the recent past, few workers have attempted to carry out work on gut microbiome analysis in oysters and mussels in marine and freshwater environments. Pierce and Ward (2019) studied the gut microbiome of oysters, *Crassostrea Virginia,* and mussel, *Mytilus edulis* in relation to temporal variation and also the influence of marine aggregate-associated microbiome communities. It is observed that these bivalves utilize a shared feeding mechanism, which may result in similar gut microbial communities throughout the seasons. As far as the gut microbial composition of these bivalves is concerned, it is noticed that throughout the year composition of microbial diversity was almost similar but temporal variation was also observed. It has been reported that though there was not much change in microbial composition throughout the year, the same get influenced by some of the extrinsic factors like diet and nutritional conditions. These studies also compared over the course of a year, the gut microbial communities of two species of bivalves, oysters, and mussels, held under identical conditions in coastal marine waters. While the mussels and oysters harbored gut microbial communities with similar compositions, but on a functional level, they exhibited species and temporal variation. These results indicate that intrinsic factors influence the bivalve microbiome, resulting in species variability, even when environmental conditions, feeding mechanisms, and particle diet are constant. Seasonal and multispecies comparisons for bivalve-associated microbial communities are rare (Pierce and Ward 2019). Tanaka et al. (2004) while working on the gut microbiome composition of gastropod mollusc, Abalone, mentioned that when different diets are used for feeding the animals, there was a change in the microbiome diversity. Furthermore, multiple studies on bivalve molluscs showed that the gut microbiome of several oyster species was influenced by life stage, host taxonomy, and environmental parameters (Trabal et al. 2012, 2014). Specifically, the gut communities of the eastern oyster (Crassostrea virginica) were not significantly different from one another when separated geographically, but they were characterized by seasonal, temperature-dependent differences (Pierce et al. 2016). In all cases, both intrinsic and extrinsic factors are likely at play, with the environment as well as the host asserting control of the microbiome composition. In the aquatic environment, one of the least studied parameters is the influence of diet. For sessile suspension-feeding bivalves such as oysters and mussels, direct interactions with the suspended particle diet could affect their gut microbiomes. In particular, the presence of particle aggregations could impact the types and numbers of bacteria that are captured and ingested by bivalves. Pierce and Ward (2019) reported that the most abundant phyla in oysters, mussels, and other bivalves were the Proteobacteria, Tenericutes, Verrucomicrobia, Bacteroidetes, Cyanobacteria, Planctomycetes, Actinobacteria, Firmicutes, Cladithrix, and Fusobacteria. Overall, bivalves contained significantly higher abundances of Verrucomicrobia and Tenericutes but significantly fewer Bacteroidetes than marine aggregates. Temporally, marine-aggregate communities change little at the phylum level. Bivalve samples demonstrated seasonal variation in the proportion of abundant phyla. Greater temporal variability could be seen at lower bacterial taxonomic levels within each of the sample types.

12.2.8.1 Impact of Thermal Stress

Li et al. (2019) carried out studies on the effect of thermal stress on gut microbiome composition in the mussel *Mytilus galloprovincialis.* It is reported that exposure to a high temperature of 27 °C significantly reduced the survival of animals associated with increased microbial diversity of the gut. The main reason speculated was the presence of a relative abundance of Vibrio and Arcobacter in live Mytilus in response to high temperature. An increase in the levels of such pathogenic bacteria might increase host susceptibility to disease and contribute greatly to mortality. The findings of this study provide knowledge on ecological adaptation for Mytilus and host–bacteria interaction during temperature stress. Previous pieces of evidence also showed that exposure to a high water temperature of 31 °C altered the gut microbiome of *Mytilus coruscus* by facilitating the proliferation of opportunistic bacteria in adult mussels (Li et al. 2018). Therefore, the dysbiosis of gut microbiome may reflect the health status of the host. Li et al. (2019) observed three dominant phyla of microbiota viz., Bacteroidetes, Proteobacteria, and Verrucomicrobia in gut samples, which accounted for 87.2–95.9% of total bacteria present at normal temperature. At high temperatures of 27 °C, the relative abundance of Chlamydiae, Epsilonbacteraeota, Planctomycetes, and Proteobacteria has been noticed in gut microbiome composition. A significant reduction in the relative abundance of Bacteroidetes, Spirochaetes, and Verrucomicrobia was observed in warm acclimated mussels compared to control mussels. Temperature stress has been reported as a key factor in summer mortality syndrome (SMS) of marine bivalves by a number of workers (Myrand et al. 2000; Xiao et al. 2005; Malham et al. 2009). It was reported that warm temperatures deteriorated health conditions in oysters which is associated with decreased microbial diversity (Lokmer and Wegner 2015). It remains to be established whether the variation of microbial community diversity is caused by the mussel's physiological responses to heat stress or a direct effect of temperature.

12.2.8.2 Freshwater Mussel

In freshwater mussels, despite recent evidence suggesting that the gut microbiome plays a role in nutrient digestion (Mioduchowska et al. 2020; McCauley et al. 2021), data on how the microbiome varies across species and environmental conditions, and the functional role of the microbiome within host mussel is still largely unexplored. Weingarten et al. (2019) carried out studies on the gut microbiome composition of freshwater mussels in different host species collected from different sites of the river and from different environmental situations. The freshwater mussel species studied were *Cyclonaias asperata, Fusconaia cerina, Lampsilis ornata,* and *Obovaria unicolor.* The analysis of gut microbial composition data indicated that there was a great influence of environmental variations as well as hot species on the gut microbiome and the dominant microbiomes observed were from the members of the group Clostridiales and nitrogen-fixing bacteria *Methylocystis* sp. The other groups of microbiota observed in the gut of these animals belong to the phylum Proteobacteria, Firmicutes, Planctomycetes, Bacteroidetes, Actinobacteria, Verrucomicrobia, Fusobacteria and there were other bacterial groups also which could not be classified. These findings suggested that selective retention of bacterial taxa by the host and both host species and the environment are important in determining mussel gut microbiome composition. Compared to marine oysters and mussels there is relatively scanty research into the gut microbiome composition of freshwater bivalves. It has been reported that the different physico-chemical conditions of the freshwater environment could generate different selection pressures for the recruitment of bacterial species (Atkinson and Vaughn 2015; Vaughn et al. 2008). Recent research has also shown that differences in bacterial composition between clams, oysters, and mussels, which mirrored differences in the rates of organic matter processing, could indicate a strong functional role of the microbiome in host metabolism. Weingarten et al. (2019) also mentioned that the gut bacterial community of freshwater mussels would differ significantly from the freely suspended bacteria in the water column as a result of the host species recruitment of potentially beneficial taxa. It was also observed that freshwater mussels harbor microbiota that is significantly different in diversity, composition, and structure than those on freely suspended seston and there are significant differences in the relative abundances of different bacterial taxa between co-occurring mussel species. Furthermore, it was reported that the site also plays a significant role in determining the composition of the gut microbiome within three of the four mussel species. McCauley et al. (2021) investigated gut microbiomes of six freshwater mussels, viz., *Lampsilis ornata, Obovaria unicolor, Elliptio arca, Fusconaia cerina, Cyclonaias asperata,* and *Tritogonia* verrucosa in relation to temporal variations in their river habitat environment for a period of 3 years. During the study period, they examined the taxonomic, phylogenetic, and inferred functional facets of microbiome diversity for all six species of mussels. It has been reported that significant differences have been observed for 2 years in gut microbial composition in five of the six species. The major bacterial classes observed were Betaproteobacteria, Cyanobacteria, Actinobacteria, Spartobacteria, Firmicutes, Fusobacteria, Gammaproteobacteria, Alphaproteobacteria, Planctomycetia, and Clostridia. The relative abundance of bacteria such as Proteobacteria, Firmicutes, and Planctomycetia was seen throughout the seasons during 3 years period in all the species of mussel. However, not all species that exhibited a temporally variable microbiome were functionally distinct across years, indicating functional redundancy within the mussel gut microbiome. Furthermore, it has been reported that biosynthesis pathways showed temporal variation involved in the degradation process, while pathways involved in cellular metabolism were stable. There was no evidence for phylosymbiosis across any facet of microbiome biodiversity. These results indicated that temporal variation is an important factor in the assembly of the gut microbiomes of freshwater mussels and provides further support that the mussel gut microbiome is involved in host development and activity.

12.3 Conclusions

The impact of global warming on animal and plant biodiversity is well known. Consequently, the impact of climate change on the gut microbiome of fish and shellfish is bound to be there and getting momentum, and in the recent past, considerable research work has been carried out in such niche areas of science. It is well known that the impact of climate change on the gut microbiome could lead to possible dysbiosis with strong consequences on hosts and ecosystems. These changes may be due to a number of environmental factors particularly rising temperatures, changes in water salinities, PH, turbidity, depletion of oxygen levels, pollutants, etc. that may lead to an increase in disease prevalence in the habitat ecosystems. Hence, understanding the factors affecting host–microbiome interactions with cultivable organisms would improve predictions on biodiversity responses to climate change. Several environmental factors which act as stressors can disrupt the strong relationship between hosts and their associated microbiomes leading to dysbiosis and enhancing the chances of host diseases. Hosts and their associated microbiomes form a coherent biological entity termed "holobionts"—that must be studied together to properly understand biological systems and how they will be impacted by climate change. This is particularly crucial for habitat-forming organisms because impacts on these holobionts can affect entire ecosystems. There are several reports indicating that the capacity of organisms to adapt to new habitat conditions basically depends on their plasticity attributes, of which their gut commensal microbiome species might be an essential impact factor. Particularly in organisms, which inhabit the aquatic environment, the complex and dynamic microbiomes have significant effects on health and development. However, an understanding of the relative contribution of internal sorting and colonization processes of the microbiome in the gut system is still not clear and research investigation is required in this particular area on a priority basis. Besides environmental factors, pollutants in the form of contaminants and toxins present in the environment may also influence the fish gut microbiota, and also other kinds of chemicals such as pesticides, heavy metals, and antibiotics can also induce gut microbiome dysbiosis associated with changes in the intestinal mucus layer and inflammation in the digestive tract in fish, thereby reducing the ability to absorb nutrients.

A number of workers have reported that host-specific intrinsic factors also influence the microbial composition of the digestive tract in the case of fish. Intrinsic factors such as phylogeny, genetics and sex, age/life stages, and diet/feeding habits also determine microbial composition among fish. However, there is still some controversy that whether host genetics has got any influence on gut microbiome composition in fish and this particular area needs further investigation. Many times environment plays an important role in shaping the gut microbiome composition suggesting that environmental factors can overcome differences in host genetics. Besides environmental factors, changes in the diet composition also influence microbial profiles in the gut tract of several fish. Extreme dietary changes, such as fasting bring changes in the gastrointestinal tract structure which affects reduced nutrient uptake, due to changes observed in the gut microbiome composition.

Furthermore, depletion of nutrients induces changes in gut microbiome composition to favor bacterial species/communities that use more diverse energy sources and are capable of survival under limited nutrient conditions. In an aquatic ecosystem, the food chain or trophic levels of natural food web configuration are always get influenced by several environmental factors. The food web configuration and available food density at different water levels influence the microbial composition of the gut system of the animals inhabiting different trophic levels of the ecosystem. However, more investigations are required to establish the relationship between the influences of trophic levels on the gut microbial composition of finfish and shellfish inhabiting different environmental ecosystems. Seasonal changes in the environment particularly in freshwater ecosystems have a greater impact on gut microbiome composition in fish. Considerable literature that is available has highlighted seasonal variations as important parameters for fish gut microbial composition. Further research is required to determine the effects of seasonal variations on the gut microbiome of fish inhabiting different ecosystems.

Several studies have been carried out to investigate whether there is any difference in the gut microbial composition in the fish that is caught from the naturally inhabiting environment and those caught from captive farming systems. In the recent past due to improvements in aquaculture farming technologies fish produced in captive conditions are of better quality with good growth. Due to a controlled environment, balanced feed diet, and preventative measures in controlling infectious diseases, there is substantial improvement in the quality production of fish at the farm gate. However, knowledge of the gut microbiome in wild fish requires more attention to provide a baseline for comparative purposes to better understand the effects of captive rearing.

For the maintenance of healthy microbiome composition in the digestive tract of fish, water quality is the most important parameter irrespective of whether the fish lives in natural waters or is cultivated in captive conditions. Due to anthropogenic and other activities, inland water bodies, as well as seas and oceans all over the world, are getting heavily polluted and this kind of environmental impact on the microbial composition of fish becomes a great concern for the fisheries sector. Hence, a comparison of the water quality and host gut microbiome is expected to improve the understanding of environmental influence. Besides anthropogenic factors, other fluctuating physico-chemical parameters such as salinity, temperature, pH, oxygen, CO_2 levels, etc. also have a great influence on the gut microbial composition of fish and other animals. An increase in the environmental temperature seriously affects various physiological activities of fish, particularly feeding, growth, immunity, and disease resistance capacity. It has been reported that these changes are closely related to the destruction of intestinal barrier function, the change in intestinal microbiome composition, and metabolic dysfunction. Therefore, it is necessary to explore the changes in the intestinal microbiome under heat stress to understand the regulatory mechanism of heat stress and protect the intestinal health of fish from the negative effects of rising water temperature. Two types of colonization of fish gut microbiomes have been reported in the literature. One type is the stochastic or neutral organization of microbes and the second one is a deterministic or non-neutral assemblage of microbiota which may be dependent on host selective pressures or active dispersal by the host and host–microbe and microbe–microbe interactions.

As far as shellfish are concerned particularly while dealing with shrimp, it has been reported that due to the poor quality of water and the presence of pollutants or contaminants, there have been several incidences of the outbreak of diseases in aquaculture farming, and shrimp mortality both in wild and culture water bodies. In such deteriorated ecosystems the microbiome composition in surrounding waters and also in the gut of cultivable organisms has been reported to be of poor quality and the population of pathogenic microbiota was found to be predominant. More research investigations are needed to confirm these findings. Another area we need to pay attention to is how to build up a healthy gut microbiome in such animals so that outbreaks of diseases and mortality can be reduced in aquaculture farming. The microbial composition of the gut tract and other body parts of some of the commercial shrimp collected from wild/natural waters has been analyzed by a number of workers. Comparing the differences in intestinal bacteria of wild-caught and domesticated and cultured shrimp can reveal changes in intestinal bacteria composition as shaped by aquaculture farming protocols. These changes have implications for shrimp health and consequently, the application of this knowledge may improve the survival rates of shrimp under domestication.

References

Alberdi A, Aizpurua O, Bohmann K, Zepeda-Mendoza ML, Gilbert MTP (2016) Do vertebrate gut metagenomes confer rapid ecological adaptation? Trends Ecol Evol 31:689–699. https://doi.org/10.1016/j.tree.2016.06.008

Al-Harbi AH, Naim Uddin M (2004) Seasonal variation in the intestinal bacterial flora of hybrid tilapia (Oreochromis niloticus × Oreochromis aureus) cultured in earthen ponds in Saudi Arabia. Aquaculture 229:37–44. https://doi.org/10.1016/S0044-8486(03)00388-0

Atkinson CL, Vaughn CC (2015) Biogeochemical hotspots: temporal and spatial scaling of the impact of freshwater mussels on ecosystem function. Freshw Biol 60:563–574

Bakke-McKellep AM, Penn MH, Salas PM, Refstie S, Sperstad S, Landsverk T et al (2007) Effects of dietary soybean meal, inulin, and oxytetracycline on intestinal microbiota and epithelial cell stress, apoptosis and proliferation in the teleost Atlantic salmon (*Salmo salar* L.). Br J Nutr 97: 699–713. https://doi.org/10.1017/S0007114507381397

Bano N, deRae Smith A, Bennett W, Vasquez L, Hollibaugh JT (2007) Dominance of Mycoplasma in the guts of the long-jawed mud sucker, *Gillichthys mirabilis*, from five California salt marshes. Environ Microbiol 9:2636–2641. https://doi.org/10.1111/j.1462-2920.2007.01381.x

Bass D, Stentiford GD, Wang H-C, Koskella B, Tyler CR (2019) The pathobiome in animal and plant diseases. Trends Ecol Evol 34(11):996–1008. https://doi.org/10.1016/j.tree.2019.07.012

Bates JM, Akerlund J, Mittge E, Guillemin K (2007) Intestinal alkaline phosphatase detoxifies lipopolysaccharide and prevents inflammation of zebrafish in response to the gut microbiota. Cell Host Microbe 2:371–382

Bellard C, Bertelsmeier C, Leadley P, Thuiller W, Courchamp F (2012) Impacts of climate change on the future of biodiversity. Ecol Lett 15:365–377. https://doi.org/10.1111/j.1461-0248.2011.01736.x

Bestion E, Jacob S, Zinger L, Di GL, Richard M, White J et al (2017) Climate warming reduces gut microbiota diversity in a vertebrate ectotherm. Nat Ecol Evol 1:161. https://doi.org/10.1038/s41559-017-0161

Bledsoe JW, Waldbieser GC, Swanson KS, Peterson BC, Small BC (2018) Comparison of channel catfish and blue catfish gut microbiota assemblages show minimal effects of host genetics on microbial structure and inferred function. Front Microbiol 9:1073. https://doi.org/10.3389/fmicb.2018.01073

Bolnick DI, Snowberg LK, Hirsch PE, Lauber CL, Org E, Parks B et al (2014) Individual diet has sex-dependent effects on vertebrate gut microbiota. Nat Commun 5:ncomms5500. https://doi.org/10.1038/ncomms5500

Bordenstein SR, Theis KR (2015) Host biology in light of the microbiome: ten principles of holobionts and hologenomes. PLoS Biol 13(8):e1002226. https://doi.org/10.1371/journal.pbio.1002226

Bruce TJ, Neiger RD, Brown ML (2018) Gut histology, immunology and the intestinal microbiota of rainbow trout, Oncorhynchus mykiss (Walbaum), fed process variants of soybean meal. Aquacult Res 49:492–504. https://doi.org/10.1111/are.13480

Bucio A, Hartemink R, Schrama JW, Verreth J, Rombouts FM (2006) Presence of lactobacilli in the intestinal content of freshwater fish from a river and from a farm with a recirculation system. Food Microbiol 23:476–482. https://doi.org/10.1016/j.fm.2005.06.001

Burns AR, Stephens WZ, Stagaman K, Wong S, Rawls JF, Guillemin K, Bohannan BJ (2016) Contribution of neutral processes to the assembly of gut microbial communities in the zebrafish over host development. ISME J 10:655–664. https://doi.org/10.1038/ismej.2015.142

Burokas A, Moloney RD, Dinan TG, Cryan JF (2015) Chapter One. Microbiota regulation of the mammalian gut-brain axis. In: Advances in applied microbiology. Academic Press, Cambridge, MA, pp 1–62. https://doi.org/10.1016/bs.aambs.2015.02.001

Butt RL, Volkoff H (2019) Gut microbiota and energy homeostasis in fish. Front Endocrinol 10:6–8

Cabello FC (2006) Heavy use of prophylactic antibiotics in aquaculture: a growing problem for human and animal health and for the environment. Environ Microbiol 8:1137–1144. https://doi.org/10.1111/j.1462-2920.2006.01054.x

Cai Y, Benno Y, Nakase T, Oh TK (1998) Specific probiotic characterization of Weissella hellenica DS-12 isolated from founder intestine. J Gen Appl Microbiol 44:311–316

Cardona E et al (2016) Bacterial community characterization of water and intestine of the shrimp *Litopenaeus stylirostris* in a bioflocs system. BMC Microbiol 16:157. https://doi.org/10.1186/s12866-016-0770-z

Castex M, Daniels C, Chim L (2014) Probiotic applications in crustaceans. In: Aquaculture nutrition. Gut health, probiotics and prebiotics, pp 290–327

Ceccaldi HJ (1989) Anatomy and physiology of digestive tract of Crustaceans Decapods reared in aquaculture. In: Advances in tropical aquaculture, Tahiti, Aquacop, IFREME, Actes de Colloque, vol 9, pp 243–259

Chaiyapechaia S, Rungrassamee W, Suriyachay I, Kuncharin Y, Klanchui A et al (2011) Bacterial community associated with the intestinal tract of *P. monodon* in commercial farms. Microb Ecol 63:938–953

Chaiyapechara S et al (2012) Bacterial community associated with the intestinal tract of P. monodon in commercial farms. Microb Ecol 63:938–953. https://doi.org/10.1007/s00248-011-9936-2

Chen CY, Chen PC, Weng FC, Shaw GT, Wang D (2017) Habitat and indigenous gut microbes contribute to the plasticity of the gut microbiome in oriental river prawn during rapid environmental change. PloS One 12(7):e0181427. https://doi.org/10.1371/journal.pone.0181427

Chessman SE, Guillemin K (2007) We know you are in there: conversing with the indigenous gut microbiota. Res Microbiol 158:2–9. https://doi.org/10.1016/j.resmic.2006.10.005

Cheung MK et al (2015) Rapid change of microbiota diversity in the gut but not the hepatopancreas during gonadal development of the new shrimp model Neocaridina denticulata. Marine Biotechnol 17:811–819

Clements KD, Pasch IBY, Moran D, Turner SJ (2007) Clostridia dominate 16S rRNA gene libraries prepared from the hindgut of temperate marine herbivorous fishes. Mar Biol 150:1431–1440. https://doi.org/10.1007/s00227-006-0443-9

Clements KD, Angert ER, Montgomery WL, Choat JH (2014) Intestinal microbiota in fishes: what's known and what's not. Mol Ecol 23:1891–1898. https://doi.org/10.1111/mec.12699

Cordero H, Guardiola FA, Tapia-Paniagua ST, Cuesta A, Meseguer J, Balebona MC et al (2015) Modulation of immunity and gut microbiota after dietary administration of alginate encapsulated Shewanella putrefaciens to gilthead sea bream (*Sparus aurata* L.). Fish Shellfish Immunol 45:608–618. https://doi.org/10.1016/j.fsi.2015.05.010

Dai W et al (2019) Gut eukaryotic disease-discriminatory taxa are indicative of Pacific white shrimp (*Litopenaeus vannamei*) white faces syndrome. Aquaculture 506:154–160

Dai W et al (2020) Shrimp disease progression increases the gut bacterial network complexity and abundances of keystone taxa. Aquaculture 517:734802

De Silva SS, Doris S (2009) Climate change and aquaculture: potential impacts, adaptation and mitigation. In: Cochrane K, de Young C, Soto D, Bahri T (eds) Climate change implication for fisheries and aquaculture. Overview of current scientific knowledge FAO Fisheries and Aquaculture Technical Paper 530. FAO, Rome, pp 151–212

Defoirdt T, Sorgeloos P, Bossier P (2011) Alternatives to antibiotics for the control of bacterial disease in aquaculture. Curr Opin Microbiol 14:251–258. https://doi.org/10.1016/j.mib.2011.03.004

Dehler CE, Secombes CJ, Martin SAM (2017) 16 Environmental and physiological factors shape the gut microbiota of Atlantic salmon parr (*Salmo salar* L.). Aquaculture 467:149–157. https://doi.org/10.1016/j.aquaculture

Desai AR, Links MG, Collins SA, Mansfield GS, Drew MD, Van Kessel AG, Hill JE (2012) Effects of plant-based diets on the distal gut microbiome of rainbow trout (Oncorhynchus mykiss). Aquaculture 350–353:134–142. https://doi.org/10.1016/j.aquaculture

Dhanasiri AK, Brunvold L, Brinchmann MF, Korsnes K, Bergh O, Kiron V (2011) Changes in the intestinal microbiota of wild Atlantic cod *Gadus morhua* L. upon captive rearing. Microb Ecol 61:20–30. https://doi.org/10.1007/s00248-010-9673-y

Di Maiuta N, Schwarzentruber P, Schenker M, Schoelkopf J (2013) Microbial population dynamics in the faeces of wood-eating loricariid catfishes. Lett Appl Microbiol 56:401–407

Diwan AD, Harke SN, Gopalkrishna, Panche AN (2021) Aquaculture industry prospective from the gut microbiome of fish and shellfish: an overview. J Anim Physiol Anim Nutr 106:441–469. https://doi.org/10.1111/jpn.13619

Egan S, Harder T, Burke C, Steinberg P, Kjelleberg S, Thomas T (2013) The seaweed holobiont: understanding seaweed –bacteria interactions. FEMS Microbiol Rev 37:462–476. https://doi.org/10.1111/1574-6976.12011

Egerton S, Culloty S, Whooley J, Stanton C, Ross RP (2018) The gut microbiota of marine fish. Front Microbiol 9:1–17

Eichmiller JJ, Hamilton MJ, Staley C, Sadowsky MJ, Sorensen PW (2016) Environment shapes the faecal microbiome of invasive carp species. Microbiome 4:44. https://doi.org/10.1186/s40168-016-0190-1

Feng Q, Chen W-D, Wang Y-D (2018) Gut microbiota: an integral moderator in health and disease. Front Microbiol 9:151. https://doi.org/10.3389/fmicb.2018.00151

Fidopiastis PM, Bezdek DJ, Horn MH, Kandel JS (2006) Characterizing the resident, fermentative microbial consortium in the hindgut of the temperate-zone herbivorous fish, Hermosilla azurea (Teleostei: Kyphosidae). Mar Biol 148:631–642. https://doi.org/10.1007/s00227-005-0106-2

Fishelson L, Montgomery WL, Myrberg AA Jr (1985) A unique symbiosis in the gut of tropical herbivorous surgeonfish (Acanthuridae: Teleostei) from the Red Sea. Science 229:49–51. https://doi.org/10.1126/science.229.4708.49

Folke C, Carpenter S, Walker B, Scheffer M, Elmqvist T, Gunderson LH, Holling CS (1998) Regime shifts, resilience and biodiversity in ecosystem management. Annu Rev Ecol Syst 35: 557–581. https://doi.org/10.1146/annurev.ecolsys.35.021103.105711

Fournier V, Huelvan C, Desbruyeres E (2004) Incorporation of a mixture of plant feedstuffs as substitute for fish meal in diets of juvenile turbot (*Psetta maxima*). Aquaculture 236:451–465. https://doi.org/10.1016/j.aquaculture.2004.01.035
Franchini P, Fruciano C, Frickey T, Jones JC, Meyer A (2014) The gut microbial community of Midas cichlid fish in repeatedly evolved limnetic-benthic species pairs. PloS One 9:e95027
Garnier M, Labreuche Y, Garcia C, Robert M, Nicolas JL (2007) Evidence for the involvement of pathogenic bacteria in summer mortalities of the pacific oyster *Crassostrea gigas*. Microb Ecol 53:187–196
Gatesoupe FJ (1999) The use of probiotics in aquaculture. Aquaculture 180:147–165. https://doi.org/10.1016/S0044-8486(99)00187-8
Giri SS, Yun S, Jun JW, Kim HJ, Kim SG, Kang JW et al (2018) Therapeutic effect of intestinal autochthonous *Lactobacillus reuteri* P16 against waterborne lead toxicity in *Cyprinus carpio*. Front Immunol 9:1824. https://doi.org/10.3389/fimmu.01824
Givens CE (2012) A fish tale: comparison of the gut microbiome of 15 fish species and the influence of diet and temperature on its composition. PhD Thesis, University of Georgia, Athens, GA, USA
Givens CE, Ransom B, Bano N, Hollibaugh JT (2014) Comparison of the gut microbiomes of 12 bony fish and 3 shark species. Mar Ecol Prog Ser 518:209–223
Gomez GD, Balcazar JL (2008) A review on the interactions between gut microbiota and innate immunity of fish. FEMS Immunol Med Microbiol 52:145–154. https://doi.org/10.1111/j.1574-695X.2007.00343.x
Goncalves AT, Gallardo-Escárate C (2017) Microbiome dynamic modulation through functional diets based on pre-and probiotics (manna oligosaccharides and Saccharomyces cerevisiae) in juvenile rainbow trout (Oncorhynchus mykiss). J Appl Microbiol 122:1333–1347. https://doi.org/10.1111/jam.13437
Granados CF et al (2017) Microbiome of Pacific White leg shrimp reveals differential bacterial community composition between wild, aquacultured and AHPND/EMS outbreak conditions. Sci Rep 7(1):1–15
Granados CF, Gallardo-Becerra L, Leonardo-Reza M, Ochoa-Romo JP, Ochoa-Leyva A (2018) A meta-analysis reveals the environmental and host factors shaping the structure and function of the shrimp microbiota. PeerJ 6:e5382. https://doi.org/10.7717/peerj.5382
Green TJ, Smullen R, Barnes AC (2013) Dietary soybean protein concentrate-induced intestinal disorder in marine farmed Atlantic salmon, Salmo salar is associated with alterations in gut microbiota. Vet Microbiol 166:286–292. https://doi.org/10.1016/j.vetmic.2013.05.009
Hagi T, Tanaka D, Iwamura Y, Hoshino T (2004) Diversity and seasonal changes in lactic acid bacteria in the intestinal tract of cultured freshwater fish. Aquaculture 234:335–346. https://doi.org/10.1016/j.aquaculture.2004.01.018
Hakim JA et al (2015) An abundance of Epsilonproteobacteria revealed in the gut microbiome of the laboratory cultured sea urchin *Lytechinus variegates*. Front Microbiol 6:1047
Hamer HM, Jonkers DMAE, Bast A, Vanhoutvin SALW, Fischer MAJG, Kodde A, Troost FJ, Venema K, Brummer RJM (2009) Butyrate modulates oxidative stress in the colonic mucosa of healthy humans. Clin Nutr 28:88–93
Hansen G, Olafsen J (1999) Bacterial interactions in early life stages of marine cold water fish. Microb Ecol 38:1–26. https://doi.org/10.1007/s002489900158
Hansen G, Strøm E, Olafsen J (1992) Effect of different holding regimens on the intestinal microflora of herring (*Clupea harengus*) larvae. Appl Environ Microbiol 58:461–470
Harley CDG, Anderson KM, Demes KW, Jorve JP, Kordas RL, Coyle TA, Graham MH (2012) Effects of climate change on global seaweed communities. J Phycol 48:1064–1078. https://doi.org/10.1111/j.1529-8817.2012.01224.x
Harvell CD, Mitchell CE, Ward JR, Altizer S, Dobson AP, Ostfeld RS, Samuel MD (2002) Climate warming and disease risks for terrestrial and marine biota. Science 296:2158–2162. https://doi.org/10.1126/science.1063699

Hassenrück C, Reinwald H, Kunzmann A, Tiedemann I, Gärdes A (2021) Effects of thermal stress on the gut microbiome of juvenile milkfish (*Chanos chanos*). Microorganisms 9:5. https://doi.org/10.3390/microorganisms9010005

Hennersdorf P, Kleinertz S, Theisen S, Abdul-Aziz MA, Mrotzek G, Palm HW, Saluz HP (2016) Microbial diversity and parasitic load in tropical fish of different environmental conditions. PloS One 11:e0151594. https://doi.org/10.1371/journal.pone.0151594

Holben W, Williams P, Saarinen M, Särkilahti L, Apajalahti J (2002) Phylogenetic analysis of intestinal microflora indicates a novel mycoplasma phylotype in farmed and wild salmon. Microb Ecol 44:175–185. https://doi.org/10.1007/s00248-002-1011-6

Holt CC, van der Giezen M, Daniels CL et al (2020) Spatial and temporal axes impact ecology of the gut microbiome in juvenile European lobster (*Homarus gammarus*). ISME J 14:531–543. https://doi.org/10.1038/s41396-019-0546-1

Holt CC, Bass D, Stentiford GD, van der Giezen M (2021) Understanding the role of the shrimp gut microbiome in health and disease. J Invertebr Pathol 186:107387. https://doi.org/10.1016/j.jip.2020.107387

Hooper LV, Gordon JI (2001) Commensal host-bacterial relationships in the gut. Science 292: 1115–1118. https://doi.org/10.1126/science.1058709

Hovda MB, Fontanillas R, McGurk C, Obach A, Rosnes JT (2012) Seasonal variations in the intestinal microbiota of farmed Atlantic salmon (*Salmo salar* L.). Aquacult Res 43:154–159. https://doi.org/10.1111/j.1365-2109.2011.02805.x

Huang Z, Li X, Wang L, Shao Z (2014) Changes in the intestinal bacterial community during the growth of white shrimp, *Litopenaeus vannamei*. Aquacult Res 47(6):1737–1746

Huang L et al (2020a) The bacteria from large-sized bioflocs are more associated with the shrimp gut microbiota in culture system. Aquaculture 523:735159

Huang Z et al (2020b) Micro ecological Koch's postulates reveal that intestinal microbiota dysbiosis contributes to shrimp white faces syndrome. Microbiome 8(1):1–13

Huyben D, Sun L, Moccia R, Kiessling A, Dicksved J, Lundh T (2018) Dietary live yeast and increased water temperature influence the gut microbiota of rainbow trout. J Appl Microbiol 124:1377–1392. https://doi.org/10.1111/jam.13738

Jarchum I, Pamer EG (2011) Regulation of innate and adaptive immunity by the commensal microbiota. Curr Opin Immunol 23:353–360

Jayasree L, Janakiram P, Madhavi R (2006) Characterization of Vibrio spp. associated with diseased shrimp from culture ponds of Andhra Pradesh (India). J World Aquacult Soc 37(4): 523–532

Johnson KV, Foster KR (2018) Why does the microbiome affect behavior? Nat Rev Microbiol 16 (10):647–655. https://doi.org/10.1038/s41579-018-0014-3. PMID: 29691482

Johnson CN, Barnes S, Ogle J, Grimes DJ (2008) Microbial community analysis of water, foregut, and hindgut during growth of pacific white shrimp, *Litopenaeus vannamei,* in closed-system aquaculture (Vol 39, pp 251, 2008). J World Aquacult Soc 39:440–440

Joshi J et al (2014) Variation in Vibrio parahaemolyticus isolates from a single Thai shrimp farm experiencing an outbreak of acute hepatopancreatic necrosis disease (AHPND). Aquaculture 428-429:297–302

Kalia VC, Prakash J, Koul S, Ray S (2017) Simple and rapid method for detecting biofilm forming bacteria. Indian J Microbiol 57:109–111. https://doi.org/10.1007/s12088-016-0616-2

Kamada N, Seo SU, Chen GY, Núñez G (2013) Role of the gut microbiota in immunity and inflammatory disease. Nat Rev Immunol 13:321–335. https://doi.org/10.1038/nri3430

Kan H, Zhao F, Zhang XX, Ren H, Gao S (2015) Correlations of gut microbial community shift with hepatic damage and growth inhibition of *Carassius auratus* induced by pentachlorophenol exposure. Environ Sci Technol 49:11894–11902. https://doi.org/10.1021/acs.est.5b02990

Kanther M, Rawls JF (2010) Host-microbe interactions in the developing zebra fish. Curr Opin Immunol 22:10–19. https://doi.org/10.1016/j.coi.2010.01.00692

Kashinskaya EN, Simonov EP, Kabilov MR, Izvekova GI, Andree KB, Solovyev MM (2018) Diet and other environmental factors shape the bacterial communities of fish at in a eutrophic lake. J Appl Microbiol 125:1626–1641. https://doi.org/10.1111/jam.14064

Khosravi S, Rahimnejad S, Herault M, Fournier V, Lee CR, Dio Bui HT et al (2015) Effects of protein hydrolysates supplementation in low fish meal diets on growth performance, innate immunity and disease resistance of red sea bream Pagrus major. Fish Shellfish Immunol 45: 858–868. https://doi.org/10.1016/j.fsi.2015.05.039

Krogdahl Å, Penn M, Thorsen J, Refstie S, Bakke AM (2010) Important antinutrients in plant feedstuffs for aquaculture: an update on recent findings regarding responses in salmonids. Aquacult Res 41:333–344. https://doi.org/10.1111/j.1365-2109.2009.02426.x

Landsman A, St-Pierre B, Rosales-Leija M, Brown M, Gibbons W (2019) Impact of aquaculture practices on intestinal bacterial profiles of Pacific White leg Shrimp *Litopenaeus vannamei*. Microorganisms 7:93. https://doi.org/10.3390/microorganisms7040093

Larsen AM, Mohammed HH, Arias CR (2014) Comparison of DNA extraction protocols for the analysis of gut microbiota in fishes. FEMS Microbiol Lett 362:fnu031. https://doi.org/10.1093/female/fnu031

Lavilla-Pitogo CR, Leaño EM, Paner MG (1998) Mortalities of pond-cultured juvenile shrimp, Penaeus monodon, associated with dominance of luminescent vibrio in the rearing environment. Aquaculture 164(1–4):337–349

Lee CT, Chen IT, Yang YT, Ko TP, Huang YT, Huang JY, Huang MF, Lin SJ, Chen CY, Lin SS, Lightner DV, Wang HC, Wang AH-J, Wang HC, Hor LI, Lo CF (2015) The opportunistic marine pathogen Vibrio parahaemolyticus becomes virulent by acquiring a plasmid that expresses a deadly toxin. Proc Natl Acad Sci U S A 112:10798–10803. https://doi.org/10.1073/pnas.1503129112

Ley RE, Peterson DA, Gordon JI (2006) Ecological and evolutionary forces shaping microbial diversity in the human intestine. Cell 124:837–848. https://doi.org/10.1016/j.cell.2006.02.017

Li X, Yu Y, Feng W, Yan Q, Gong Y (2012) Host species as a strong determinant of the intestinal microbiota of fish larvae. J Microbiol 50:29–37. https://doi.org/10.1007/s12275-012-1340-1

Li X, Yan Q, Xie S, Hu W, Yu Y, Hu Z (2013) Gut microbiota contributes to the growth of fast-growing transgenic common carp (*Cyprinus carpio* L.). PloS One 8:e64577. https://doi.org/10.1371/journal.pone.0064577

Li J, Ni J, Li J, Wang C, Li X, Wu S et al (2014) Comparative study on gastrointestinal microbiota of eight fish species with different feeding habits. J Appl Microbiol 117:1750–1760. https://doi.org/10.1111/jam.12663

Li X, Zhou L, Yu Y, Ni J, Xu W, Yan Q (2017) Composition of gut microbiota in the gibel carp (*Carassius auratus gibelio*) varies with host development. Microb Ecol 74:239–249. https://doi.org/10.1007/s00248-016-0924-4

Li YF, Yang N, Liang X, Yoshida A, Osatomi K, Power D et al (2018) Elevated seawater temperatures decrease microbial diversity in the gut of Mytilus coruscus. Front Physiol 9:839. https://doi.org/10.3389/fphys.2018.00839

Li Y-F, Xu J-K, Chen Y-W, Ding W-Y, Shao A-Q, Liang X, Zhu Y-T, Yang J-L (2019) Characterization of gut microbiome in the Mussel Mytilus galloprovincialis in response to thermal stress. Front Physiol 10:1086. https://doi.org/10.3389/fphys.2019.01086

Liston J (1956) Quantitative variations in the bacterial flora of flatfish. Microbiology 15:305–314. https://doi.org/10.1099/00221287-15-2-305

Liu PC, Lee KK (2002) Cysteine protease is a major exotoxin of pathogenic luminous Vibrio harveyi in the tiger prawn, Penaeus monodon. Lett Appl Microbiol 28(6):428–430

Liu H, Wang L, Liu M, Wang B, Jiang K et al (2011) The intestinal microbial diversity in Chinese shrimp (*Fenneropenaeus chinensis*) as determined by PCR-DGGE and clone library analyses. Aquaculture 317:32–36

Liu H, Guo X, Gooneratne R, Lai R, Zeng C, Zhan F et al (2016a) The gut microbiome and degradation enzyme activity of wild freshwater fishes influenced by their trophic levels. Sci Rep 6:24340. https://doi.org/10.1038/srep24340

Liu Y, Yao Y, Li H, Qiao F, Wu J, Du ZY et al (2016b) Influence of endogenous and exogenous estrogenic endocrine on intestinal microbiota in zebra fish. PloS One 11:e0163895. https://doi.org/10.1371/journal.pone.0163895

Llewellyn MS, Boutin S, Hoseinifar SH, Derome N (2014) Teleost microbiomes: The state of the art in their characterization, manipulation and importance in aquaculture and fisheries. Front Microbiol 5:1

Llewellyn MS, McGinnity P, Dionne M, Letourneau J, Thonier F, Carvalho GR, Creer S, Derome N (2016) The biogeography of the Atlantic salmon (*Salmo salar)* gut microbiome. ISME J 10: 1280–1284. https://doi.org/10.1038/ismej.2015.189

Lokmer A, Wegner KM (2015) Haemolymph microbiome of Pacific oysters in response to temperature, temperature stress and infection. ISME J 9:670–682. https://doi.org/10.1038/ismej

Macmillan JR, Santucci T (1990) Seasonal trends in intestinal bacterial flora of farm-raised channel catfish. J Aquat Anim Health 2:217–222. https://doi.org/10.1577/1548-8667(1990)0022.3.CO;2

Malham SK, Cotter E, O'Keeffe S, Lynch S, Culloty SC, King JW et al (2009) Summer mortality of the Pacific oyster, Crassostrea gigas, in the Irish Sea: the influence of temperature and nutrients on health and survival. Aquaculture 287:128–138. https://doi.org/10.1016/j.aquaculture.2008.10.006

Manilal A et al (2010) Virulence of vibrio isolated from diseased black tiger shrimp, Penaeus monodon, Fabricius. J World Aquacult Soc 41(3):332–343

Martins R, Fernandez N, Beiras R, Vasconcelos V (2007) Toxicity assessment of crude and partially purified extracts of marine Synechocystis and Synechococcus cyanobacteria strains in marine invertebrates. Toxicon 50:791–799. https://doi.org/10.1016/j.toxico

McCauley M, Chiarello M, Atkinson CL, Jackson CR (2021) Gut microbiomes of freshwater mussels (Unionidae) are taxonomically and phylogenetically variable across years but remain functionally stable. Microorganisms 9(2):411. https://doi.org/10.3390/microorganisms9020411

McFall-Ngai M et al (2013) Animals in a bacterial world, a new imperative for the life sciences. Proc Natl Acad Sci U S A 110:3229–3236. https://doi.org/10.1073/pnas.1218525110

Meng XL, Li S, Qin CB, Zhu ZX, Hu WP, Yang LP et al (2018) Intestinal microbiota and lipid metabolism responses in the common carp (*Cyprinus carpio* L.) following copper exposure. Ecotoxicol Environ Saf 160:257–264. https://doi.org/10.1016/j.ecoenv.2018.05.050

Mente E, Eleni N, Efthimia A, Martin SAM, Kormas KA (2018) Core versus diet-associated and postprandial bacterial communities of the rainbow trout (*Oncorhynchus mykiss*) midgut and faeces. Biol Open 7:bio034397. https://doi.org/10.1242/bio.034397

Meziti A, Alban R, Eleni M, Konstantinos Ar K (2010) Temporal shifts of the Norway lobster (*Nephrops norvegicus*) gut bacterial communities. FEMS Microbiol Ecol 74(2):472–484. https://doi.org/10.1111/j.1574-6941.2010.00964.x

Meziti A, Mente E, Kormas KA (2012) Gut bacteria associated with different diets in reared *Nephrops norvegicus*. Syst Appl Microbiol 35:473–482

Mioduchowska M, Zajac K, Bartoszek K, Madanecki P, Kur J, Zajac T (2020) 16S rRNA gene-based metagenomic analysis of the gut microbial community associated with the DUI species *Unio crassus* (Bivalvia: Unionidae). J Zoolog Syst Evol Res 58:615–623

Miyake S, Ngugi DK, Stingl U (2015) Diet strongly influences the gut microbiota of surgeon fishes. Mol Ecol 24:656–672. https://doi.org/10.1111/mec.13050

Mohd Nosi MZ, Syed Jamil Fadaak SNE, Muhammad MDD, Iehata S (2018) Assessment of gut microbiota in different developmental stages of Malaysian Mahseer (*Tor tambroides*). Aquacult Res 49:2977–2987. https://doi.org/10.1111/are.13757

Mouchet MA, Bouvier C, Bouvier T, Troussellier M, Escalas A, Mouillot D (2012) Genetic difference but functional similarity among fish gut bacterial communities through molecular and biochemical fingerprints. FEMS Microbiol Ecol 79:568–580. https://doi.org/10.1111/j.1574-6941.2011.01241.x

Mountfort DO, Campbell J, Clements KD (2002) Hindgut fermentation in three species of marine herbivorous fish. Appl Environ Microbiol 68:1374–1380. https://doi.org/10.1128/AEM.68.3.1374-1380.2002

Myrand B, Guderley H, Himmelman JH (2000) Reproduction and summer mortality of blue mussels Mytilus edulis in the Magdalen Islands, southern Gulf of St. Lawrence. Mar Ecol Prog Ser 197:193–207. https://doi.org/10.3354/meps197193

Nagothu US, Muralidhar M, Kumaran M, Muniyandi B, Umesh NR, Krishna KS, Silva SD (2012) Climate change and shrimp farming in Andhra Pradesh, India: socio-economics and vulnerability. Energy Environ Res 2(2):137–143. https://doi.org/10.5539/eer

Nam H, Ha M, Bae O, Lee Y (2002) Effect of Weissella confusa strain PL9001 on the adherence and growth of *Helicobacter pylori*. Appl Environ Microbiol 68:4642–4645

Navarrete P, Mardones P, Opazo R, Espejo R, Romero J (2008) Oxytetracycline treatment reduces bacterial diversity of intestinal microbiota of Atlantic salmon. J Aquat Anim Health 20:177–183. https://doi.org/10.1577/H07-043.1

Navarrete P, Magne F, Araneda C, Fuentes P, Barros L, Opazo R et al (2012) PCRTTGE analysis of 16S rRNA from rainbow trout (*Oncorhynchus mykiss*) gut microbiota reveals host-specific communities of active bacteria. PloS One 7:e31335. https://doi.org/10.1371/journal.pone.0031335

Naviner M, Giraud E, Le Bris H, Armand F, Mangion C, Ganiere J (2006) Seasonal variability of intestinal microbiota in rainbow trout (Oncorhynchus mykiss), with a particular attention to Aeromonas spp. as candidate indicator of antimicrobial resistance. Rev Méd Vét 157:597–602

Nayak SK (2010) Role of gastrointestinal microbiota in fish. Aquacult Res 41:1553–1573. https://doi.org/10.1111/j.1365-2109.2010.02546.x

Nelson TM, Rogers TL, Brown MV (2013) The gut bacterial community of mammals from marine and terrestrial habitats. PloS One 8:e83655. https://doi.org/10.1371/journal.pone.0083655

Neuman C, Hatje E, Zarkasi KZ, Smullen R, Bowman JP, Katouli M (2016) The effect of diet and environmental temperature on the faecal microbiota of farmed Tasmanian Atlantic Salmon (*Salmo salar* L.). Aquacult Res 47:660–672

Olson JB, Thacker RW, Gochfeld DJ (2014) Molecular community profiling reveals impacts of time, space, and disease status on the bacterial community associated with the Caribbean sponge *Aplysina cauliformis*. FEMS Microbiol Ecol 87:268–279. https://doi.org/10.1111/1574-6941.12222

Omont A, Regina EG, Eduardo QG, Cristina EF, Roberto HH, Alberto PR (2020) Digestive microbiota of shrimp *Penaeus vannamei* and oyster *Crassostrea gigas* co-cultured in integrated multi-trophic aquaculture system. Aquaculture 521:735059. https://doi.org/10.1016/j.aquaculture.2020.735059

Oxley APA, Shipton W, Owens L, McKay D (2002) Bacterial flora from the gut of the wild and cultured banana prawn, *Penaeus merguiensis*. J Appl Microbiol 93:214–223

Parris DJ, Brooker RM, Morgan MA, Dixson DL, Stewart FJ (2016) Whole gut microbiome composition of damselfish and cardinal fish before and after reef settlement. PeerJ 4:e2412. https://doi.org/10.7717/peerj.2412

Penn MH, Bendiksen EÅ, Campbell P, Krogdahl Å (2011) High level of dietary pea protein concentrate induces enteropathy in Atlantic salmon (*Salmo salar* L.). Aquaculture 310:267–273. https://doi.org/10.1016/j.aquaculture.2010.10.040

Perry WB, Lindsay E, Payne CJ, Brodie C, Kazlauskaite R (2020) The role of the gut microbiome in sustainable teleost aquaculture. Proc R Soc B Biol Sci 287:20200184

Pierce ML, Ward JE (2019) Gut microbiomes of the eastern oyster (Crassostrea virginica) and the blue mussel (Mytilus edulis): temporal variation and the influence of marine aggregate-associated microbial communities. mSphere 4:e00730–e00719. https://doi.org/10.1128/mSphere.00730-19

Pierce ML, Ward JE, Holohan BA, Zhao X, Hicks RE (2016) The influence of site and season on the gut and pallial fluid microbial communities of the Eastern oyster, Crassostrea virginica

(Bivalve, Ostreidae): community level physiological profiling and genetic structure. *Hydrobiologia* 765:97–113. https://doi.org/10.1007/s10750-015-2405-z

Prayitno SB, Sarwan S (2015) 2021The diversity of gut bacteria associated with milkfish (*Chanos chanos* Forsksal) from Northern coast of Central Java, Indonesia. Proc Environ Sci 23:375–384

Qiu Z, Coleman MA, Provost E, Campbell AH, Kelaher BP, Dalton SJ, Thomas T, Steinberg PD, Marzinelli EM (2019) Future climate change is predicted to affect the microbiome and condition of habitat-forming kelp. Proc R Soc B 286:20181887. https://doi.org/10.1098/rspb.2018.1887

Ransom BL (2008) Intestinal microbial community composition of six Actinopterygii fish species in the Southeastern United States. University of Georgia, Athens, GA

Rasheeda MK, Rangamaran VR, Srinivasan S, Ramaiah SK, Gunasekaran R, Jaypal S, Gopal D, Ramalingam K (2017) Comparative profiling of microbial community of three economically important fishes reared in sea cages under tropical offshore environment. Mar Genom 34:57–65

Rawls JF, Samuel BS, Gordon JI (2004) Gnotobiotic zebrafish reveal evolutionarily conserved responses to the gut microbiota. Proc Natl Acad Sci U S A 101:4596–4601. https://doi.org/10.1073/pnas.0400706101

Rawls JF, Mahowald MA, Ley RE, Gordon JI (2006) Reciprocal gut microbiota transplants from zebrafish and mice to germ-free recipients reveal host habitat selection. Cell 127:423–433. https://doi.org/10.1016/j.cell.2006.08.043

Ray AK, Ghosh K, Ringø E (2012) Enzyme-producing bacteria isolated from the fish gut: a review. Aquacult Nutr 18:465–492. https://doi.org/10.1111/j.1365-2095.2012.00943.x

Rieu A, Aoudia N, Jego G, Chluba J, Yousfi N, Briandet R, Deschamps J, Gasquet B, Monedero V, Garrido C, Guzzo J (2016) The biofilm mode of life boosts the anti-inflammatory properties of Lactobacillus. Cell Microbiol 16:1836–1853. https://doi.org/10.1111/cmi.12331

Rimmer D, Wiebe W (1987) Fermentative microbial digestion in herbivorous fishes. J Fish Biol 31: 229–236. https://doi.org/10.1111/j.1095-8649.1987.tb05228.x

Ringo E, Sperstad S, Myklebust R, Refstie S, Krogdahl A (2006) Characterisation of the microbiota associated with intestine of Atlantic cod (*Gadus morhua* L.) the effect of fish meal, standard soybean meal and a bioprocessed soybean meal. Aquaculture 261:829–841. https://doi.org/10.1016/j.aquaculture.2006.06.030

Ringø E, Zhou Z, Vecino JLG, Wadsworth S, Romero J, Krogdahl A, Olsen RE, Dimitroglou A, Foey A, Davies S, Owen M, Lauzon HL, Martinsen LL, De Schryver P, Bossier P, Sperstad S, Merrifield DL (2016) Effect of dietary components on the gut microbiota of aquatic animals. A never-ending story? Aquacult Nutr 22:219–282. https://doi.org/10.1111/anu.12346

Roeselers G, Mittge EK, Stephens WZ, Parichy DM, Cavanaugh CM, Guillemin K, Rawls JF (2011) Evidence for a core gut microbiota in the zebra fish. ISME J 5:1595–1608

Rosenberg E, Falkovitz L (2004) The Vibrio shiloi/Oculina patagonica model system of coral bleaching. Annu Rev Microbial 58:143–159. https://doi.org/10.1146/annurev.micro.58.030603.123610

Rosenberg E, Koren O, Reshef L, Efrony R, Zilber RI (2007) The role of microorganisms in coral health, disease and evolution. Nat Rev Microbiol 5:355–362. https://doi.org/10.1038/nrmicro1635

Rungrassamee W, Klanchui A, Chaiyapechara S, Maibunkaew S, Tangphatsornruang S, Jiravanichpaisal P, Karoonuthaisiri N (2013) Bacterial population in intestines of the black tiger shrimp (*Penaeus monodon*) under different growth stages. PLoS One 8(4):e60802. https://doi.org/10.1371/journal.pone.0060802

Rungrassamee W et al (2014) Characterization of intestinal bacteria in wild and domesticated adult black tiger shrimp *(Penaeus monodon*). PloS One 9:e91853

Rungrassamee W et al (2016) Bacterial dynamics in intestines of the black tiger shrimp and the Pacific white shrimp during *Vibrio harveyi* exposure. J Invertebr Pathol 133:12–19

Saha S, Roy RN, Sen SK, Ray AK (2006) Characterization of cellulase-producing bacteria from the digestive tract of tilapia, *Oreochromis mossambica* (Peters) and grass carp, *Ctenopharyngodon idella* (Valenciennes). Aquacult Res 37:380–388. https://doi.org/10.1111/j.1365-2109.2006.01442.x

Sahlmann C, Sutherland BJ, Kortner TM, Koop BF, Krogdahl Å, Bakke AM (2013) Early response of gene expression in the distal intestine of Atlantic salmon (*Salmo salar* L.) during the development of soybean meal induced enteritis. Fish Shellfish Immunol 34:599–609. https://doi.org/10.1016/j.fsi.2012.11.031

Sanchez LM, Cheng AT, Warner CJ, Townsley L, Peach KC, Navarro G, Shikuma NJ, Bray WM, Riener RM, Yildiz FH, Linington RG (2016) Biofilm formation and detachment in gram-negative pathogens is modulated by select bile acids. PloS One 11:e0149603. https://doi.org/10.1371/journal.pone.0149603

Schluter J, Foster KR (2012) The evolution of mutualism in gut microbiota via host epithelial selection. PLoS Biol 10:e1001424

Schmidt VT, Smith KF, Melvin DW, Amaral-Zettler LA (2015) Community assembly of a euryhaline fish microbiome during salinity acclimation. Mol Ecol 24:2537–2550. https://doi.org/10.1111/mec.13177

Scott KP, Gratz SW, Sheridan PO, Flint HJ, Duncan SH (2013) The influence of diet on the gut microbiota. Pharmacol Res 69:52–60. https://doi.org/10.1016/j.phrs

Seibert CH, Pinto AR (2012) Challenges in shrimp aquaculture due to viral diseases: distribution and biology of the five major penaeid viruses and interventions to avoid viral incidence and dispersion. Braz J Microbiol 43(3):857–864

Sekirov I et al (2010) Gut microbiota in health and disease. Physiol Rev 90(3):859–904

Semova I, Carten JD, Stombaugh J, Mackey LC, Knight R, Farber SA, Rawls JF (2014) Microbiota regulate intestinal absorption and metabolism of fatty acids in the zebrafish. Cell Host Microbe 12:277–288

Sogin ML, Morrison HG, Huber JA, Mark Welch D, Huse SM et al (2006) Microbial diversity in the deep sea and the underexplored "rare biosphere". Proc Natl Acad Sci U S A 103:12115–12120

Soriano EL, Ramirez DT, Araujo DR, Gómez-Gil B, Castro LI, Sánchez CG (2018) Effect of temperature and dietary lipid proportion on gut microbiota in yellowtail kingfish *Seriola lalandi* juveniles. Aquaculture 497:269–277

Soto-Rodriguez SA, Gomez-Gil B, Lozano-Olvera R, Betancourt-Lozano M, Morales-Covarrubias MS (2015) Field and experimental evidence of *Vibrio parahaemolyticus* as the causative agent of acute hepatopancreatic necrosis disease of cultured shrimp (*Litopenaeus vannamei*) in Northwestern Mexico. Appl Environ Microbiol 81:1689–1699

Stentiford GD, Neil DM, Peeler EJ, Shields JD, Small HJ, Flegel TW, Vlak JM, Jones B, Morado F, Moss S, Lotz J, Bartholomay L, Behringer DC, Hauton C, Lightner DV (2012) Disease will limit future food supply from the global crustacean fishery and aquaculture sectors. J Invertebr Pathol 110(2):141–157

Sugita H, Ito Y (2006) Identification of intestinal bacteria from Japanese flounder (*Paralichthys olivaceus*) and their ability to digest chitin. Lett Appl Microbiol 43(3):336–342

Sullam KE, Essinger SD, Lozupone CA, O'Connor MP, Rosen GL, Knight R, Kilham SS, Russell JA (2012) Environmental and ecological factors that shape the gut bacterial communities of fish: a meta-analysis. Mol Ecol 21:3363–3378. https://doi.org/10.1111/j.1365-294X.2012.05552.x

Sullam KE, Rubin BE, Dalton CM, Kilham SS, Flecker AS, Russell JA (2015) Divergence across diet, time and populations rules out parallel evolution in the gut microbiomes of Trinidadian guppies. ISME J 9:1508–1522. https://doi.org/10.1038/ismej.2014.231

Svanevik CS, Lunestad BT (2011) Characterisation of the microbiota of Atlantic mackerel (*Scomber scombrus*). Int J Food Microbiol 151:164–170. https://doi.org/10.1016/j.ijfoodmicro.2011.08.016

Sylvain FE, Cheaib B, Llewellyn M, Correia TG, Fagundes DB, Val AL, Derome N (2016) pH drop impacts differentially skin and gut microbiota of the Amazonian fish tambaqui (*Colossoma macropomum*). Sci Rep 6:32032. https://doi.org/10.1038/srep32032

Tanaka R, Ootsubo M, Sawabe T, Ezura Y, Tajima K (2004) Biodiversity and in situ abundance of gut microflora of abalone (Haliotis discus hannai) determined by culture-independent techniques. Aquaculture 241:453–463. https://doi.org/10.1016/j.aquaculture

Thangamani V, Rajendran N (2016) Total heterotrophic bacterial load in the gut of detritus fishes: a case study of Pichavaram Mangrove environment, Southeast coast, India. Curr World Environ 11:778–783

Trabal N, Mazón-Suástegui JM, Vázquez-Juárez R, Asencio-Valle F, Morales-Bojórquez E, Romero J (2012) Molecular analysis of bacterial microbiota associated with oysters (*Crassostrea gigas* and *Crassostrea corteziensis*) in different growth phases at two cultivation sites. Microb Ecol 64:555–569. https://doi.org/10.1007/s00248-012-0039-5

Trabal N, Mazón-Suástegui JM, Vázquez-Juárez R, Ascencio-Valle F, Romero J (2014) Changes in the composition and diversity of the bacterial microbiota associated with oysters (*Crassostrea corteziensis, Crassostrea gigas and Crassostrea sikamea*) during commercial production. FEMS Microbiol Ecol 88:69–83. https://doi.org/10.1111/1574-6941.12270

Tran L et al (2013) Determination of the infectious nature of the agent of acute hepatopancreatic necrosis syndrome affecting penaeid shrimp. Dis Aquat Organ 105(1):45–55

Tremaroli V, Bäckhed F (2012) Functional interactions between the gut microbiota and host metabolism. Nature 489:242–249. https://doi.org/10.1038/nature11552

Tsuchiya C, Sakata T, Sugita H (2008) Novel ecological niche of *Cetobacterium somerae*, an anaerobic bacterium in the intestinal tracts of freshwater fish. Lett Appl Microbiol 46:43–48

Turnbaugh PJ, Hamady M, Yatsunenko T, Cantarel BL, Duncan A, Ley RE, Sogin ML, Jones WJ, Roe BA, Affourtit JP, Egholm M, Henrissat B, Tzeng TD, Pao YY, Chen PC, Weng FC, Jean WD, Wang D (2015) Effects of host phylogeny and habitats on gut microbiomes of Oriental River Prawn (*Macrobrachium nipponense*). PloS One 10(7):e0132860. PMID: 26168244; PubMed Central PMCID: PMC4500556. https://doi.org/10.1371/journal.pone.0132860

Tzeng TD, Pao YY, Chen PC, Weng FC, Jean WD, Wang D (2015) Effects of host phylogeny and habitats on gut microbiomes of Oriental River Prawn (Macrobrachium nipponense). PLoS One 10(7):e0132860. https://doi.org/10.1371/journal.pone.0132860

Uchii K, Matsui K, Yonekura R, Tani K, Kenzaka T, Nasu M et al (2006) Genetic and physiological characterization of the intestinal bacterial microbiota of bluegill (*Lepomis macrochirus*) with three different feeding habits. Microb Ecol 51:277–284. https://doi.org/10.1007/s00248-006-9018-z

Van Kessel MAHJ, Dutilh BE, Neveling K, Kwint MP, Veltman JA, Flik G, Jetten MSM, Klaren PHM, Op den Camp HJM (2011) Pyro sequencing of 16S rRNA gene amplicon to study the microbiota in the gastrointestinal tract of carp (*Cyprinus carpio* L.). AMB Express 1:1–9

Vatsos IN (2016) Standardizing the microbiota of fish used in research. Lab Anim 51:353–364. https://doi.org/10.1177/0023677216678825

Vaughn CC, Nichols SJ, Spooner DE (2008) Community and food web ecology of freshwater mussels. J North Am Benthol Soc 27:409–423

Vergés A et al (2014) The tropicalization of temperate marine ecosystems: climate-mediated changes in herbivory and community phase shifts. Proc R Soc B 281:20140846. https://doi.org/10.1098/rspb.2014.0846

Verschuere L, Rombaut G, Sorgeloos P, Verstraete W (2000) Probiotic bacteria as biological control agents in aquaculture. Microbiol Mol Biol Rev 64:655–671. https://doi.org/10.1128/MMBR.64.4.655-671.2000

Vilchis LI, Tegner MJ, Moore JD, Friedman JD, Friedman CS, Riser KL, Robbins TT, Dayton PK (2005) Ocean warming effects on growth re production and survivorship of southern California abalone. Ecol Appl 15:469. https://doi.org/10.1890/03-5326

Wahl M, Molis M, Hobday AJ, Dudgeon SR, Neumann R, Steinberg PD, Campbell AH, Marzinelli EM, Connell SD (2015) The responses of brown microalgae to environmental change from local to global scales: direct versus ecologically mediate effects. Perspect Phycol 2:11–30. https://doi.org/10.1127/pip/2015/0019

Wang CZ et al (2014) The cellular community in the intestine of the shrimp *Penaeus penicillatus* and its culture environments. Fish Sci 80:1001–1007

Wang AR, Ran C, Ringø E, Zhou ZG (2017) Progress in fish gastrointestinal microbiota research. Rev Aquac 10:626–640. https://doi.org/10.1111/raq.12191

Weingarten EA, Atkinson CL, Jackson CR (2019) The gut microbiome of freshwater Unionidae mussels is determined by host species and is selectively retained from filtered seston. PloS One 14(11):e0224796. https://doi.org/10.1371/jounal.pone.0224796

Wu S, Wang G, Angert ER, Wang W, Li W, Zou H (2012) Composition, diversity, and origin of the bacterial community in grass carp intestine. *PloS One* 7:e30440. https://doi.org/10.1371/journal.pone.0030440

Xia JH, Lin G, Fu GH, Wan ZY, Lee M, Wang L, Liu XJ, Yue GH (2014) The intestinal microbiome of fish under starvation. BMC Genomics 15:266. https://doi.org/10.1186/1471-2164-15-266

Xiao J, Ford SE, Yang H, Zhang G, Zhang F, Guo X (2005) Studies on mass summer mortality of cultured zhikong scallops (*Chlamys farreri*) in China. Aquaculture 250:602–615. https://doi.org/10.1016/j.aquaculture.2005.05.002

Xing MX, Hou ZH, Yuan JB, Liu Y, Qu Y, Liu B (2013) Taxonomic and functional metagenomic profiling of gastrointestinal tract microbiome of the farmed adult turbot (*Scophthalmus maximus*). FEMS Microbiol Ecol 86:432–443. https://doi.org/10.1111/1574-6941.12174

Xiong J et al (2017) Integrating gut microbiota immaturity and disease-discriminatory taxa to diagnose the initiation and severity of shrimp disease. Environ Microbiol 19:1490–1501. https://doi.org/10.1111/1462-2920.13701

Yan Q, Li J, Yu Y, Wang J, He Z, Van Nostrand JD et al (2016) Environmental filtering decreases with fish development for the assembly of gut microbiota. Environ Microbiol 18:4739–4754. https://doi.org/10.1111/1462-2920.13365

Ye L, Amberg J, Chapman D, Gaikowski M, Liu WT (2014) Fish gut microbiota analysis differentiates physiology and behavior of invasive Asian carp and indigenous American fish. ISME J 8:541–551. https://doi.org/10.1038/ismej.2013.181

Ye W, Han D, Zhu X, Yang Y, Jin J, Xie S (2017) Comparative study on dietary protein requirements for juvenile and pre-adult gibel carp (Carassius auratus gibelio var. CAS III). Aquacult Nutr 23:755–765. https://doi.org/10.1111/anu.12442

Yoshimizu M, Kimura T (1976) Study on the intestinal microflora of salmonids. J Fish Dis Res 10: 243–259. https://doi.org/10.3147/jsfp.10.243

Yukgehnaish K, Kumar P, Sivachandran P, Marimuthu K, Arshad A, Paray BA, Arockiaraj J (2020) Gut microbiota metagenomics in aquaculture: factors influencing gut microbiome and its physiological role in fish. Rev Aquac 12:1903–1927

Zhang ML et al (2014) Characterization of the intestinal microbiota in Pacific white shrimp, *Litopenaeus vannamei*, fed diets with different lipid sources. Aquaculture 434:449–455. https://doi.org/10.1016/j.aquaculture

Zhang Z, Li D, Refaey MM, Xu W, Tang R, Li L (2018a) Host age affects the development of southern catfish gut bacterial community divergent from that in the food and rearing water. Front Microbiol 9:495. https://doi.org/10.3389/fmicb.2018.00495

Zhang Z, Li D, Refaey MM, Xu W (2018b) High spatial and temporal variations of microbial community along the southern catfish gastrointestinal tract: insights into dynamic food digestion. Front Microbiol 8:1531. https://doi.org/10.3389/fmicb.2017.01531

Zheng Z, Aweya JJ, Bao S, Yao D, Li S, Tran NT, Ma H, Zhang Y (2021) The microbial composition of Penaeid Shrimps' Hepatopancreas is modulated by Hemocyanin. J Immunol 207(11):2733–2743. https://doi.org/10.4049/jimmunol.2100746

Zhou C, Yang S, Ka W, Gao P, Li Y, Long R, Wang J (2022) Association of gut microbiota with metabolism in rainbow trout under acute heat stress. Front Microbiol 13:846336. https://doi.org/10.3389/fmicb.2022.846336

Zoqratt MZ et al (2018) Microbiome analysis of Pacific white shrimp gut and rearing water from Malaysia and Vietnam: implications for aquaculture research and management. PeerJ 6:e5826

SPRINGER NATURE

GPSR Compliance

The European Union's (EU) General Product Safety Regulation (GPSR) is a set of rules that requires consumer products to be safe and our obligations to ensure this.

If you have any concerns about our products, you can contact us on ProductSafety@springernature.com

In case Publisher is established outside the EU, the EU authorized representative is:

Springer Nature Customer Service Center GmbH
Europaplatz 3
69115 Heidelberg, Germany

The manufacturer's authorised representative in the EU is Springer Nature Customer Service Centre GmbH, Europaplatz 3, 69115 Heidelberg, Germany. If you have any concerns regarding our products, please contact ProductSafety@springernature.com

Printed and bound by CPI Group (UK) Ltd, Croydon, CR0 4YY

30/09/2026

02225319-0010